HARVARD HISTORICAL STUDIES • 172
Published under the auspices
of the Department of History
from the income of the
Paul Revere Frothingham Bequest
Robert Louis Stroock Fund
Henry Warren Torrey Fund

CONFLUENCE

The Nature of Technology and the
Remaking of the Rhône

SARA B. PRITCHARD

Harvard University Press
Cambridge, Massachusetts
London, England
2011

Copyright © 2011 by the President and Fellows of Harvard College
All rights reserved
Printed in the United States of America

Library of Congress Cataloging-in-Publication Data

Pritchard, Sara B., 1972–
Confluence : the nature of technology and the remaking of the
Rhône / Sara B. Pritchard.
    p. cm.
Includes bibliographical references and index.
ISBN 978-0-674-04965-9 (alk. paper)
1. River engineering—Rhône River Watershed (Switzerland and
France) 2. Rhône River (Switzerland and France) I. Title.
TC472.R6P75 2011
333.91'620944—dc22   2010039268

For Ron

CONTENTS

List of Maps and Figures ... ix

Prologue ... xi

Introduction: Nature, Technology, and History ... 1

1. Envisioning a New Rhône ... 28
2. Imagining the Nation's River ... 55
3. Postwar Transformations ... 78
4. Local Responses ... 132
5. Rethinking the Nation ... 163
6. Rethinking the Rhône ... 193
7. A New Modern ... 212

Conclusion: Legacies of the Rhône ... 240

List of Abbreviations ... 253

Notes ... 255

Bibliography ... 339

Acknowledgments ... 353

Index ... 359

MAPS AND FIGURES

Maps

P.1. A transnational river   xii
P.2. The French Rhône   xiii
1.1. The CNR's completed projects   47
3.1. The Rhône formula   81
3.2. Donzère-Mondragon   86
3.3. Nuclear and industrial development in the Rhône valley   114
4.1. Donzère-Mondragon's counter-canal network   155
5.1. The BRL's irrigation system in southern France   175
7.1. Developing the upper Rhône   217

Figures

1.1. The multipurpose Rhône   50
2.1. Inaugurating Donzère-Mondragon   56
2.2. Interior of André Blondel hydroelectric plant   69
2.3. Side view of Donzère-Mondragon   71
3.1. Building Donzère-Mondragon: the tailrace   88
3.2. Building Donzère-Mondragon: Blondel hydroelectric plant   89
3.3. Building Donzère-Mondragon: reservoir dam   93
4.1. Challenging Donzère-Mondragon   139

PROLOGUE

In the summer of 1946, officials from the recently founded Commissariat Général au Plan traveled down the French reach of the Rhône River.[1] Widespread devastation across France meant politicians and bureaucrats would be planning for the repairs of basic infrastructure and the rebuilding of key industries. Without a doubt, postwar reconstruction, both physical and symbolic, was the government's highest priority, but such efforts required a detailed assessment of existing conditions across the country. The Commissariat Général au Plan (Planning Commission, usually referred to as the Plan), then headed by Jean Monnet and dominated by economic and technical elites, had already begun to oversee coordinated state planning in January 1946. They would formulate the first Plan de Modernisation et d'Equipement (1947–1952) the following year. These experts targeted fundamental sectors of the economy, outlining general goals such as the revival and expansion of energy production as well as specific projects for reconstruction, while attempting to avoid the extremes of either unbridled liberalism or authoritarian *dirigisme* (state intervention).[2] As it turned out, the Rhône would play a crucial role in the reconstruction and ultimately modernization of France after World War II. Indeed, the vital importance of the Rhône to France's political economy helps explain why the government turned first to this crucial waterway, along with the Seine, after the war.[3]

That summer, Plan officials followed the river southwest and then south over its three-hundred-mile course, from the Swiss border to the Mediterranean Sea (see Maps P.1 and P.2). The Rhône poured from Lake Léman into French territory at Geneva and, although only ten to fifteen meters wide in most places with modest volume, it nonetheless maintained a forceful flow through the lower Alps. The river rushed through

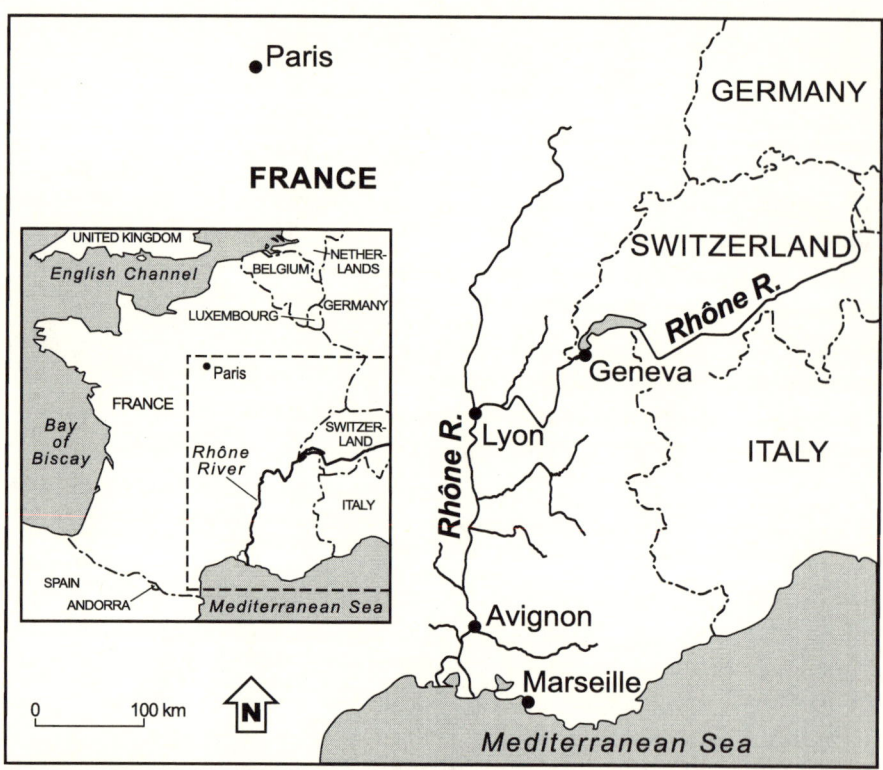

Map P.1. *A transnational river.* The Rhône's source is the Furka glacier, high in the Swiss Alps. The river enters French territory as it exits Lake Léman at Geneva. (Map by Joseph W. Stoll, Syracuse University Cartographic Laboratory.)

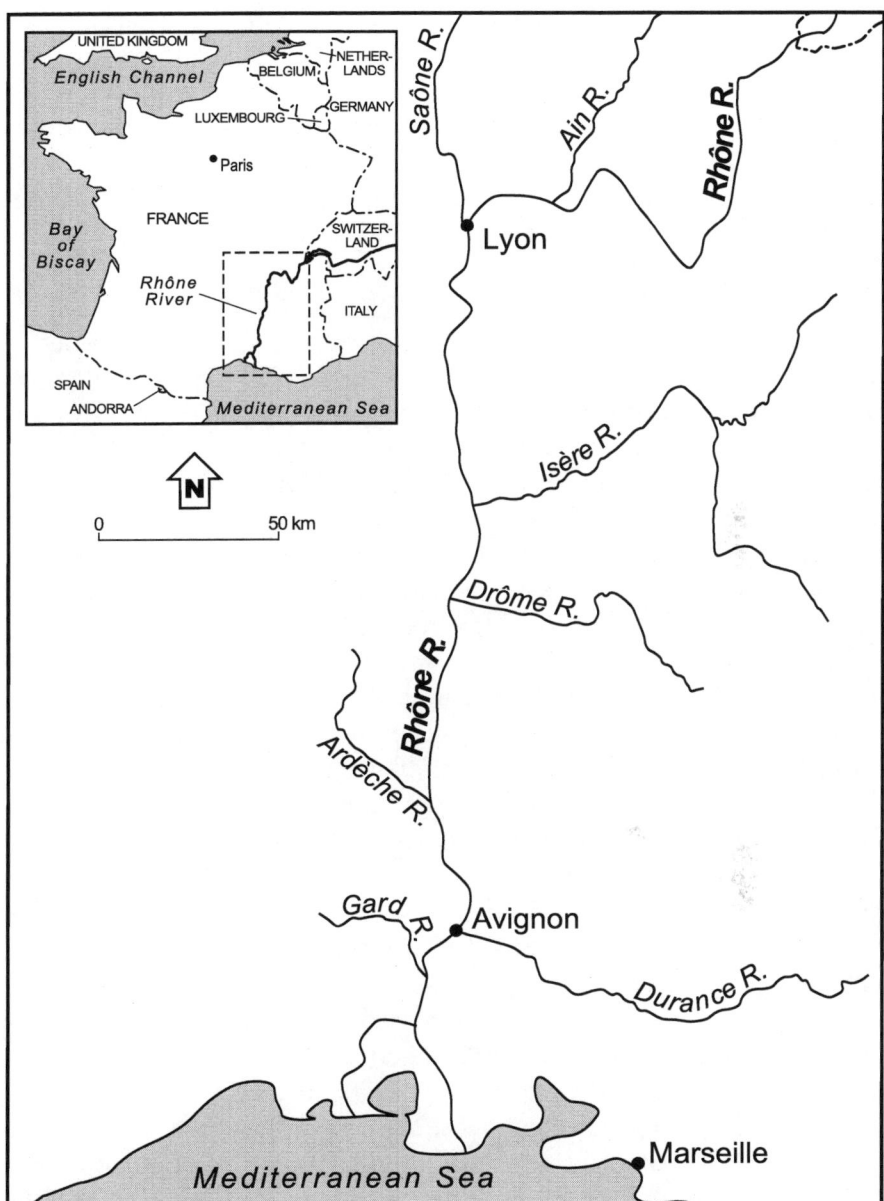

Map P.2. *The French Rhône*. Several rivers contribute to the Rhône's considerable flow, especially south of Lyon. (Map by Joseph W. Stoll, Syracuse University Cartographic Laboratory.)

narrow valleys and craggy gorges with cascades of water marking the steepness of its course. At Sault-Brénaz, it dropped two meters in one-tenth of a kilometer.[4] The river's rocky, precipitous flow made the upper Rhône, the reach of the river between the Franco-Swiss border and Lyon, nearly impossible to navigate. For this reason, river travel, especially by large commercial vessels, was limited until the Rhône reached Lyon. Meanwhile, the officials observed weathered villages perched halfway up steep hillsides to avoid flooding; downhill fields and pastures were not so fortunate.

While the features of the upper Rhône valley may have constrained navigation and farming, they afforded other possibilities. Several small, late-nineteenth-century hydroelectric plants harnessed the river's flow, while an enormous, high-chute dam at Génissiat was finally nearing completion after the occupation of northern France had brought construction to a standstill. Rumors circulated that valiant members of the Resistance, or perhaps frustrated locals, had thrown the bodies of German soldiers into the still damp concrete of the monumental work site, forever entombing them along the upper Rhône.[5]

After winding one hundred mountainous miles through the French lower Alps, Plan officials followed the river as it veered south near Lyon, the former capital of Roman Gaul. Here the Rhône appeared more subdued. Its bed widened, and its waters flowed more gently as the valley flattened. These attributes yielded opportunities for farmers in the city's fertile hinterlands. Also on the outskirts of Lyon, the officials passed by Jonage, a multipurpose project from the 1890s that channeled the river northeast of the city. It produced energy and facilitated navigation while reducing floods that had plagued the ancient metropolis for over two millennia.

Lyon's two rivers, the Rhône and the Saône, had attracted the Romans and helped spur the founding of "Lugudunum" in 43 BCE. Centuries later, textile mills tapped the rivers, literally fueling the city's international reputation as a producer of fine silks. Yet periodic flooding of the Rhône and the Saône threatened Lyon, a commercial crossroads originally established on a thin slice of land between their banks, now called the "Peninsula" (Presqu'île). Embankments eventually were built lining the Rhône as it passed through the city center, in an attempt to keep the river from spilling onto cobblestone streets and bustling squares, but many building walls were scarred as testament to floodwaters past. Still, for once, the river's threats were overshadowed: the immediate effects of war and occupation, which had destroyed essential infrastructure, crippled the economy, sent thousands of citizens to work in Germany, and forced France's remaining population to survive on meager rations, were impossible to ignore.

As the officials continued south, they watched the Rhône's hydrology become not only more complex and imposing but also more promising. The merging of the Rhône and the Saône in Lyon, when coupled with the hydraulic and geologic contexts of the central Rhône valley, began to change the river's ecological character. Between Lyon and Orange, a series of narrow gorges and expansive plains punctuated the river valley. Where a gorge channeled the Rhône, it created a strong, fast surge or "faucet" *(robinet)* of water, which then fanned out and slowed down as the river ran through the plain downstream. As the Rhône streamed through the central valley, its volume was augmented by the Isère, Drôme, and Ardèche rivers. Moreover, the Rhône still had far to fall as it continued toward Avignon. In the 331 kilometers between Lyon and the Mediterranean, the river dropped 162 meters before finally reaching sea level.[6] It took the Rhine and Danube rivers two and five times that distance, respectively, to make similar drops.[7] Although the Rhône's descent would not have impressed American engineers familiar with some of the rivers of the U.S. West, this river flowed through an ancient, densely populated region with a long history of agriculture, commerce, industry, and urbanization. The Rhône had both facilitated and endangered these endeavors probably for as long as humans had sought to live with it.

As Plan officials surveyed the coursing waters of the central Rhône, they began to understand the river's ominous nickname: "furious bull." The Rhône was remarkable under normal conditions, but a confluence of factors, including spring runoff and converging storms, could result in extraordinary floods when the river channeled nearly six times its "average" flow, rising five or even six meters above low-water levels and stretching for miles beyond its banks. During thousand-year floods, the central Rhône valley essentially became an inland sea.

And yet, as with other rivers, the Rhône's destructive capacity also defined its immense potential, a point that did not escape those seeking to rebuild France after World War II. In addition to the river's hydroelectric possibilities, the rich alluvial soil of its floodplain, abetted by the climate of southeastern France, made the central Rhône valley ideal for certain crops. In this agricultural Eden, the officials glimpsed extensive, ancient irrigation networks that tapped the river, siphoning some of the Rhône's waters for farms and vineyards stretching inland from its banks. Locals called this region "the empire of the sun" *(l'empire du soleil)* and pinned their futures on the rhyming formula *"L'eau plus le soleil égale la prospérité"* ("water plus sun equals prosperity").[8]

Plan officials also observed how local communities and other groups used, managed, and reshaped the river in other ways. In the not-so-distant past, teams of men and horses had hauled commodity-laden barges upstream during the Rhône's limited navigation season. Now diesel- and coal-powered barges moved up the central Rhône with far greater ease than had their forebears. Battles, bombs, and sabotage may have damaged many of the country's highways and bridges, but rivers like the Rhône continued to flow through the country's war-torn landscape, serving as vital thoroughfares, despite the desperate, last-ditch efforts of retreating German soldiers.[9] The officials also noted the villages clustered along the edge of the central Rhône, their water and sanitation systems literally connecting the river with its cities.[10] Because of the Rhône's propensity for flooding, many families in the river's vast lowlands built elevated houses and tied boats to porches. Meanwhile, Plan officials watched fishers wade into the Rhône, hoping the river's bounty would supplement their still limited food supplies. Many stretches were fast and dangerous, but experienced fishers chose sites away from the Rhône's main channel.

Although the Rhône spilled into its expansive floodplain with predictable unpredictability, the river usually maintained its regular course into the lower Rhône valley. After it curved around the walled papal city of Avignon, passing under what remained of the city's famous bridge, the Durance River and soon the Gard joined the Rhône. By Beaucaire, a fortressed village in the heart of Provence known for its international fairs during the Renaissance, the Rhône reached an average volume three times that of the Seine and twice that of the famous château-surrounded Loire. The river's flow here rivaled that of the Po and even the formidable Rhine.[11]

The Rhône's waters began to widen and slow as they approached the Mediterranean. During these final miles, the vast Rhône delta was in constant flux. The river meandered, with its flow dividing into *le grand Rhône, le petit Rhône,* and numerous minor channels. South of Arles, the Rhône fed the Camargue, brackish marshes that were home to feral horses, flamingos, and the river's namesake bulls, as well as rice paddies and a natural reserve. Finally, after its five-hundred-mile journey from high in the Swiss Alps, the Rhône gushed into the Mediterranean west of Marseille. In a single year, France's most powerful river channeled 54 billion cubic meters of water and 21 million cubic meters of sediment.[12]

If these Plan officials had returned to the Rhône in 1986—just forty years later—they might not have recognized the landscape of the river valley, even though their own efforts were central to the river's remaking. Seven hydroelectric dams, two nuclear reactors, and a dozen recreational facilities transformed the one hundred miles of the upper Rhône alone. The changes along the river's central and lower reaches were even more striking. Lyon's population had grown, and its long industrial history had taken new forms. The city's southern suburbs were now home to numerous riverside petrochemical plants. Twelve more hydroelectric plants harnessed the two hundred miles of the Rhône between Lyon and the Mediterranean. Here long diversion canals sent the vast majority of the Rhône's flow through turbines of enormous hydroelectric plants. Ten or fifteen miles downstream from the point of diversion, these canals rejoined the main channel of the river, whose intense faucet had been reduced to a mere trickle, only to be rerouted again. The Rhône was still a braided river, but channelization and industrialization yielded a series of radically simplified figure eights. Meanwhile, France's atomic agencies had built four nuclear power plants and processing facilities in the shadow of these hydroelectric complexes. They drew some of the river's flow into their cooling systems and consumed vast quantities of Rhône-produced electricity. Near the shores of the Mediterranean, the industrial site of Fos sprawled over former wetlands. Oil refineries and container ships were now part of the ecology of the lower Rhône. In total, by 1986, the three-hundred-mile French Rhône boasted nineteen hydroelectric plants, nineteen dams, fourteen navigation canals, six nuclear facilities, and several large industrial centers.[13] These were perhaps the most dramatic and visible changes, but Plan officials would have found the valley's agriculture revolutionized as well. Agricultural experts and farmers had sought to boost productivity by rationalizing farms. New irrigation systems brought more of the Rhône's water to these agroindustrial complexes, sometimes located dozens of miles from the river. Nature preserves and recreational areas set aside other reaches of the Rhône for tourism, leisure, and environmental protection. The officials might also have noticed that few fishers ventured into the river anymore. Then again, there was little reason to do so. There weren't many fish.

This book examines how the modern Rhône and, in the process, modern France came to be.

CONFLUENCE

INTRODUCTION:
NATURE, TECHNOLOGY, AND HISTORY

> And by the Rhône, France became France.
> —*Gabriel Hanotaux,*
> *French statesman and historian*
> *(1853–1944)*

France's Rhône River experienced dramatic, perhaps even stunning, changes between the end of World War II and the late twentieth century. Yet to perceive the Rhône of 1946 as natural, and the recent Rhône as technological, is not only historically inaccurate but also conceptually problematic. Whether viewed from a horse-drawn barge, a retreating German tank, or the observation deck of a nuclear power plant, the Rhône was and remains an envirotechnical landscape. It is a river that has been repeatedly remade by politicians, scientific experts, and ordinary people. Its transformations have taken place through the workings of both formal institutions and informal practices, and hydrologic processes beyond humanity's complete control. This convergence of nonhuman and human factors—this reblending of ecological and technological systems—defines the Rhône as an envirotechnical landscape.

Geographers emphasize that such a landscape is, by definition, a hybrid space of both nature and culture, what leading scholars in science and technology studies (STS) call "naturecultures" or "nature-culture."[1] The Rhône's history underscores the critical role diverse technologies have played in both shaping and being shaped by that landscape. As David Nye states, "technology is not alien to landscape, but integral to it."[2] At the same time, historians of technology and other STS scholars have stressed that these "technologies" are, in fact, sociotechnical artifacts and systems. The Rhône thus exemplifies the complex and dynamic amalgamation of what historical actors and scholarly analysts usually call more simply nature, technology, and society. Yet the history of the river's transformations since World War II also reveals how various groups and institutions vied to redefine and remake the Rhône. They sought to manage the river's natural resources—from its flow to its fish—and its ecology as a whole, through the construction and operation of technologies designed to suit their shifting and often competing goals for France. Or, to paraphrase Amita Baviskar, the Rhône River has long been a "landscape" saturated with power.[3]

This book traces the multifaceted and contested histories of the Rhône's remaking since 1945: the ways various (human) constituencies appropriated, enrolled, and transformed the "same" river to pursue aims that were sometimes political, other times economic, still others cultural, and most often wholly intertwined. For simplicity's sake, I will usually refer to "the Rhône's history," even though this phrase obscures the river's very multiplicity, both within and throughout historical periods, and the intense negotiations over its management and place in the political, cultural, and environmental histories of contemporary France.[4] Yet, despite the numerous cultural representations and material forms of the Rhône over the past sixty-five years, it has remained a singular river. In fact, the Rhône's physical and symbolic centrality to diverse goals and its location within multiple, often overlapping systems has exacerbated the conflicts over it.

As its title already suggests, this book's argument develops through three interconnected levels of analysis—descriptive, historical, and conceptual—that integrate insights from the history of technology and environmental history while situating the Rhône's transformations within the history of modern France. The Rhône is at the center of each layer of this narrative.

The first level of analysis describes how various social groups interacted with the Rhône over time, what environmental historians call human-natural relations. I trace the histories of the river's transformations, showing how politicians, technical elites, and citizens have understood and ultimately remade it since 1945. Their particular representations and redesigns of the Rhône reflected diverse, changing, and often competing goals for France. For example, during the interwar years, the French state approved the multipurpose development of the Rhône, striving to balance national and regional demands. World War II and its legacy, however, altered the context of the river's development after

1945. Postwar reconstruction, dominated by demands for energy, therefore recast the interpretation and realization of multipurpose development.[5] Yet these objectives were not fixed, either. Decolonization, European integration, and a resurgence in regionalism altered both development goals and the political geography of the remade Rhône in subsequent decades. Tracing the ongoing transformation of the Rhône between World War II and the late twentieth century thus illuminates how the river's remaking both embodied and enacted these changes in French society, culture, and political economy.

In describing the changes the Rhône underwent, this study also explores the relationship between technological development and environmental management on the one hand and political identities and state building on the other. This second level of analysis thus moves beyond recounting the particular redesign of the Rhône at various historical moments—diverting a certain amount of its flow for hydroelectric generation or channelizing its course—to analyzing the historical processes that shaped its remaking in particular ways and not in others. A surprising array of groups used the Rhône to enact postwar reconstruction, both culturally and materially, by associating ideas about nature and technology with French national identity while pursuing literal nation building through the construction of large-scale technological systems and intensive river management. As the imperatives of reconstruction faded, the specific meanings and objectives of state building shifted, but the Rhône's importance to these new frameworks and programs remained. The Rhône's history therefore demonstrates the centrality of technological change and environmental management to the political and cultural histories of modern France, bolstering the claim of some recent historians of technology and environmental historians that they have crucial contributions to make to so-called mainstream history. As scholars in both specialties have begun to show, technology studies, environmental history, and their intersection help expose the material contexts of key historical processes such as nation building while challenging common assumptions about the nature of historical agency.

Finally, in writing this history of the Rhône's transformations since 1945, I have developed several analytic tools to help conceptualize nature, technology, and their relationship *within* and *as* history. This is the book's third and final level of analysis. The concepts of envirotechnical systems and envirotechnical regimes, by calling attention to both actors' and analysts' categories, and the ecological dimensions of "technological" artifacts, combine central insights from environmental history and technology studies. As I discuss in more depth below, these concepts demonstrate how technological objects and systems are productive sites of inquiry for environmental historians while drawing attention to nonhuman nature within studies of technology. This theoretical framework emerged from my study of the Rhône but has, as I hope this book shows, wider relevance, perhaps not only within academia but beyond.

"Confluence" provides, then, a fruitful metaphor derived from river hydrology. Historically, nature and technology, both discursively and materially, are critical to understanding the politics and culture of contemporary France. Analytically, bridging environmental history and technology studies strengthens each field while enriching our understanding of the past, precisely because that past is at once social, technological, and natural.[6]

Rivers in/as History

In order to trace the history of the Rhône's repeated transformations, the book's first level of analysis, I examine the institutions, goals, and power structures behind its "development" *(aménagement)* since World War II. The French state may have created the Compagnie Nationale du Rhône (CNR, or Rhône River Authority) to develop and manage the river in 1921, but the agency's officials still had to collaborate with agricultural engineers from the Génie Rural (part of the state's agricultural bureaucracy), navigation experts from the Corps des Ponts et Chaussées (the state's engineering corps for "bridges and highways," but also canals, harbors, and railroads), and representatives from other state agencies. For instance, when France developed an independent nuclear power program after the war, this commitment altered the existing institutional matrix and vision for the river. Yet not just state experts and other elites participated in the river's remaking. Locals had ambiguous relationships with the Rhône, just as they had complex interactions with the large projects eventually built in their midst. The following chapters examine how and why these various groups envisioned, designed, negotiated, realized, and at times modified the Rhône's ongoing reconstruction between 1945 and the end of the twentieth century. Energy, navigation, and agriculture reshaped the river during the postwar era in both complementary and conflicting ways, and as the emergence of nuclear power suggests, definitions of development were certainly not static. Agriculture carried less weight until the early 1960s, while environmentalism challenged large-scale industrial development beginning in the late 1960s. All of these objectives were deeply politicized; even the notion of a "development project" became contentious.[7]

In writing the history of the Rhône's repeated remaking, I have employed methods from the history of technology and science and technology studies more broadly. In particular, this study opens the "black box" of river management technologies to examine the contested process of their design, development, and use. Subsequent chapters analyze numerous debates over the Rhône and its possible futures, tracking how invested groups and their goals changed over time. Drawing also on insights from environmental history, unpacking technologies of river management illuminates that field's central concern: the historical interactions between human and nonhuman nature.[8] As the Rhône's history shows, technology—particularly the process of technological development—offers valuable perspectives on human-natural relations, not only those at a given historical moment but also those envisaged and prescribed for the future.

Too often, however, scholars outside STS have not adequately analyzed the contested process of technological development. As historians of technology have ably demonstrated, to present only what happened—"the path taken"—creates a sense of inevitability, perhaps even a teleological representation of the past. Put another way, to study and recount only what was actually built risks naturalizing technological change. Given this book's focus, this is an especially apt but loaded metaphor that captures the depoliticization of technology, particularly for those related to environmental management.[9] As I discuss below, valuable techniques from the history of technology instead help to uncover the historical and cultural contingencies of technological change. At the same time, environmental historians remind STS scholars that not all such contingencies emanate from humanity, another crucial point to which I will return. Overall, approaches from science and technology studies emphasize who participated in debates over technological change, how these groups' views may have differed, and what historical forces may have strengthened the position of some constituencies while undermining others. Together, these insights call into question the idea that the ultimate outcome was either inevitable or necessarily "the best" one, a conclusion with vital political implications for humans and nonhumans alike.

The environmental "impacts" of the Rhône's transformations since 1945 are, then, only one part (indeed, not the primary focus) of this history. I do not doubt the importance, both historical and contemporary, of ecological changes associated with twentieth-century river development: from shifts in species composition to chemical and thermal pollution, to name just two. But the historical analysis of this book focuses instead on the objectives, design, negotiation, and materialization of river management in France since World War II. In some ways, this history is more abstract than a simple recounting of the environmental consequences of postwar river development. In fact, I discuss below some of the assumptions embedded within such narratives of environmental impacts, including the tidy separation of nature and technology and the unproblematic representation of environmental knowledge. A fine-grained historical analysis of technological development—one that places cultural assumptions, social relations, and power in the foreground—serves as a critical reminder of what otherwise could have been, including alternative ecological possibilities. Indeed, the ultimate environmental changes were closely linked to the particular redesigns of the Rhône. If the river had been managed in other ways, then the environmental implications might have been different.[10]

Furthermore, as rivers have garnered increasing attention from historians, two approaches to writing river histories have emerged. One examines a river's history for its own sake. The other explores how political questions, economic debates, cultural ideals, and social struggles invariably become interwoven with rivers and their management. In other words, these studies investigate the constitutive relationship between rivers and their remaking, and established historical processes such as state building, industrialization, and identity formation.[11] This brings us to the book's second level of analysis.

Nature, Technology, and Nation

As the opening vignette of this book suggested, France faced a severe, multifaceted crisis after World War II. Materially, parts of the country lay in ruins. Many roads, bridges, and other essential infrastructure had been destroyed by combat and sabotage. Numerous factories, even entire industries, faced similar problems, while energy shortages stymied basic repairs, let alone the revival of the nation's lagging economy. But it was not just the landscape and economy that had been affected: France's citizenry had suffered through years of food shortages; in fact, bread rations were not fully lifted until 1949. And culturally, the country had once again been defeated by Germany, for the third time in seventy years. Collaborating with Germans during the war had allowed many French citizens to "get by" (what was called "*système d*," shorthand for *se débrouiller*), but it was a shameful chapter in the country's history, one that would be scrutinized and purged with violence after the war. Moreover, the deportation of Jews from France had betrayed the nation's revolutionary ideals, with tragic consequences. These humiliating experiences tarnished France's supposed *grandeur* (greatness) and rendered suspect the authority and legitimacy of the government itself. The Third Republic (1870–1940) may have been the first stable republican form of government since the revolution of 1789, but it had once again failed to fend off the German military, eventually leading to the country's occupation. Then the wartime Vichy regime had epitomized collaboration at the highest (or lowest) level.[12]

It was under these daunting conditions that France's leaders and citizens sought to reconstruct their nation, both conceptually and literally. Benedict Anderson's influential *Imagined Communities* has offered one powerful way to consider how they did so. In examining how national communities have been imagined, Anderson describes nations *becoming*—that is, the ways national communities have been conceptualized, constructed, and eventually maintained by particular groups at specific historical moments, rather than being either self-evident or self-reproducing. In other words, Anderson denaturalizes the modern nation-state. Other scholars such as Peter Sahlins have built on Anderson's important work.[13]

Yet nations, whether one focuses on their cultural dimensions or bureaucratic institutionalization, are *not only* imagined. Anderson's metaphor has proved illuminating and productive among a number of disciplines, but as a social constructivist framework, it risks the implication, however erroneous, that "imagined" (national) communities are somehow not "real."[14] Yet, as Anderson and subsequent scholars have shown, the seemingly abstract ideologies of inclusion (and by definition, exclusion) can have serious, tangible, even horrific consequences for those excluded from a given community. National "imaginings" can be—and have been—materialized through laws, policies, practices, and, as I discuss below, technological artifacts and environmental management strategies. Furthermore, as James C. Scott's influential study *Seeing Like a State* persuasively shows, these materializations have often had profound implications for both humans and nonhumans.[15]

Nations may be imagined, then, but they also *take place*. This metaphor operates on two levels. First, as Anderson and Sahlins demonstrate, the formulation and maintenance of nation-states take enormous work. Therefore, examining how, in what ways, and through what means this political reproduction occurs is as important as the particular eventual outcome. It is also worth noting here that STS scholars might reframe the books, newspapers, and other forms of print-capitalism central to Anderson's national communities as technopolitics—specifically, as technopolitics of the emergent nation-state. This concept serves to highlight the material enactment of nascent nationalism.[16] Second, as the metaphor itself connotes, nations are *situated* not only geographically but also ecologically. This second layer of meaning underlines the spatial and environmental contexts of the nation-state. While geographers and cultural historians are right to stress how "space" and "the environment" are themselves cultural constructs, they are nonetheless simultaneously material and materialized.[17] In short, nations are *both* abstract and concrete.

This notion of concreteness is useful because it suggests how a nation, or conceptions of national communities, can be instantiated in material forms. But it is particularly telling for the Rhône and its place in the (national) history of France because, as the river's dams and nuclear reactors show, concreteness is more than just a metaphor. The "concreteness" of the French nation-state literally cemented nature and technology in the fibers of national reconstruction, both materially and discursively. Or, as Wiebe Bijker succinctly puts it, "dikes and dams: thick with politics."[18]

As the Rhône's physical transformations demonstrate, environmental management and technological development helped enact the French nation-state since 1945 by reshaping the landscape and building structures in ways that regulated humans, nonhumans, and their interactions. In other words, technologies and strategies of environmental management materialized France as a nation in the territorial space declared within its borders. This argument combines Scott's analysis of how state elites and bureaucratic structures make socioenvironmental landscapes "(il)legible" with Chandra Mukerji's critique of this framework, because it tends to take the state itself for granted. Employing the concept of co-production from STS, Mukerji instead argues that "the (high-modernist) state" should be viewed as a simultaneous product of the very processes Scott analyzes. David Blackbourn and Karen O'Neill have made similar arguments for Germany and the United States, respectively.[19]

Turning to cultural constructions of the nation, the Rhône's history reveals how ideas about nature, technology, and their relationship infused conceptions of the nation in France throughout the second half of the twentieth century. Certainly, religious, linguistic, racial, and other human-focused categories of inclusion and exclusion have been defining features of France's national community since 1945, all focal points of Anderson's own analysis. Yet politicians, technical elites, writers, and citizens also invoked ideas about nature and technology to define—and redefine—France. Controlling an unruly Rhône and manifesting technical prowess proved central to French postwar *grandeur*. However, neither these ideas nor their supporters were static. By the late 1960s, critics actually inverted these arguments, using environmental preservation and the cancellation of several projects on the Rhône to defend their alternative visions of France.

As Mukerji has shown for early modern France, one cannot underestimate the symbolic power of nature and its management to the definition and legitimacy of political regimes. Gabrielle Hecht has made parallel arguments regarding the relationship of technology and politics in post-1945 France. Yet, as Hecht and other scholars have shown, it is not only a question of symbolic power. At the same time, technologies and environmental management strategies offered powerful ways to inscribe the nation into what eventually came to be seen as infrastructure and the landscape itself, sometimes to the point where it became invisible and therefore taken-for-granted. Part of this power came from its very materiality. Questioning the technopolitics of what I call envirotechnical systems was more difficult precisely because of its physical instantiations in material objects and the landscape itself.[20]

The fact that the Rhône is a transnational river does not entirely account for the ways its remaking and French nation building, both discursively and physically, have been repeatedly entwined over the second half of the twentieth century. Periodically, Switzerland and France have negotiated over the Rhône's management, but even debates completely within France have involved questions of national identity and processes of state building. Of course, the history of the Rhône's transformations does demonstrate that international and transnational developments such as the Cold War, decolonization, and European integration heightened national anxieties in France, often spurring both the rhetoric and materialization of nationalism. Yet the Rhône's transnationality alone fails to explain these ongoing connections forged between the river's remaking and specifically French nation building. Questions of political sovereignty, politics, and power were, then, relevant not only when the Rhône crossed national borders but regardless of the correspondence (or lack thereof) between political and hydrologic boundaries.[21]

Indeed, moving from the narrower question of the relationship between nation and nature to the broader framework of politics and nature provides a distinct advantage: it historicizes the nation in ways that substantially extend Anderson's important insights. Anderson's analysis of the nation and its ongoing production is brought into even sharper relief by examining alternative political geographies considered, and sometimes even enacted, at the same time as the nation-state.[22] In this way, a constructivist perspective highlights not only how and in what ways nation building has occurred but also how "the nation" itself has become a relevant and dominant, but not always sole, political unit of analysis. For instance, the Rhône's transformations since World War II reveal how a national framework was simultaneously reproduced through the river's remaking. Yet other political geographies, including the region and a transnational Europe, emerged (in fact, reemerged) after 1945, thereby challenging the apparent primacy of the nation-state during this era. As I will show, the political geography of technological change and environmental management in the Rhône valley fragmented and shifted over the second half of the twentieth century, moving from the nation-state to a multifaceted, often multilayered geography of region, nation, and Europe. These links among nature, technology, and nation were, then, dynamic on several levels, and even the most basic political unit of analysis was at issue. In short, the nation and other political geographies were never a priori to the Rhône's remakings but were simultaneously articulated, reproduced, and maintained—or conversely, challenged—through that very process.[23]

That diverse groups in France framed the Rhône's transformations in regional, national, and European terms, sometimes simultaneously, thus shifts our attention from the nation-state to a broader, potentially far more inclusive connection between politics and nature, thereby suggesting that the nation may have been inadequately problematized.[24] Echoing elements of Samer Alatout's notion of bio-territoriality, while extending it beyond a focus on human populations, I emphasize a dynamic relationship between nonhuman nature and political territorialities. To be clear, my goal is certainly not a return to environmental or perhaps hydraulic determinism, thus risking the naturalization of the nation or any other political geography. Neither do I assume that the link between politics and nature is relevant only to "environmental" politics, nor do I simply extend Foucault's notion of governmentality to environmental management—what several recent scholars have variously called environmentality, ecogovernmentality, and ecogovernance. Rather, widening our analytic lens from nature and nation to politics and the environment serves to highlight how various groups in France implicated nature, both discursively and materially, in the formation and maintenance of political territorialities, regardless of their particular geographic framing. In other words, this framework opens up not only the politics of nature but also the nature of politics.[25]

Toward Envirotechnical Analysis

As the Rhône's history demonstrates, nature, technology, and nation were linked *historically*, thus raising important questions about how scholars have conceptualized them and their interrelationships. Envirotechnical analysis places these important issues in the foreground.

As the term itself suggests, *envirotechnical analysis* reflects the recent confluence of two fields, environmental history and the history of technology, building on important insights from both. One of environmental historians' primary contributions is their insistence that not all historical processes emanate from humanity. This argument is typically, if problematically, framed as the "agency" of nature.[26] The concept of the envirotechnical emphasizes the "nature" of technology, or the ways nonhuman nature affords material constraints to technological development and use, ultimately partly constituting "technology" itself.[27] Environmental historians also analyze anthropogenic environmental change, and envirotechnical analysis builds on this long-standing interest. Yet, as William Cronon and others have shown, wilderness was a founding myth of not only the United States but also U.S. environmental history.[28] Only more recently have many American environmental historians shifted their attention to cities, factories, and other landscapes that clearly bear the mark of humanity.[29] Envirotechnical analysis thus reflects American environmental historians' growing interest in managed environments and obviously "technological" processes such as industrialization, topics that European scholars began studying much earlier than their American counterparts.[30] When combined with the cultural turn in history, U.S. environmental historians have finally become less wedded to the idea of "Nature" over the past two decades.[31] Perhaps no concept better captures this turn than the envirotechnical.

Envirotechnical analysis, while it also builds on important work in the history of technology and in science studies more broadly, at first glance may appear to reverse recent trends within these fields. Most historians of technology "black box" the environment, treating it as an unproblematic, ahistorical backdrop to studies of technological change, implying that nature and technology are entirely distinct, and that environmental factors and ecological processes play no role in technological development.[32] These arguments, whether implicit or explicit, signal these historians' rejection over the past three decades of "technological determinism," a term often embracing two ideas: technological change is autonomous, and societal development is determined by technology. By renouncing technological determinism, historians of technology have echoed larger developments within STS that maintain that "nature" is actually the product of scientific and technological work, not the self-evident explanation of that work. These arguments are central to constructivist models of technology, which have dominated the field since the early 1980s.

In some ways, then, examining the envirotechnical seeks both to parallel and to refine the sociotechnical turn of the 1980s and 1990s, which yielded such prominent approaches as systems theory, the social construction of technology, and actor-network theory (ANT).[33] Arguably, one could integrate a critical conception of nature into these frameworks; ANT does just that, while revising Thomas Hughes's systems theory also has potential.[34] Yet the term "envirotechnical" calls attention to the entangled web of nature and technology just as early work on the sociotechnical stressed the inextricable ties between society and technology, thereby challenging these categories as distinct.

Highlighting the envirotechnical is precisely the objective of research that began to crystallize in the late 1990s in the work of a small but growing group of scholars—what has since become known as "envirotech."[35] Two sets of studies have emerged thus far. One set has focused on the historical production of organisms and landscapes, from domesticated wheat and industrial chickens to managed forests. These studies emphasized how such "biological" organisms are in fact *envirotechnical objects,* or material hybrids of species and technologies. They also suggested how ecosystems such as forests or rivers are *envirotechnical landscapes* (physical hybrids of ecological and technological systems).[36] The second set has highlighted the cultural meanings and representations of nature, technology, and their relationship. These studies have examined the cultural specificity of these ideas, how they have changed over time, and the political work they have performed for the groups that mobilized them. For example, as I will show, advocates of hydroelectric development after World War II framed the Rhône as a hydraulic object to help justify their mission.[37] Scholars working at the intersection of environmental history and the history of technology have therefore begun to develop envirotechnical analysis on two levels: the study of historical interactions between ecological and technological systems, both materially and discursively. Less common are studies integrating materialist and cultural analyses within a single case.

This preliminary scholarship has begun to demonstrate the value of an envirotechnical perspective, and from this collective work a few insights have already emerged.[38] First, these scholars have exposed the often invisible (and sometimes masked) "nature" of technology, a conclusion with vital implications for the biological and ecological entities implicated by a given technology.[39] Second, although practitioners in this subfield have generally not framed their studies in these terms, they have in fact examined the "life cycle" of technology—the contested process of technological design, development, use, and disposal—through the lens of environmental history: they have begun to explore technological objects and systems as *artifacts* of human-natural relations. However, drawing on valuable perspectives from technology studies, this scholarship has also suggested that these relations are partly constituted through technological development and use. Technological objects and systems are therefore also *mediators* of human-natural interactions: technologies shape the experiences that individuals, social groups, and communities can have with nonhuman nature.[40] These insights therefore push Arthur McEvoy's argument that "technology is the tangible instrument . . . it is the point of interaction between the human and the natural," so as to frame technology, society, and nature as mutually constitutive.[41] Integrating sociotechnical and envirotechnical analyses, then, suggests how human-natural relations are embedded within the sociotechnical and vice versa, providing further evidence that these approaches are complementary.[42]

To date, there has been relatively little discussion of the analytic premises, methodology, or theoretical implications of envirotechnical analysis. General consensus has yet to emerge around what envirotechnical analysis means (or should mean) or what an envirotechnical approach entails (or should entail). Furthermore, tensions between core tenets of STS and environmental history have been left largely unaddressed. The main question to be addressed, let alone theorized, is how scholars can treat *both* nature and technology critically without resorting to the determinism, reductionism, or realism of early work in both fields.[43]

Before turning to that central tension, it is worth noting that most environmental historians and STS scholars do share several important intellectual commitments. For one, constructivist frameworks are central to both fields. For example, as scholars in technology studies have moved away from technological determinism, they have adopted several theoretical approaches, but the majority has asserted and continues to emphasize the social, cultural, and political *shaping* of technology. Meanwhile, the social or cultural construction of nature, which helped undermine the idea of a transcendent "Nature," has become an important premise of most environmental histories in the form of two recognitions: first, that anthropogenic forces have profoundly affected "nonhuman" nature; second and perhaps more important, that the category itself is historically and culturally contingent. Environmental historians have highlighted, then, the human shaping of nature, both literally and discursively.[44]

Both fields also share a strong interest in the role of the nonhuman as not only an object of scholarly inquiry but also a force in shaping historical and contemporary worlds. Although scholars in technology studies are still committed to exposing the sociopolitical contingencies of technological development, more now acknowledge the constraints that artifacts and systems present. They assert that technological objects and networks cannot be reduced *only* to their social, cultural, or political dimensions. In other words, these critics insist that the materiality of technology also matters.[45] Meanwhile, environmental historians maintain that human interactions with the environment constitute a vital part of the historical record that scholars should seek to recount and understand. Yet nature is not simply a passive object of history. Rather, one of the field's foundational arguments is that nature is an agent, actor, "active player," or "force" in that very history. For these reasons, most environmental historians would probably sympathize with Bruno Latour's critique of social constructivist approaches because they favor a priori the social and endorse his notion of "actants" because it is amenable to the inclusion of nonhuman nature.[46] Scholars of STS and environmental history share a materialist orientation, then, although environmental historians particularly highlight "natural" materiality.

There are significant analytic affinities between environmental history and STS, but here scholars in the two fields generally part ways, along several fronts. For one, science studies scholars have developed more critical stances on environmental knowledge and change, at least as presented by many environmental historians, for several reasons. Environmental historians often rely on scientific and technical documents in order to historicize and assess environmental change. Although there is certainly variation in how environmental historians treat these records, they are frequently not scrutinized and contextualized in the same way as other historical documents. For many environmental historians, the fact that that these documents are "scientific" or "technical" grants them a privileged epistemological status.[47] In contrast, STS scholars analyze knowledge and the process of knowledge making to a far greater extent than most environmental historians. Even the apparently self-evident concepts of "the environment" or "environmental change" are historically and conceptually problematic within STS.[48]

Science studies scholars are also more likely to question claims about the historical agency of nonhuman nature than are many environmental historians. For instance, invoking agency is a powerful move for both historical actors and analysts. One must be wary of accepting actors' assertions of agency at face value. "Nature's" agency might actually be a strategic argument made by certain (human) groups for particular ends. In the case of the Rhône, attributions of the river's power helped justify its "taming." Consequently, it is important to avoid conflating historical actors' and analysts' invocations of agency, especially with regard to what political work might be gained from such claims (see Chapter 3). Overall, most environmental historians subscribe to a weaker notion of constructivism that shapes their treatment of technoscientific evidence and their understanding of and claims about the agency of nonhuman nature.

Although the stronger form of constructivism within STS offers important insights and significant cautions to scholars in environmental history, the latter also provides valuable perspectives for those working in STS. In fact, environmental history helps expose one of the main limits of social constructivist frameworks within STS. Analyzing the Rhône's floods illustrates this point. As a bumper sticker might say, "floods happen." And as Chapter 3 demonstrates for the Rhône, the category of a "flood" (or conversely, "normal" conditions) and the rankings of hundred- or thousand-year floods are contingent. They depend on assumptions about a river and how it will flow, assumptions that may say more about that culture than the nature it purports to represent. The categorization and hierarchy of floods also depend on the knowledge, practices, and tools employed in the calculation of a river's "average" and "extreme" flows. Scholars in both STS and environmental history therefore share a view of floods as cultural constructs. Moreover, floods, like many "natural" events, are usually the products of *both* "natural" and "cultural" processes. For instance, the construction of levees might reduce flooding in certain areas but exacerbate it in others. In fact, as Timothy Mitchell shows in his rich analysis of Egypt's malaria epidemic in 1942, it may be impossible to ascertain where "natural" processes end and "cultural" ones begin.[49] Indeed, if the Rhône's history is any indication, historical actors may attempt to distinguish between "natural" and "cultural" causes in order either to reduce their political, financial, and legal responsibility for damages or to bolster their position as claimants.[50]

Yet, despite the many cultural and historical contingencies of the category "flood," and the confusing, inextricable matrix of "natural" and "cultural" factors that might be mobilized by actors to serve their own interests, rivers will run high on occasion, and human ideas and social processes are not the only things at work in what comes to be called a flood. One of environmental historians' real strengths, and perhaps the field's unique contribution to sociohistorical analysis, is their insistence that we hold onto this point: nonhuman nature may be profoundly mediated and constructed, both literally and metaphorically, but it is not wholly reducible to culture. Nature and cultural meanings of nature are utterly and problematically entwined, but they are not synonymous. The cultural turn within history and related disciplines has led many scholars to abandon the idea of nature that exists outside of culture; hence the prolific references to "nature" in the humanities and certain social sciences.[51] Most environmental historians, however, are unwilling to concede this argument; in fact, several leading scholars have fought vehemently against it.[52]

As constructivist (versus social constructivist) models have gained currency within STS over the past decade, these two fields appear to be converging in complementary ways, as illustrated by the emergence of "envirotech." A number of science studies scholars have expressed renewed interest in the technoscientific mediation of social practices, while most environmental historians continue to insist that nature cannot and should not be reduced to a cultural construct.[53] The challenge, then, is to uphold both fields' chief contributions—nature, like science and technology, is constructed (STS) *and* nature is not only constructed (environmental history)—at the same time. In other words, the trick is to open two black boxes, nature and technology, *simultaneously* without resorting to determinism or realism, or relinquishing the argument that nonhuman nature presents constraints to both technological and historical possibilities.[54]

In seeking to refine how scholars conceptualize technology, nature, and the relationship between them, technology studies offers at least one useful technique that practitioners in this specialty have used to frame the relationship between two pivotal conceptual pairs within the field: technology and culture, and technology and politics. This is the idea of co- or mutual construction—what, within STS, is broadly called the idiom of co-production.[55] However, a crucial difference between these two sets of binaries, and the dichotomy of technology and nature, is that the established binaries often get mapped onto a material/culture divide: technology is "obviously" material, while culture and politics are "evidently" social. Yet, as Hecht and others have shown, it is precisely these dichotomies and their associations that STS scholars challenge. As Hecht argues, rather than buying into the material/culture divide, scholars should explore the social and cultural dimensions of technologies while examining the ways the cultural and discursive can become materialized.[56] Rethinking nature and technology, as envirotechnical analysis does, risks another false dichotomy of materiality and culture or, even worse, falling back on realism, such that nature serves as unproblematically real while technology might be viewed as at once material and cultural. Combining contributions from technology studies and environmental history instead suggests that both technology and nature are at once material and discursive. Moreover, this kind of analysis conveys the need for a deeper historicization and interrogation of the categories of "nature" and "technology" themselves.

In fact, the history of the Rhône's transformations demonstrates that diverse groups strategically categorized the river and represented its multiplicity (or singularity). They used ideas of "nature," "technology," and "society" and formulated relationships among them tactically, sometimes holding apart these concepts, other times merging them. Tensions also emerged between actors' frameworks and practices on the ground—and in the water. To allow for the Rhône's multiplicity, both historically and analytically, I have developed the concepts of envirotechnical systems and envirotechnical regimes, which I elaborate below. They are clearly analyst's categories, not those of historical actors, although, as we will see, certain groups did move at times to "unite technology and nature" (see the Conclusion). Nonetheless, these concepts offer a productive framework for the Rhône's multifaceted, contested past by drawing attention to the strategic invocation, maintenance, and erasure of these categories and their material instantiations over time.

Conceptual Tools

In writing the history of the Rhône's transformations, I have developed several conceptual tools to help analyze the cultural and physical connections among nature, technology, and nation forged by historical actors. These tools are rooted in the history of technology and environmental history and integrate valuable questions, methods, and approaches from each field. This theoretical framework is the book's third and final level of analysis.

First, I turn to my historical actors and their perspectives. How did *they* define "nature" and "technology" and think about their relationship? As numerous environmental historians and historians of technology have shown, ideas of nature and technology are contingent. If scholars use static, monolithic definitions to gauge the past, they risk ignoring, if not misrepresenting, its historical and cultural specificities. Furthermore, it is not only unproductive but also problematic to ask "Where does nature end and technology begin?"[57] To pose that question presumes a universal, ahistorical definition of what nature and technology are (and by extension, what they are not). The question and its answer can only reflect the perspective of the questioner, not that of the historical subject. Instead, scholars need to investigate and analyze these very concepts rather than taking them as self-evident, at once employing these categories to help frame their own analyses while being mindful of the definitions of our historical actors and the often strategic ways they invoked these concepts.[58] After all, their ideas and assumptions informed the decisions they made.[59] By remaining sensitive to both actors' and analysts' categories, we can begin to understand "nature" and "technology" through the eyes of our historical subjects.[60]

My second analytic tool is the concept of *envirotechnical systems*.[61] I define this term as the historically and culturally specific configurations of intertwined "ecological" and "technological" systems, which may be composed of artifacts, practices, people, institutions, and ecologies. I have chosen "enviro" (shorthand for "environment") rather than "eco" (for "ecosystem") for two reasons. "Environment" tends to connote surroundings that may not be limited to "natural" features. As this point suggests, I use "environment" to emphasize that human societies shape the vast majority of ecosystems, either directly or indirectly. This is certainly presently true, as it has been for most of human history.[62] For this study's purposes, the Rhône's environment consists of the river and its watershed but extends beyond these boundaries when decisions and actions outside the river's hydraulic system affected or were affected by it. My use of "technical" (shorthand for "technological") follows recent work in the social and historical studies of technology that posits an inclusive definition of technology: the knowledge, skills, and tools of making or doing something—a definition more closely aligned with the French *technique*, which seamlessly combines technical knowledge, practices, and objects, than the Anglo-American, artifact-centered "technology."[63] Several relevant technologies in this history include hydroelectric plants, irrigation networks, the practices locals and engineers used to operate these artifacts, and perhaps most controversially, even the Rhône itself.[64] Drawing on the influential work of Thomas Hughes and subsequent scholarship in the history of technology, "systems" describes a network of interrelated and interacting parts of a larger whole.[65] For my purposes, envirotechnical systems include what both actors and analysts might usually label "nature" and "technology."[66] Although it may not be explicit in the term itself, envirotechnical systems therefore encompass not only "nature" and "technology" but also all of the social, cultural, and political dimensions of "technology" that historians and sociologists of technology have ably explored over the past three decades.[67] I have consciously chosen *envirotechnical* systems over alternatives such as "technoenvironmental" because the term acknowledges that nonhuman nature, however altered physically and mediated discursively, did come first.[68]

Envirotechnical systems are, therefore, inextricably embedded environments and technologies that continually reshape individual parts of the system and the whole. Environmental conditions constrain (but do not determine) technological possibilities.[69] For instance, engineers hoped to maximize hydroelectric generation after World War II, but they had to work within both the Rhône's fluctuating hydraulic regime and their knowledge of it. As Chapter 3 shows, the hydroelectric plants' designs reflected these constraints. Moreover, new opportunities and new challenges for the development and operation of those technologies emerged as "environmental" (in fact, envirotechnical) conditions shifted. For example, Chapter 7 traces how existing nuclear facilities on the upper Rhône mediated the design of hydroelectric projects during the 1980s because the reactors required the river's flow to be within a specific thermal range. This example suggests how the process of technological development and use may result, in turn, in new "environmental" (again, envirotechnical) contexts, both intended and unintended.

Time, place, culture, ecology, and technology delineate the specific features of a particular envirotechnical system. In this study, France, World War II, and river ecology are all relevant (but not exclusive) factors to understanding the Rhône as a set of envirotechnical systems. Let me emphasize that the present-day Rhône does not "have" an envirotechnical system. It *is* one. And to be more accurate, it is several overlapping systems.[70] This concept thus extends Richard White's notion of "the organic machine" beyond either the "natural" dimensions of "technology" or the "technological" features of "nature" to challenge the very boundaries between both these categories and artifacts. To better understand the envirotechnical character of the Rhône today, I discuss pre-1945 envirotechnical systems in Chapter 1, while subsequent chapters trace the emergence of and modifications to these systems over the second half of the twentieth century.

I refer to envirotechnical systems in the plural for several reasons. First, the plural underscores the diverse and historically contingent forms such systems may take. Both medieval mills and hydroelectric dams tapped the Rhône's flow to produce energy, but the specific features of these envirotechnical systems clearly differed. Second, the plural suggests that several systems may coexist or compete within, or throughout, scales and societies. The development of France's nuclear power program, including the completion of several reactors on the central Rhône during the postwar era, created an envirotechnical system oriented for the production of atomic energy and weapons-grade plutonium. Yet it actually relied on a preexisting system built to generate hydroelectricity, in at least two ways: hydroelectric plants produced ample supplies of electricity while regulating the Rhône's flow, thereby assuring more constant supplies of water for the reactors' cooling processes. But in turn, these new nuclear facilities ended up mediating the management of the hydroelectric plants. Because these two envirotechnical systems literally intersected, as part of the Rhône's flow passed through turbines and reactors alike, engineers had to negotiate the sometimes conflicting demands of both. As these points suggest, the plural "systems" accentuates the concept's flexibility and therefore utility to include quite diverse ecological and technological systems. Consequently, it offers a framework within which to explore their similarities and differences.[71]

In elaborating the concept of envirotechnical systems, I develop two corollaries. The first is the idea of *technology as natural*.[72] Since technology is usually associated with humanity, this corollary challenges the tidy divide between nature and culture. It also questions the supposition that the development and use of technology replace nature. As Edmund Russell has succinctly characterized it, the popular assumption (at least before the rise of biotech) has been that nature must be dead to be technology.[73] This idea presumed that technology is entirely distinct from and opposed to nature.[74] Consider, however, as historians of technology do, the design and development of any given technological artifact. It was created from organic and nonorganic materials, even if they have already been radically transformed. Earth and valuable minerals eventually became, say, the concrete walls and turbines of hydroelectric plants dotting the Rhône valley. Production of those artifacts also involved natural resources and ecological processes. Manufacturing concrete and the components of hydroelectric plants required electricity, which had to come from somewhere, whether rivers, coal, or uranium. The production process itself tends to mask how these technological artifacts are connected to and thus dependent on the environment, both in terms of their material substance and the processes by which they came into existence. Ultimately, "technological" artifacts are envirotechnical ones.[75]

One might likewise analyze "technological" systems as a whole. They are embedded within, while simultaneously reshaping, their "environments."[76] By definition, hydroelectricity depends on water. Yet hydroelectric production involves remaking aquatic environments, often radically, to better suit them to energy generation. The systemic character of envirotechnical networks also has the potential to alter the spatial dimensions of human and nonhuman relations. Electricity produced in the Rhône valley, whether hydroelectric or nuclear-generated, can be transferred over long distances. Growing energy demands in eastern Europe and the creation of transnational energy networks, for instance, might have, then, significant consequences for the management of rivers in western Europe like the Rhône.[77]

Framing technology as natural seeks to highlight the manner in which technological artifacts and systems are thoroughly implicated in/by/with ecologies. (In fact, the difficulty in describing this "relationship," not to mention that problematic term itself, alludes to the ways this binary permeates language and thinking.) As such, these technologies are part of the environment, even though they may be substantial transformations of nature and contribute to significant environmental change. Seeing technology as natural therefore illuminates the dependency of technology, and thus humanity, on the natural world, which the processes of technological development, production, and use and the artifact itself often obscure. While historians of technology now see technology as thoroughly political, social, and cultural, I argue that we must also perceive its environmental dimensions—without returning, however, to either technological or environmental determinism. The idea of envirotechnical systems provides a way to illuminate those dimensions.[78]

My second corollary inverts the first. In addition to seeing technology as natural, I suggest that we think about *nature as technological*.[79] This challenges the argument that technology is what nature is not, and fundamental Western assumptions about the separation of nature from culture. Rather, humans can, have, and do appropriate nonhuman nature, whether Hegel's "first" or "second" nature, as technology; that is, humans use biological organisms, environmental features, and ecological processes in order to make or do things.[80] The fertility of a river's floodplain increases agricultural yields, so farmers around the world have developed practices to take advantage of its benefits. As the Rhône's history suggests, a river's flow may be harnessed to turn waterwheels, spin turbines, and cool nuclear reactors. Thus, natural resources and ecological processes can simultaneously be technological. As a result, the corollary expands "technology" further, not to mention who (or even what) might be included as a technological actor.[81]

This idea of nature as technological can also be more abstract. As Chapters 2 and 3 demonstrate, the postwar French state fostered the hydroelectric development of the Rhône. The river, redesigned to maximize energy production, thereby served state objectives of reconstruction, industrialization, and modernization. This example indicates the way state officials used and ultimately reshaped the river for political and economic aims. The managed Rhône essentially became a technology of the postwar state, serving as the envirotechnical means to these ends. As other chapters show, the Rhône played parallel roles in the emergence of the atomic age and regional economic development in France.[82]

My third and final tool builds on the concept of envirotechnical systems while focusing on the social relations and power structures behind their development, production, and operation. Here I propose the idea of *envirotechnical regimes:* the institutions, people, ideologies, technologies, and landscapes that together define, justify, build, and maintain a particular envirotechnical system as normative. If the notion of envirotechnical systems is largely descriptive—that is, it *describes* the particular features of embedded ecological and technological systems—then envirotechnical regimes are the *prescriptive, instrumental* formulation and use of those systems by groups or institutions for specific ends. The plural again emphasizes that multiple regimes are possible within and throughout periods and cultures.

I employ the regime metaphor, drawing both on Hecht's explanation of its effectiveness in her discussion of technopolitical regimes and Michelle Murphy's insightful "regimes of (im)perceptibility," to call attention to the systemic, prescriptive, and thus profoundly political character of envirotechnical systems. After all, "regime" implies resistance from within and without.[83] Moreover, some of the power of an envirotechnical regime resides in its system. The material manifestation of a regime helps to sustain—perhaps even to naturalize, both literally and metaphorically—the authority and interests of that regime, thereby making it more difficult to challenge. This final conceptual tool therefore stresses the *politics* of envirotechnical systems by highlighting the power relations behind their development, implementation, and use.[84]

The regime concept is (and should be) central to what I am calling envirotechnical analysis. Describing envirotechnical systems without paying attention to the history and politics of their formulation, production, and implementation may be insightful but is ultimately inadequate. In contrast, examining the power structures that shaped the design and realization of envirotechnical systems reveals the vested interests they serve. In fact, by neglecting the regime that sustains a given system, one risks naturalizing (and therefore depoliticizing) that very system. The concept of power not only is central to envirotechnical regimes (and by extension envirotechnical systems) but also helps foreground the tactical separation or conflation of "nature" and "technology" by people in the past. To date, scholars working at the intersection of environmental history and technology studies have tended to focus their attention on identifying envirotechnical systems. But by investigating the politics of their constitution and operation, they have an opportunity to bring all the vital insights of social, cultural, and political studies of technology to envirotechnical analysis while incorporating nature into their inquiries in a reflective, critical way.

Chapter 1 sets up the remaking of the Rhône since 1945 by examining the river's place in French culture, society, and political economy before World War II. Focusing on the Rhône's management since the seventeenth century, this chapter highlights how the river's uses and users and the role of the French state shifted, especially between 1870 and 1945. Although many politicians and engineers proposed bold visions, political, economic, and technological factors together constrained human management of the Rhône until the mid-nineteenth century. At that time, the French state began to expand its authority over rivers with significant military and commercial value. Meanwhile, long-standing disagreements among agricultural, navigation, and industrial interest groups worsened when entrepreneurs began to build hydroelectric plants along the upper Rhône in the 1870s.

After fifty years of sustained conflict, these diverse constituencies attempted to find common ground, or rather, a common river that might suit their multiple interests, even if they recreated the river in the process. To do so, they aimed to remake the Rhône into an envirotechnical system that would negotiate and ultimately reconcile various demands. The eventual result was the Compagnie Nationale du Rhône (CNR). Established by the state in 1921 and formally constituted in 1934, the CNR blended public and private participation with a mission of multipurpose development—the simultaneous pursuit of electricity, navigation, and agriculture—while aiming to meet both regional and national needs. In short, the CNR's program sought to resolve the historic conflicts among the river's users over its potential purposes. And for a few years, at least, the agency largely met these goals.

World War II was a major watershed in the history of the Rhône and its management. The war wrought massive destruction and challenged France's place in the world. In the Rhône valley, the exigencies of war and occupation put the river's transformation on hold, although, remarkably, planning continued apace. Chapter 2 traces how the Rhône's development during the postwar era attempted to regenerate France economically and culturally. It analyzes political, technical, and popular discourses about the river, showing how state elites, writers, and citizens all enrolled the waterway in nation building on both material and cultural levels. In the process, they constructed the Rhône, literally and metaphorically, as the nation's river.

State agencies and engineers attempted to realize this goal. Hydroelectric production boomed, agriculture became industrialized, and France joined the world's nuclear powers. All of these envirotechnical systems depended on, yet reshaped, the Rhône's "ecological" features. Chapter 3 examines how hydroelectricity, the atomic age, and agriculture all remade the postwar river—to a level unprecedented in scale and scope—a highly reengineered landscape of overlapping and often competing envirotechnical systems.

These projects may have transformed the Rhône, sometimes radically, but the process of technological development was quite complicated and contested, a point the final results often obscured. Many of the old conflicts among users resurfaced in new ways. In fact, there was little consensus over what postwar development meant, even for state officials. Government bureaucrats involved in agriculture, economic development, energy, navigation, and nuclear power agreed on the abstract objective of national reconstruction and modernization, yet fierce conflicts over the allocation and management of the Rhône's flow, the focal point of Chapter 3, demonstrated that these groups did not agree on the meaning of modernization. These debates hinted at deeper struggles over the redefinition of France after World War II, as the nation's leaders and citizens considered how to balance agriculture and industry, tradition and modernity, the nation and an increasingly integrated world. In the short term, however, France's desperate need for energy initially favored hydroelectric generation over all other objectives. As a result, state engineers began to redesign the Rhône primarily as a hydroelectric river to fuel the nation's reconstruction and industrialization.

These conflicts within the French state revealed fractures in the ostensible consensus over the Rhône's postwar transformation and the creation of the nation's river, but there was also opposition on the ground—the focus of Chapter 4. Local communities, both human and nonhuman, presented challenges to the ambitious postwar vision. The area's human inhabitants had ambiguous views of the river's remaking.

The environmental changes that locals attributed to the projects partly explain this ambiguity. Technical elites proved unable to predict, let alone control, the Rhône's hydrology. While expanded irrigation and subsidized electricity benefited many locals, consequences of river development threatened the livelihood of others, including some farmers and fishers, and the fish themselves. This chapter explores these local responses to the Rhône's postwar development. Together, intrastate conflict, community opposition, and environmental change all exposed just how difficult it was to construct the nation's river.

During the 1950s and 1960s, French centralization came under increasing fire. Chapter 5 examines the Rhône's continued development amid debates over the spatial distribution of political and economic power in France. In fact, the Rhône valley became a laboratory for regionalization nationally. When critics challenged the primacy of the nation-state, they offered new spatial frameworks for the river's transformation. Regionalists claimed that the Rhône would generate greater autonomy for southeastern France, especially following decolonization, while transnationalists asserted that it would secure an important role for the country within a newly unifying Europe. Even critics agreed that the destinies of the nation and these alternative political geographies were ultimately entwined. As these groups began to complicate the idea of "the nation," they also began to question "the nation's river."

By the late 1960s, then, the postwar vision had begun to weaken on several fronts. These challenges did not end the development of the Rhône, however. Instead, they offered new ideas about its remaking, including opposition to development itself. Chapter 6 traces how politicians, engineers, and citizens came to reframe, even reconsider, the meaning of the river's transformation. In a fascinating inversion of the postwar framework, advocates of new large-scale projects mobilized the region and Europe to legitimate the Rhône's continued remaking while critics actually invoked the nation for their antidevelopment cause.

Chapter 7 outlines the remaking of the Rhône between the early 1970s and mid-1980s, showing how these new perspectives guided the latest wave of river development. The design of these later projects reveals how the hydroelectric paradigm had weakened, agriculture received greater investment, and ecological and recreational objectives achieved more importance. A new modern thus began to govern the Rhône in the 1970s, as evidenced by both completed and unbuilt CNR projects. The emergence of this perspective did not mean, however, that postwar high modernism had entirely disappeared. The projects of the late twentieth century, both those carried out and those left unrealized, did embody these new concerns. Yet they did not fundamentally challenge the previous four decades' transformation of the river but ultimately reconciled late-twentieth-century environmentalism with the ambitious visions and transformative technologies of the postwar era.

The Conclusion considers the complex and often ambiguous legacies of the Rhône and its repeated remaking since 1945. Since the early 1990s, restoration has reshaped but not radically altered the envirotechnical systems built in the Rhône valley after World War II. For instance, although an ambitious scheme to modernize the link between the Rhône and Rhine rivers was cancelled in 1997, since the 1990s, France and Spain have repeatedly debated a proposal to divert part of the Rhône's flow to supply the parched region around Barcelona. A Rhône-Barcelona aqueduct would essentially extend the existing water diversion network in south-central France across the border into northeastern Spain. Such an extension would echo the ambitions and technopolitics of the postwar era that united hydrologic and technological systems in the name of the nation while putting them in the service of the transnational, European politics of the late twentieth and early twenty-first centuries. As this proposal suggests, postwar high modernism has begun to travel far beyond France. Many colonial and postcolonial governments have sought technical assistance and financial aid in developing their rivers and other water resources, encouraged by global powers, international institutions, and other influential forces. Several agencies involved in the Rhône's remaking have exported their expertise and technologies of river management around the world, especially in the global South. The envirotechnical systems formulated and built in the Rhône valley have therefore circulated well beyond the political frontiers of France. Although aspects of high modernism have been called into question in the West, this ideology has thus proven versatile transnationally and had a remarkable second life in the global South.

1

ENVISIONING A NEW RHÔNE

Between 1945 and 1986, the Rhône became the river that is now familiar to those born since World War II, but vestiges of its long, entwined human and natural histories remain today. The Romans first built embankments along the river as it flowed through Lugudunum, modern-day Lyon.[1] During the Rhône's frequent floods, residents etched dates on the walls of buildings to indicate high-water marks, inscriptions that lingered long after floodwaters had receded. The names of places throughout the Rhône valley—from Aigues-Mortes ("dead waters"), which was once located along the Mediterranean shoreline, to Les Brotteaux ("muddy waters"), a now posh Lyon neighborhood built on reclaimed wetlands—all recall the river's sometimes distant ecological past.[2]

While traces of these histories remain, the modern Rhône is undoubtedly most visible today. Huge hydroelectric dams blocking lengthy diversion canals and the gigantic cooling towers of nuclear plants along the river's banks dominate the landscape. Large pipes connect the Rhône to petrochemical factories, while homogeneous suburbs now sprawl around ancient villages nestled along the river. Expansive irrigation systems disperse some of the river's flow throughout the countryside, plotted neatly into the rectangular quilt of industrial monoculture. It took enormous effort—technological, political, and cultural work carried out by politicians and bureaucrats, engineers and hydrologists, farmers and merchants—to remake the Rhône.

Since 1945, construction of these large-scale, state-sponsored projects has substantially transformed the Rhône. These projects helped create a different river, which recombines ecological and technological systems and resulted in a new envirotechnical landscape in the Rhône valley. They ultimately helped facilitate the realization of a new France, but only after seventy-five years of intense debate over the Rhône and its future. At issue were what purposes a river should serve, which objectives should guide the Rhône's management, and who should gain from its remaking. The fact that goals often conflicted with one another only complicated these deliberations. The state's role in this process also shifted. Especially after World War II, state elites involved in the Rhône's transformation faced two new challenges. First, they had to generate designs for individual projects rather than simply formulate an abstract mission for the river's management. Second, the war and its legacy had altered the context of the Rhône and its development, forcing them to adapt their mission as well as the design of specific projects to reflect the challenging realities of postwar France.

To understand what happened after World War II, we need to return to the "former" Rhône (a phrase the CNR's hydraulic engineers used to refer to the river's historic channels) because the Rhône's postwar story flows from this much longer history of river management. After sketching out these early visions of large-scale development and exploring the role of the state before 1945, this chapter offers a brief institutional history of the agency that eventually played a major role in the river's remaking, the CNR. I also examine the philosophy that defined the CNR and became inscribed in its projects: multipurpose development, or projects that promoted energy, navigation, and agriculture simultaneously. Finally, we see how, despite the CNR's multipurpose mandate, the legacies of World Wars I and II, the agency's financial structure, and conflicts among state agencies together made hydroelectric generation the CNR's primary goal between 1945 and the early 1960s.[3] The pre-1945 era proved critical in shaping both the vision of river development after World War II and the state's role in that process.

River Management and the State in Modern France

Although many of the country's largest public works projects, including river reengineering, date to the second half of the twentieth century, the territory that became modern-day France reflects a long-standing relationship among political authority, technological development, and environmental management—or envirotechnical regimes. River development proved important to both the early modern and modern French states. In fact, historians have described the ancien régime's pursuit of *la politique de l'eau*. Large public works and civil engineering projects that remade the

French landscape, especially those pertaining to water, such as the Canal du Midi, helped consolidate royal authority into absolute monarchy.[4]

In the late seventeenth century, King Louis XIV and his influential advisor Jean-Baptiste Colbert strengthened the monarchy's jurisdiction over France's rivers. Because the early modern French state classified rivers primarily by navigability (a precedent rooted in Roman law), a combination of ecological characteristics and cultural perceptions of these features formed the basis of juridical distinctions between rivers. A river had to be sufficiently wide, deep, voluminous, and strategically located to merit the status of "navigable." This legal standard reflected the monarchy's military, political, and economic interests in controlling certain waterways. After a protracted struggle with provincial seigneurs, the king reserved what were perceived as France's most important rivers for the Crown. According to the 1669 Edict Concerning the General Regulation of Waters and Forests, "navigable and floatable" rivers belonged to the monarchy as "domainal" waterways, while nonnavigable rivers remained under the seigneurs' jurisdiction. This edict also specified that the state's domain included the waters and beds of navigable rivers. The monarchy did not control the banks of navigable waterways, but they were subject to easements and possible claims of rights-of-way. The 1669 edict therefore attempted to reverse the feudal parcelization of navigable waterways by placing management of France's major rivers under the firm control of the monarchy.

As part of his regime of territorial management and *la politique de l'eau,* Louis XIV expanded his royal administration in order to oversee his growing holdings. The administration of waters and forests (Administration des Eaux et Forêts) dated to the thirteenth century, but the Sun King increased its powers. Meanwhile, a navigation service within the Ministry of Public Works oversaw the management of all navigable waterways in a separate administration.[5] Although the 1669 edict aimed to ensure the monarchy's control of important rivers and centralize elements of river management through these two administrations, its management of domainal rivers was ultimately divided between the agencies. The state's fractured authority over France's rivers largely continued until the 1960s.

Jurisdiction over rivers became even more complicated when artificial waterways transformed the landscape of early modern France. The reign of Louis XIV witnessed significant investment in public works, and the king's ambitious canal-building campaign created new rivals to natural waterways. The Canal des Deux Mers, better known as the Canal du Midi, epitomized this trend. It linked the Atlantic Ocean with the Mediterranean, creating an alternative route to the sometimes dangerous circumnavigation of the Iberian peninsula. Such projects facilitated commerce while embodying the territorial ambitions of the French monarchy and enacting state building through the construction of public works that might integrate the country's potentially rebellious periphery.[6]

Although late-eighteenth-century revolutionaries sought to overthrow the monarchy, they too mobilized rivers in reforging the political order, in hopes that nature might provide an objective and more stable basis than the country's historic reliance on blood, title, and privilege. As Josef Konvitz argues in his study of cartography in France between 1660 and 1848, "advocates of maps in France saw nature as a valid template for human activity. They assumed that to the extent that economic activities and developments conformed to France's topography, France would become more prosperous. The political corollary of this concept was that political boundaries based upon natural divisions were in fact natural, and hence logical, rational, and stable."[7] When these revolutionaries attempted to transform the foundation of France's government and society, rivers became founding precepts of the new administrative unit of the state: the *département* (department). Surveying the nation's territory, including its rivers, helped determine the particular boundaries of these new departments. Revolutionaries therefore aimed to ground the new state in its physical territory, and particularly in its rivers.

On some level, this naturalization of the emergent French nation-state was a revolutionary break from the Old Regime.[8] The rational, the objective, and the natural—indeed, closely entwined concepts for eighteenth-century philosophes—helped guide the reorganization of sociopolitical order in France after 1789; at least, idealistic supporters of the revolution asserted this was the case. Yet there were continuities in the relationship between political order and river management before and after the French Revolution. The law of November 22–December 1, 1790 declared, for instance, that navigable and floatable rivers were part of the public domain and, consequently, could not be owned by private parties. The new government therefore seamlessly transferred navigable rivers from the domain of the prerevolutionary monarchy to its own control. The influential Code Napoléon of 1804 reaffirmed the central government's rule over navigable waterways, stipulating that roads, streets, highways, and ports in the state's charge, as well as "navigable or floatable rivers and streams ... are considered parts of the public domain," adding that they "are not susceptible to private property." While the range of waterways declared part of the public domain under the 1804 Code was not as broad as Roman law, the state nonetheless preserved exclusive rights to navigable waterways in the early nineteenth century. Moreover, its authority contrasted with both English and American jurisdictions.[9] Centralized political authorities, whether monarchical or those born of revolution, aimed to control France's major rivers.

Laws passed during the late eighteenth and early nineteenth centuries not only affirmed but also widened the state's control over navigable rivers. French water law included riparian rights, which held that those living along the banks of a river could claim privileges to use its flow. In fact, owning land along a river was the only way for individuals to hold a riparian water right. Water rights thus flowed from landed property rights. In contrast, the prior appropriation doctrine popular in the nineteenth-century U.S. West separated land ownership from the ability to claim water rights.[10] While riparian doctrine may have implied that early-nineteenth-century property owners maintained free reign over "their" water, in practice, laws, historic royal grants, and local customs together constrained the "private" appropriation of river flows in France.

Property owners maintained rights to nonnavigable waterways, but they were conditional. They had to observe local custom and balance the interest of agriculture, which was perceived as a public good, with that of private property. Furthermore, two laws passed during the early years of the revolution broadened the definition of public good, expanding the state's jurisdiction over "private" water rights. A 1790 law granted the government control over any waters that served the common good, while a 1797 decree authorized it to assure the free flow of rivers and irrigation canals. Constructing any potential obstacle or taking irrigation water, even from nonnavigable streams, therefore required state authorization. Then the Code Napoléon made illegal any act that caused a stream to flow differently from the way it had the previous summer. Because downstream parties held rights to water that flowed from lands uphill, upstream property owners were obliged to act with downstream property owners in mind. Sometimes called "natural flow theory," this restriction generally discouraged significant changes in river management practices. As a result, late-eighteenth- and early-nineteenth-century laws tended to promote continuity of use—a sort of ecological as well as economic status quo—over new development. In this respect, the early postrevolutionary state generally preserved a river's "natural state." Although all of these laws permitted property owners along a nonnavigable river to continue using its waters, they also circumscribed "private" rights.[11]

Overall, the postrevolutionary era witnessed a polarization of river management between the state and private property owners, mediated relatively little by local authorities. Because the central government appointed prefects, or heads of departments, these representatives were more closely aligned with national interests than local concerns. As Correia writes, "this system (State v. Property) left less authority for local authorities and to the local management of common property than elsewhere in Europe."[12] Expanding state oversight, however, increasingly blurred this distinction between "state" and "property."

The state's authority in river management had grown over the nineteenth century, but no single engineering corps within a particular government agency oversaw France's rivers. Impressed by the Netherlands' long tradition of hydraulic engineering, Napoléon considered establishing a special separate corps of water engineers.[13] Such a "water corps" might have unified diverse users and uses of water within one government agency. However, neither Napoléon nor any government since created a single bureau devoted to river management, let alone all kinds of water management across France. For instance, Eaux et Forêts remained part of the Ministry of Agriculture, while the Ministry of Public Works housed services affiliated with navigation. In subsequent decades, the administration of France's rivers, even those declared within the state's jurisdiction, became even more fractured as additional uses complicated the existing bureaucracy. The management of rivers may have been centralized under state jurisdiction, but authority was divided among several agencies.[14] This approach, what might be called fragmented centralization, continued until recently. It was not until 1964 that new laws were passed that created river basin–based water management agencies *(Agences de l'Eau)*. These political and legal structures, which attempted to coordinate all the uses, users, and government agencies within a major watershed, represent only a more recent chapter in a much longer history of river management.[15]

Managing the Rhône until the Early Third Republic

For two millennia, politicians and engineers repeatedly proposed large-scale river development, but their grand visions for the Rhône generally far exceeded the projects that actually came to fruition. Tacitus recounted that in 18 CE the Romans debated building a canal to connect the Moselle and Saône rivers, thereby linking the Rhine and Rhône. Charlemagne revived the proposal, but it too went nowhere. The sixteenth and seventeenth centuries marked a shift toward the construction of modest projects that had received royal approval. For example, in 1693 the king conceded to the Prince of Conti the authority to build the Canal de Pierrelatte, which has irrigated a plain on the Rhône's eastern bank south of Valence for over three centuries.[16] In the years just preceding and following the French Revolution, government officials returned to the bold aspirations of Rome and Charlemagne. They considered projects such as regulating the Rhône's mountainous stretch through the French Alps or aiming to control its unruly delta.[17] Between the late eighteenth century and 1840, engineers and state officials also debated building a lateral canal—basically an artificial channel paralleling the Rhône—stretching from Lyon to a major city near the Mediterranean, and they recommended the construction of infrastructure to promote irrigation, navigation, and manufacturing.[18] Although many projects were proposed, few ever moved beyond the planning stage.[19] Generally engineers had to be content with moderate flood control efforts, particularly around key cities such as Lyon and Avignon.[20] Political fragmentation and instability, high costs of construction, limited knowledge, and persistent reservations about altering the environment together limited the transformation of the Rhône until the mid-nineteenth century.[21] Although few projects were completed, politicians, engineers, and property owners had nonetheless already begun to envision the Rhône in terms of its potential as new envirotechnical systems.

Thousand-year floods that inundated the Rhône valley twice in less than two decades, first in 1840 and again in 1856, spurred state officials to formulate plans that were finally achieved. These devastating floods washed away homes, soil, and many of the lingering hesitations about large-scale, state-sponsored river development. In 1840, the Rhône spilled into its floodplain throughout the central and lower river valley; the floodwaters flowed miles from the river's main channel. When dikes along Lyon's eastern edge failed, neighborhoods along the Rhône's banks as well as the

Presqu'île, in the heart of the city, also flooded. As the floodwaters surged, so did popular and political support for state intervention. The 1840 flood prompted the creation of the Service du Rhône, a government agency that oversaw flood prevention measures such as dredging and diking. The establishment of this agency signaled greater state intervention in the management of the Rhône, but it did not necessarily mean that technical elites were successful in their efforts to control the river.

Indeed, the flood of 1840 would pale in comparison to that of 1856. This catastrophic flood, the worst in the Rhône's recorded history, and blamed on deforestation in the Alps, both accelerated the construction of projects that attempted to regulate the river and intensified state involvement in that process.[22] The recently self-proclaimed emperor Napoléon III expressed his fervent hope that the nation's unruly rivers might be controlled during his opening speech to the parliamentary session of 1857: "By my honor I promise that rivers, like revolution, will return to their beds and remain unable to rise during my reign."[23] By associating the floods with the political turmoil of the 1840s and 1850s that had brought him to power, Napoléon III implied that restoration of political and natural order went hand in hand. In 1858, just two years after the Rhône's second flood and a year after his parliamentary speech, Napoléon III's government approved a new law that prohibited the development of floodplains located upstream of major cities in order to prevent, or at least decrease, the severity of flooding in urban areas.[24] This law effectively curtailed most development along the river between the Franco-Swiss border and Lyon in order to protect that city. In fact, little large-scale development would take place on the upper Rhône until the mid-1960s.[25]

The state's authority over the Rhône continued to expand during the late nineteenth century. In 1878, the government passed a law proclaiming the Rhône of "public utility" *(l'utilité publique)* between Lyon and the Mediterranean, thereby granting the state supervisory authority over this stretch of the river. The Rhône had been an important navigable waterway for centuries, but this declaration widened the state's authority while also eventually broadening the river's significance beyond navigation. It conferred authority on the state to approve or reject development proposals as well as to undertake projects in the name of the public good. At the time, bureaucrats still defined "public utility" largely with respect to navigation, although the term would eventually encompass hydroelectricity, atomic energy, agriculture, and regional development. During the late 1870s and 1880s, state engineers began to dredge, channelize, and dike the Rhône in order to ensure a continuously navigable channel, including during periods of low water that posed threats to commerce and transportation.[26]

The floods of 1840 and 1856 and the subsequent laws of 1858 and 1878 together opened the floodgates of river development. During the latter third of the nineteenth century, state engineers and entrepreneurs proposed a flurry of projects, both new ideas and new versions of old plans, but few of them got off the ground. Nevertheless, the widespread aspiration to dam and dredge the Rhône revealed growing interest in the river's transformation, rising state ambitions, and increasing state authority. These in turn reflected the commitment of the recently founded Third Republic (1870–1940) to expanding and improving public infrastructure. They also illustrated the government's goals to modernize rural France and to foster nation building within its own borders amid growing European nationalism.[27] The second half of the nineteenth century thus marked the beginning of a distinct shift in the scale and scope of river development in the Rhône valley, a trend that would intensify dramatically after 1945.

One River, Many Uses

The state's dredging and channelization of the Rhône during the 1870s and 1880s indicated a renewed commitment to river navigation during this period. The interest in developing the Rhône as well as France's other major rivers to facilitate navigation emerged partly from the new possibilities created by steam power.[28] No longer did teams of horses and men have to drag barges upstream for days or even weeks. Steam-powered boats sharply reduced the time necessary for travel while lengthening the river's historically short season of navigability.[29]

This resurgence in river navigation also resulted from growing conflict over the transportation of commercial goods. A canal-building campaign during the first third of the nineteenth century temporarily regenerated commerce along France's rivers and artificial waterways, but it was the country's railroad system that boomed during the rest of the century.[30] After small railroads were built near Paris during the 1840s, construction exploded under Napoléon III's Second Empire (1852–1870); and as railroad networks expanded, they took an increasing share of the market in transporting goods across and beyond France. The

Paris-Lyon-Marseille (PLM) railroad, created in 1857, undercut commercial navigation on the Rhône.[31] Chambers of commerce and politicians in towns along the river lamented their loss of power. Local leaders and small industrialists advocated state support for dredging, canalization, and diking projects that might help them regain their historic hold on commercial transportation.

Some politicians, rural engineers, and representatives from agricultural syndicates also hoped that diverting some of the Rhône's flow would improve the valley's agricultural prospects. Farmers in economic crisis pinned hopes on the Rhône as a source of agricultural renewal. Agricultural engineers aimed to expand irrigation networks south of Lyon, their vision dovetailing with the Third Republic's modernizing agenda targeting rural France.[32] Yet river reengineering drew a mixed reaction from locals. Farmers may have welcomed agricultural investment, but they were wary of intervention from outsiders, especially state officials. In their eyes, government involvement usually entailed wresting power from communities and pursuing national rather than local agendas.[33]

Conflicts soon emerged between proponents of navigation and those of agriculture. In general, the regional metropoles of Lyon and Marseille favored navigation, while smaller towns between Valence and Marseille preferred irrigation, hoping that water would yield agricultural wealth from the rich land of the Midi.[34] Each group saw the other as a threat to its own interests. Debates over the compatibility of agriculture and navigation grew even more complicated in the late nineteenth century, when hydroelectricity became a third variable in the development equation. Tapping the Rhône for energy was certainly not new. However, beginning in 1871, several industrialists began to construct small, private hydroelectric plants along the upper Rhône, the mountainous reach of the river between Geneva and Lyon. Additional proposals quickly followed suit.[35]

This second industrial revolution of the late nineteenth and early twentieth centuries created vast new opportunities—and potential new problems—for the Rhône's diverse users. The development of electricity and particularly of alternating current technology transformed the spatial possibilities of river-based energy. Because electrical power was now easily transferred over long distances, the site of production could be located far from the site of consumption, an uncoupling that had significant implications. Consumers of Rhône-produced electricity no longer needed to be near the river. In fact, they could be located even hundreds of miles away. By dramatically expanding the potential groups interested in the river's energy, electricity only intensified battles over the Rhône and its remaking.[36]

By the end of the nineteenth century, debates over the Rhône's hydroelectric potential focused less on *whether* it should be pursued than on *who* should profit. Parisians and northern industrialists had begun to stake their claims to the river's energy, much to the dismay of those living in the Rhône valley. But even if the river's energy remained within the valley, various constituencies vied for a greater share of it. Amid these heated debates, some groups began to formulate a potential compromise: they argued that the sale of energy produced by the river's hydroelectric dams could subsidize financially marginal, multipurpose projects. In other words, ample profits from electricity sales might help pay for improved navigation and agriculture. This solution, then, might satisfy multiple constituencies and reduce the disagreements among farmers, merchants, and industrialists that had historically slowed, if not prevented, the river's development.[37] By the turn of the twentieth century, then, agricultural and navigation advocates no longer viewed hydroelectricity as a third, competing objective for the river but instead regarded it as a way to subsidize irrigation or commercial projects that were not profitable on their own. To achieve such a balance, of course, experts had to design an envirotechnical system that met all of these goals.

State engineers implemented this new multipurpose philosophy in their design of Jonage, a project completed northeast of Lyon in 1892. Development proponents soon pointed to Jonage as a model for the future transformation of the entire river. Unlike the first hydroelectric plants on the upper Rhône, Jonage's design combined energy generation with improved navigation. The steep gorges of the upper Rhône offered ready-made walls for a large reservoir and high-chute dam that could produce ample electricity, but they rendered locks difficult and expensive to build. The Rhône's course near Lyon was much flatter. Numerous towns scattered along the river's banks also made a high-chute dam and reservoir system undesirable, if not impossible, since they would have been inundated. The project's supporters hoped too that Jonage would ultimately improve Lyon's flood management system by channeling the river's flow near the city through two beds, one natural and one artificial.

For these reasons, Jonage's key feature was an 18.8-kilometer (11.25-mile) diversion canal, a much shorter, modified version of the earlier lateral canal proposal, which would have required construction of an additional channel paralleling the Rhône. A dam with a hydroelectric plant regulated the flow of water into this diversion canal and the original riverbed while producing electricity. In order to ensure continued navigability, the canal also had two locks. Almost twelve miles downstream of the point of diversion, it once again merged with the Rhône into a single channel.[38] Jonage demonstrated the successful implementation of two key concepts in river management: multipurpose development and the diversion approach. This project would provide an important precedent during subsequent debates over the Rhône's transformation.

Consolidating Multipurpose Development, 1899–1921

While designers and users alike viewed Jonage as a success, the project did not lead to the immediate multipurpose development of the entire French Rhône. Nor did government officials seize on the opportunity to carry out this mission. Most of the impetus for development instead came from the private sector, as entrepreneurs proposed a variety of projects between the late nineteenth century and World War I; but their proposals worried local leaders who feared that a few industrialists might reap the river's benefits at their communities' expense.

Representatives from twenty-seven chambers of commerce in southeastern France responded to this flurry of proposals by meeting in Lyon during the summer of 1899. A proclamation from the chamber of commerce of Aubenas requesting that "the government study immediately the agricultural and industrial development of the Rhône permitting irrigation, constant navigation, and utilization of the river's potential energy" encapsulated the participants' goals as well as their growing commitment to multipurpose development. The conference's delegates referenced Jonage, pointing to the project as proof that multipurpose development could work. Over the next two years, representatives put forth numerous proposals that varied in detail but were all multipurpose in intent.[39]

It had taken nearly three decades for local and regional leaders to finally reach a consensus around the concept of multipurpose river development. But the idea was one thing, obtaining funding and actually building projects quite another. The Office des Transports, created at the 1899 meeting, was authorized to study the Rhône's remaking and how these projects might be financed. Its final report, which ended up favoring industry over navigation and agriculture, was an enormous setback for proponents of multipurpose development, not to mention commercial and farming interest groups. In addition, although constituencies in the Rhône valley had rallied around the ideal of multipurpose development, rivalries continued to plague their negotiations.[40] The fact that the

Rhône traversed a dozen different departments only complicated deliberations further.

In light of this stalemate, representatives from communities along the Rhône called for the creation of an interdepartmental body to mediate among industry, navigation, agriculture, and other interest groups and among local, regional, and national authorities. In 1900, the Conseil Général (general council) of the department of the Rhône, along with ten other Conseils Généraux in the Rhône valley, established a new commission headed by Léon Perrier, an influential senator from the department of the Isère. He played a key role in fostering what became known as regional development throughout France, but some observers credited him with even wider influence. In 1963, Pierre Salenc, a state agricultural engineer, reported that U.S. President Franklin Roosevelt had dubbed Perrier "father of the Tennessee Valley Authority" (TVA) because the eventual authorization of the multipurpose development of the Rhône predated the TVA by more than a decade.[41]

When Perrier's interdepartmental commission held its first meeting in Lyon in May 1901, it appeared that the redesign of the Rhône might be undertaken by a governmental body that was at once centralized and decentralized, bringing together leaders from many different local and regional governments but rejecting national oversight. Representatives from departments along the Rhône demanded that they retain the authority to decide the river's future. In effect, the commission defended locals' exclusive right to define a new envirotechnical regime and design its envirotechnical system.

Parisian officials disagreed and challenged the idea of multipurpose development, revealing the depths of the divisions between regional and national visions of the Rhône. Furthermore, the establishment of Perrier's interdepartmental commission prompted the Ministries of Agriculture and Public Works to create a new interministerial, rather than interdepartmental, commission to study the Rhône's future.[42] By doing so, they attempted to transfer power from communities in the Rhône valley to government ministries in Paris. Composed of national-level politicians and bureaucrats, this commission also rejected multipurpose development, concluding that it was impossible to achieve. Rather, "each of these [individual] interests [e.g., navigation, agriculture, and hydroelectricity] must be envisioned separately."[43]

Parisian officials had responded paradoxically, then, to regional plans for the Rhône. On the one hand, national leaders had asserted their authority and intervened, thereby undermining local plans for constructing a multipurpose envirotechnical system to meet area needs. On the other hand, the interministerial commission concluded that the state should play no role in developing the Rhône, at least with respect to a "private" issue such as hydroelectricity. Only agriculture and navigation fell within the state's "public" domain. Or so it argued. By asserting that "the State must not take any initiative on this issue [of hydroelectricity]," the commission essentially opened the door to private hydroelectric development. These conclusions distressed citizens and politicians from departments along the Rhône who had put aside their differences to unite behind the idea of multipurpose development. Moreover, by rejecting mixed development and placing hydroelectricity in private hands, the commission quashed the concept of using energy profits to subsidize agricultural and navigation projects that might not have been economically feasible on their own.[44]

Despite this setback, the valley's leaders persisted in their pursuit of a multipurpose, regional Rhône. A series of congresses composed of three hundred representatives from various constituencies interested in the river's development met in Lyon, Marseille, Paris, and Grenoble between June 1918 and June 1919. Because these meetings brought diverse positions to the negotiating table, they predictably exposed persistent disagreements between national and regional visions of how the Rhône should be developed, who should be in charge of development, and the distribution of the projects' benefits. Édouard Herriot, a leading figure in the Radical Party and future prime minister, defended local perspectives while arguing that the state should take up the urgent cause of the Rhône's multipurpose development, and he hoped to convene all interested groups to participate collectively in decision making. Yet Herriot demanded that national officials be excluded from the discussion, avowing that the right to decide the river's future should be reserved exclusively for Rhodanians and their political representatives.

Against Herriot's objections, the Rhône's development remained on the national stage with national players. These officials continued to reject an active role for the state in river development. In fact, they intended to parcel the Rhône into five sections and delegate each one to a different private company. Perrier vehemently opposed this proposal. Instead, he advocated the integral and comprehensive development of the entire French Rhône, asserting that the river should be managed in the "general interest," with responsibility conceded to a single organization representing all interested parties. Although the central government would have some influence over this body through its technical experts and financing by guaranteeing loans, Perrier wanted to keep decision making along the river rather than in the offices of Parisian bureaucrats or industrialists, let alone divvied up among various private companies. In Perrier's view, the people of the river valley should oversee the future of "their" entire river.[45]

Perrier's strategy reflected a peculiar blend of regionalism and statism. He advocated an active role for the state in the management of the Rhône, but one that assumed the central government would delegate control of the river to a collective body of which it was a minority, ceding most of the decision-making power to local and regional authorities.[46] This complex form of democratic representation would offer an alternative to both centralized state control and liberalism. Although laudable in many ways, Perrier ultimately failed to eliminate conflict among interest groups or to outline a clear formula for fair representation. His regional statism crystallized, rather than mitigated, the decades-old conflicts among constituencies in the Rhône valley, and between Parisians and Rhodanians.

While many in the valley sympathized with Perrier's and Herriot's critiques of Parisians' designs on the Rhône, representatives from local government also expressed concerns about high project costs and acknowledged that a significant contribution from the central government would probably be required if any projects were to be built. As a result, the interdepartmental congresses of 1918 and 1919 eventually agreed to send more electricity to Paris and the department of the Seine than Rhône valley residents themselves would receive; but in return, the state would provide substantial financial aid by guaranteeing loans, which would gradually be paid back through the sale of electricity generated by the projects on completion. The members of the final congress may have settled on these concessions, but they remained committed to multipurpose development and representational, locally based decision making. They simply hoped their pragmatism would ensure completion of the projects, not an inconsequential consideration given that disagreements had stymied action for a half century.[47]

Somewhat ironically, just as the interdepartmental congresses arrived at this compromise, state officials began to change their policy with respect to hydropower and the government's role in hydroelectric development. In the fall of 1919, a few short months after the final Grenoble congress, these officials recommended the rapid expansion of hydroelectric production and placed such projects under the state's purview. This remarkable shift can be explained by several factors. Rising energy demands sparked more interest in hydroelectricity. Electricity consumption increased by over a third in the decade between 1896 and 1906, while bureaucrats in Paris and the department of the Seine, both homes to powerful state officials and ministries, began to advocate the expansion of electricity production throughout France. Private projects on the upper Rhône also demonstrated the success of hydroelectric development, including energy distribution over long distances. In addition, by then, Jonage had a nearly three-decade record proving the compatibility of projects pursuing both electricity and navigation.[48]

The outbreak of World War I also changed state policy toward hydroelectricity and the state's role in hydroelectric development. The country's heavy dependence on coal, located primarily in Germany or war-torn and occupied regions of northern France, increased political and popular support for hydroelectric development. Expanding France's capacity for hydropower might not only fuel the country's wartime plants but also assure its energy independence in the future, an argument that would be echoed during the 1970s when oil crises led to the expansion of France's nuclear power program.[49] These wartime imperatives served to revive long-standing concerns about France's energy supplies and its capacity for energy self-sufficiency.

Soon after the outbreak of World War I, then, state officials began to reconsider their position on hydroelectric development and to pursue a more active, direct role for the government in producing hydropower, what was called *l'houille blanche* (white coal). The state's new energy policy targeted France's rivers, including the Rhône, which engineers and politicians hoped to transform into a national technology that could produce vital energy for France. This is a theme that would be echoed even more intensely after World War II, as the next chapter will show. But at this point, the Great War already bolstered a national commitment to hydroelectricity and increased the state's role in cultivating this energy source while assuaging some of the tensions among interest groups in the Rhône valley, and between those in Paris and the Rhône valley, since all could agree—at least temporarily—on the need for increased energy production. Still, the central question of who should benefit from this energy in the long term remained. In wartime, it was evident that military and industrial needs should be met first, but after the war, whom should these new sources of energy serve?[50] The old debate had simply taken on a new guise.

Given these wartime imperatives, in 1917 the government created an extraparliamentary commission composed of government representatives and scientific experts to discuss the current state and future possibilities of hydroelectricity in France. Its conclusions, written by commission member Perrier, were undeniably statist: the central government should serve the public interest through the promotion of hydroelectric development. Soon the commission's recommendation became law. On October 16, 1919, parliamentary deputies legislated that only the state could grant hydroelectric concessions, thereby rendering hydroelectricity a national resource in the public domain. Any waterway, regardless of its navigability, was now subject to the state's jurisdiction when a developer sought to produce hydroelectricity. If the state approved the proposal, the company would receive its concession for seventy-five years. The 1919 law outlined a uniform system of state-decreed concessions to permit hydroelectric development across all rivers nationwide, effectively extending the state's authority beyond navigable rivers to any that might produce electricity.[51] This law demonstrated how World War I fundamentally altered the relationship between the state and economy in France, promoting limited statist intervention in key industries, including the energy sector.[52]

The state, along with expanding its jurisdictional authority, had also adopted a more favorable stance toward hydroelectricity. Another part of the 1919 law stated that developers did not actually need to own the land on which proposed projects would be built. It did guarantee that property owners would receive indemnities, but by separating property ownership from the proposal of projects, the state helped pave the way for "outsiders" to pursue hydropower. In so doing, the state exacerbated existing tensions between local communities and "external" groups hoping to tap local resources for individual profit, albeit often in the name of the public good. The 1919 law had significant consequences for the nation's rivers, particularly those located in the French Alps: hydroelectric development had begun there in the early 1870s, but proposals proliferated after World War I.[53]

With the state's promotion of hydroelectricity, the multipurpose development of the Rhône finally fell into place as local, regional, and national visions at last coincided. On May 27, 1921, the National Assembly and Senate approved a new law whose first article stated that "the development of the Rhône between the Swiss border and the [Mediterranean] Sea will be completed according to three objectives: (1) use of hydraulic power; (2) navigation; [and] (3) irrigation, drainage, and other agricultural uses." As I will discuss below, the particular order of these goals was significant. In addition, it is perhaps surprising that flood management was not explicitly included. The law's second article mandated that "all of the projects that will be built to develop the Rhône will be the object of a unique concession made to a group of collectivities." After a half century of debate, Perrier's hopes for multipurpose river development and a representational body governing its management had been realized.

France was by no means the only country to invest in river reengineering or to choose multipurpose development during the early twentieth century. Electrification and hydroelectric development were central to Soviet economic programs as early as the first years of the Russian Revolution.[54] An integrated agency had overseen a stretch of the Ruhr River at the turn of the century.[55] During the 1920s and 1930s, Italy, Great Britain, Spain, and the United States all created similar agencies.[56] In America, the Great Depression catalyzed ambitious New Deal programs, and rivers were some of their earliest targets. While the Columbia River's Grand Coulee Dam was one of the first sites of multipurpose river management in the United States, the TVA is probably the most famous example.[57] Created in 1933, the TVA sought to promote flood control, electrification, rural development, and conservation in Appalachia while creating thousands of jobs for unemployed men. After World War II, French politicians and journalists often referred to the CNR as "the French TVA," though the law approving the multipurpose development of the Rhône and the creation of the CNR predated the TVA by twelve years. However, the CNR ended up losing most of the interwar era to conflicts over its representational composition, a costly continuation of the earlier fights. Then the outbreak of World War II stymied the agency's construction efforts. Consequently, the TVA's program moved from blueprint to electrical line more quickly and with greater fame than the CNR's.[58]

Creating the CNR, 1921–1934

Passage of the 1921 law was an essential, but ultimately incomplete, first step in the multipurpose development of the Rhône. As suggested earlier, this law outlined broad objectives for the river's remaking—energy, navigation, and agriculture simultaneously—and issued authority to the new representational body. The law also decreed that the CNR would be composed of groups known as *actionnaires* (shareholders). The law did not, however, specify their relative representation. The process of determining the *actionnaires*' composition reopened many of the conflicts among interest groups. Notwithstanding the extensive discussions of the previous five decades, nine more years were spent negotiating the CNR's composition.

Industrialists, particularly those in the electricity business, who were still the most assertive in their efforts to gain control, even attempted to overturn the 1921 law. Both of its central tenets, multipurpose development and representational decision-making, threatened their power. Eventually forced to accept these principles, they then tried to garner greater representation within the CNR. Perrier led the fight against the forceful industrialists. By the end of the nine years of debate, he had managed to negotiate equal representation for four categories of *actionnaires:* (1) the city of Paris and the department of the Seine; (2) local and regional political bodies, including towns, departments, and chambers of commerce along the Rhône; (3) the PLM railroad; and (4) industrialists, including electricity producers. The agency, then, would be a mixed-economic company: an organization composed of both public and private interests.[59] The public parties—the municipal government of Paris, the Seine department, and regional governments along the Rhône—formed half of the CNR's organization. The two private parties, the PLM and industrialists, made up the other half.

The Conseil d'Etat (Council of State) approved the CNR's statutes in July 1930; the CNR held its first assembly in Lyon on May 27, 1933; and on June 5, 1934, the newly created CNR and the Ministry of Public Works signed an agreement formally granting the CNR a seventy-five-year concession to develop and manage the Rhône. This contract also required that the CNR put together within six months a technical committee *(comité technique)* that was to have a consultative role and would include scientific and technical experts from relevant state agencies as well as those within CNR. By early 1935, the new committee had formulated its first programmatic vision for the river's transformation.[60] Subsequent agency leaders and experts implemented most of this committee's preliminary plan over the next fifty-one years (see Map 1.1).

As the timeline suggests, though the state approved the development of the Rhône in 1921, the CNR was not officially granted the concession

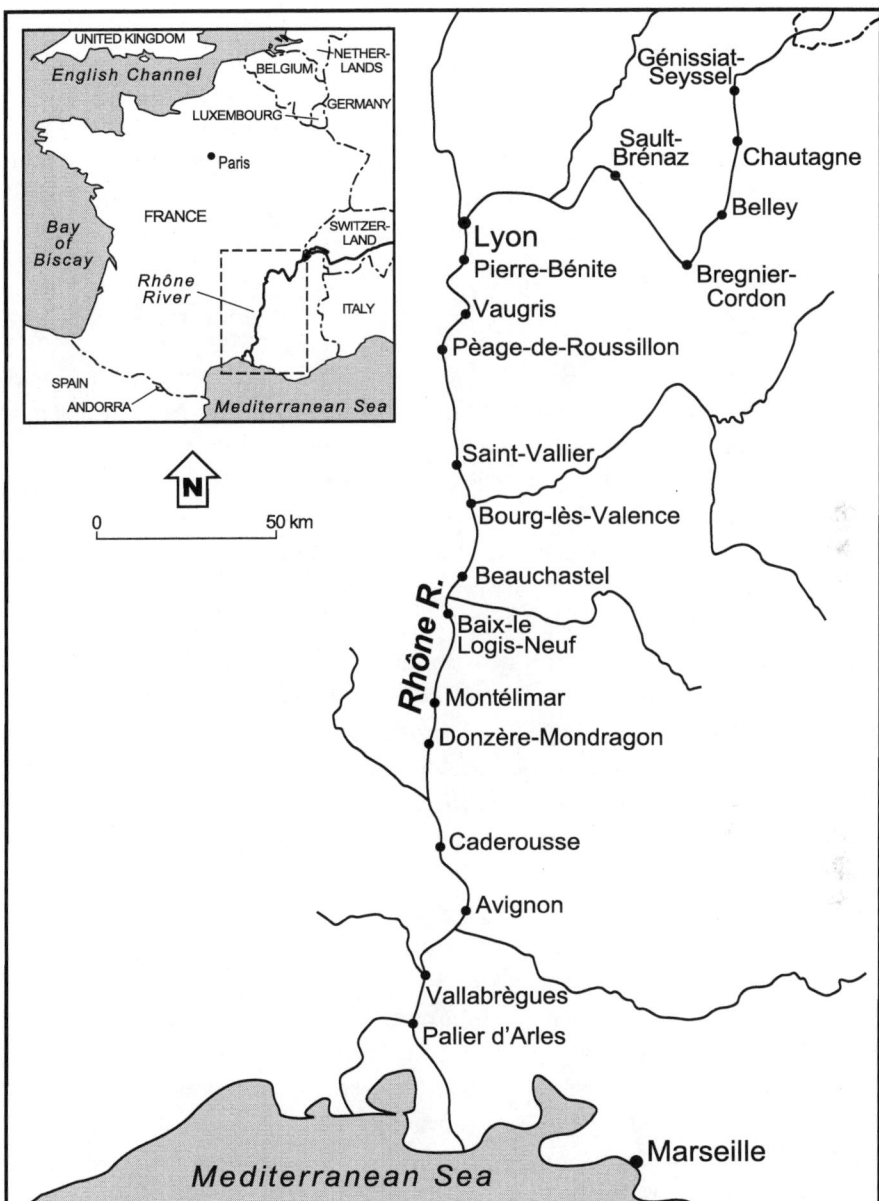

Map 1.1. *The CNR's completed projects.* The projects built by the CNR between the late 1930s and 1986. The location and features of these projects are extremely similar to what the agency's technical committee first proposed in 1935. (Map by Joseph W. Stoll, Syracuse University Cartographic Laboratory.)

until 1934. Persistent disagreements, local opposition, pressure from industrialists, the financial crises and political instability of the 1920s and 1930s, and hesitancy within factions of the government about the creation of a mixed-economic company all contributed to the substantial delays.[61] The part public, part private status of the CNR during its early years ultimately reflected the changing relationship between the French state and economy during the interwar era.[62]

The Postwar Emergence of the CNR's Hydroelectric Paradigm

Only during the late 1930s, then, were the CNR's workers finally able to begin building its first two projects: a new port in southern Lyon, named after Herriot, and a high-chute dam at Génissiat along the upper Rhône. With the benefit of hindsight, it is clear that the CNR's timing could not have been worse. The onset of World War II slowed construction on both projects, as northern France became a war zone and then fell under German occupation. Even more important, the war and its legacy had major consequences for the CNR's organizational structure, its mission, its relationship to the central government, and therefore the river itself.

The wartime Vichy government (1940–1944) was relatively short-lived, but the effects of its policies on the CNR and the Rhône were not. In December 1940, Vichy officials changed the composition of the CNR's board of directors *(conseil d'administration),* reducing the number of members from forty to just nine: four representatives from the state, four designated by the CNR's *actionnaires,* and the president of the *conseil d'administration,* who was appointed by the state. In effect, the Vichy government ensured that the state controlled a simple majority of the CNR's highest administration.[63]

The Vichy government also reshaped the agency's mission by making agricultural development, a wider goal for the regime, its highest priority. In the process, Vichy aggravated existing tensions among France's engineering communities, which tended to see themselves in a clear hierarchy. The CNR's elite engineers, most of whom were trained at the prestigious Ecole Polytechnique and Ecole des Ponts et Chaussées, were disdainful of agricultural engineers. Génie Rural engineers normally oversaw rural projects. It was indeed an insult, then, when the CNR was ordered to undertake a large agricultural project in the Crau region of Provence, particularly under the supervision of the Génie Rural.[64] Competition and resentment between the agencies continued after the war, affecting negotiations over the Rhône's transformation.⁶⁵ The end of World War II reversed some Vichy directives, such as the temporary importance placed on agriculture, but the centralizing influence of the wartime government, including its administrative reorganization of the CNR, remained.

The new government of the Fourth Republic (1946–1958) attempted to break ties with its opprobrious predecessor, but it shared with the wartime regime a functionalist view of the Rhône. For the postwar state, the economic, political, and cultural reconstruction of France depended on the material and symbolic reconstruction of the river. The government incorporated the CNR's third project, Donzère-Mondragon, into its broader agenda for postwar reconstruction. In December 1945, the Communist minister of industrial production, Marcel Paul, wrote to Léon Perrier, now CNR president, "prescrib[ing] the CNR to take, without delay, all preliminary measures, which will allow Donzère-Mondragon to be undertaken very quickly at the beginning of 1946 so that the work will be carried out in full force and the project will be finished in record speed."⁶⁶

Paul's charge reflected postwar efforts to rebuild a decimated nation. In particular, France was desperate for energy after World War II. As the opening line of a December 1945 newsreel put it, "today's problem is electricity."⁶⁷ Coal was in short supply, and combat and sabotage had taken a heavy toll on the country's industries and infrastructure alike. Rebuilding them would prove impossible without adequate and consistent supplies of energy.⁶⁸ Even in 1949 another newsreel called hydroelectric development "fundamental for France."⁶⁹ But the government's postwar obsession with energy also fit within a much older narrative that lamented France's lack of key minerals, including coal.⁷⁰ Moreover, the government turned to technology as an icon of *grandeur* after the war.⁷¹ These factors, both pragmatic and ideological, together help explain why state officials ranked energy production so high among postwar government priorities. The CNR may have had a multipurpose mandate, but the postwar state placed industrial aims above other objectives, and of all industrial goals, energy mattered most.

France's natural resources help account for the CNR's uneven development of the Rhône, which initially prioritized hydroelectricity. The country may have had little coal, but it was rich in *l'houille blanche,* as advocates of hydroelectric development were quick to point out. Of all the country's rivers, the Rhône was one of the most promising because of its volume and steep course.[72] Government officials proclaimed that the Rhône was "ready" for hydroelectric exploitation.[73] One state report even portended that "along the length of the Rhône valley will circulate a continuous current, a veritable river of electrical energy."[74] Obviously playing on the double meaning of "current," these descriptions helped justify the CNR's development of the river and particularly to legitimate its orientation toward hydroelectric production during the postwar era.[75]

Other more subtle rhetorical strategies suggested the way state officials homed in on the Rhône's hydroelectric potential. Starting with the 1921 law authorizing the river's transformation, most documents listed hydroelectricity before navigation and finally agriculture. Preliminary sketches of Donzère-Mondragon first referenced the project's energy potential, then cited benefits accrued to navigation, and only then alluded to its agricultural dimensions under the heading, "irrigation, drainage, and diverse improvements" (see Figure 1.1).

Furthermore, government documents often referred not to the river's "development" *(l'aménagement du Rhône)* but to its "hydroelectric development" *(l'aménagement hydroélectrique),* thereby obscuring the

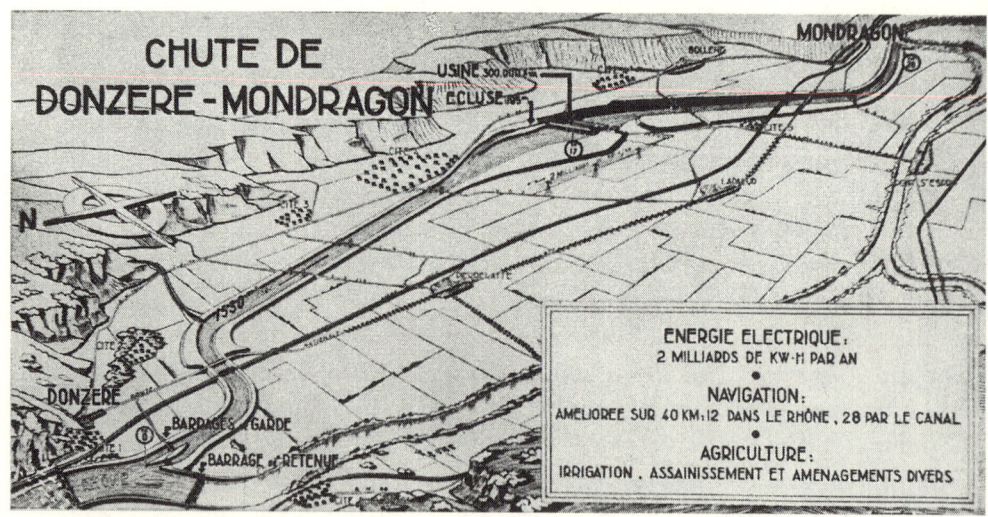

Figure 1.1. *The multipurpose Rhône.* By law, the CNR was required to pursue multipurpose development. As this CNR-produced image from 1952 shows, electricity was usually listed first, followed by navigation and then agriculture. These objectives and their relative importance did, however, shift over time. (Courtesy of Compagnie Nationale du Rhône.)

CNR's other required objectives. The CNR's administration frequently made this slip as well. Occasionally, the bias was blatant. As one government report from 1946 stated simply, the CNR's program "is weighted toward electrical development."[76]

France's politicians and bureaucrats sanctioned the CNR's program, including its orientation to hydroelectric generation, because they associated energy so strongly with France's reconstruction after World War II. With such an ample energy supply just waiting to be tapped, state officials cast electricity as the cornerstone of the nation's future. In the words of the same 1946 government report, "France needs electricity." Politicians maintained that a host of diverse groups regionally and nationally, including "electrical-chemical and electro-metallurgical industries, farmers, Parisians, and the railroads," could all benefit from increased electricity production.[77] One 1949 newsreel claimed that hydroelectric development of the Rhône would "assure the independence of the French economy."[78] The fact that this proclamation was made in a film ostensibly about French and German dam building suggests just how central hydroelectricity had become to conceptions of the nation's postwar economy.

Other government reports from the era asserted that electricity production and consumption "have almost become the indices for quality of life and the level of activity in diverse countries."[79] One writer even quoted from a 1914 "Electricity Celebration" that "the richness of a nation is evaluated no longer in gold, but in kilowatts."[80] Such claims made clear that France needed to increase its supply of electricity in order to keep up with the rest of the modern industrial world. Or as one government publication declared in 1947, "it is not an exaggeration to say that the prosperity of our country in the years to come depends in large part on the efforts that we make in this area [of hydroelectric development]."[81] After World War II, electricity had become not only an inalienable right but also a key marker of civilization itself. Together these concerns, both the practical and the ideological, help account for the emergence of the CNR's hydroelectric paradigm after the war.

Yet despite this intense discourse linking hydropower and reconstruction, conflicts over the Rhône's uses remained. Even after the state had authorized the river's development and the creation of the CNR, not all Rhône valley residents sanctioned the agency's now official mission for the river. Agricultural constituencies protested as the CNR's first projects got under way. In response, the agency's general director, R. Giguet, defended his agency to the head of the Génie Rural, arguing that far from ignoring agricultural objectives, it was only with sufficient hydroelectric production that the projects' other objectives, including farming, could be financed.[82] A few years later, however, the CNR's *conseil d'administration* did admit that the agency's projects were "still very imprecise" with regard to agricultural matters.[83] The concerns of agricultural interest groups may well have been warranted. State reports also acknowledged skirmishes between the so-called electricity and navigation camps, alluding to the lingering division among communities and perhaps within the engineering profession, but these documents generally supported the CNR's position.[84]

National politicians might have been committed to maximizing electricity generation, but questions remained about the best way to achieve this goal. The eventual solution—nationalizing electricity industries in 1946—affected the CNR's organizational structure and presumably the design of its projects.[85] When the CNR was created during the interwar era, the railroad and electricity industries, which together comprised half of the CNR's *actionnaires*, were private, but by 1946, both had become public. After nationalization, then, the CNR had no private representatives. Indeed, after considerable debate, state officials had decided that they would not incorporate the CNR into the newly nationalized electrical agency, Electricité de France (EDF). In fact, in hopes of preserving the CNR's autonomy, agency leaders and their supporters had argued that the CNR's multipurpose mandate fundamentally differentiated it from the EDF's exclusive focus on electricity generation. The multipurpose development of the Rhône, for which Perrier, Herriot, and others had fought so fiercely, thus justified the CNR's continued existence. The CNR may have successfully weathered nationalization, but this process nonetheless tightened its ties to the central government. One-fourth of the CNR's *actionnaires* were from local and regional governments, but three-fourths were now representatives of state agencies who answered to Paris.[86]

After 1946, then, the CNR embodied France's postwar turn toward nationalization rather than the mixed-economic approach of the interwar era. National prerogatives increasingly framed the CNR's activities, shaping the agency's mission and thus its projects. For instance, the EDF exerted considerable pressure on its rival agency, particularly between 1946 and 1962.[87] In part, this pressure derived from the financial constitution of the CNR. The state guaranteed loans to the CNR for the construction of its projects. But in order to pay off these loans, the CNR had to sell the electricity it generated to the newly nationalized EDF. Of course, the EDF's and CNR's interests in the negotiations of electricity rates were in direct conflict with one another. While the EDF attempted to buy the CNR's electricity for as little as possible, CNR leaders sought to negotiate high rates.

The government's postwar reconstruction program further fortified the close alliance between the state and the CNR after 1945. In the war's aftermath, the Commissariat Général au Plan, staffed by an elite corps of technical, economic, and political experts, designed a program by which the French state coordinated reconstruction. The state promoted industrialization through its Plan de Modernisation et d'Equipement (Plan for Modernization and Equipment) by targeting certain industries for not only restoration but also modernization. Every few years, the government issued a new, multiyear "Plan," both outlining general economic priorities for the nation and identifying particular projects.[88] The CNR's initial postwar project, Donzère-Mondragon, became one of the centerpieces of the First Plan (1947–1952). Incorporated into the rubric of national reconstruction and *planification,* the CNR and its projects became increasingly entangled with the central government and national policy-making after World War II.

The educational and professional backgrounds of the CNR's elites only strengthened connections between the agency and the central government. Although the CNR was a relatively new organization, it was part of a long tradition of science, technology, and engineering in France that attested to the marriage between science and the state. The CNR's engineers and administration shared strong intellectual ties with other state elites because of their education and professional training. Its personnel spanned the hierarchy of France's engineering system. Most entry-level engineers came from one of three tracks: engineers from Arts et Métiers; *centraliens,* primarily from Ecole Centrale de Lyon; and graduates of the University of Grenoble who specialized in hydraulic and electrical engineering. Engineers from the Travaux Publics de l'Etat or Ecole Supérieure d'Electricité held positions in middle management, such as heads of project work sites or lower-level departments within the CNR's central administration. Chiefships of *directions* (divisions) within the CNR were reserved for graduates of the Ecole Polytechnique, the necessary gateway for engineers who joined the elite Corps des Ponts et Chaussées. The CNR's general director, technical director, and director of research and construction all had the distinguished status of "X-Ponts." France's strong state engineering tradition thoroughly shaped the CNR's technical and administrative elite.[89]

By the late 1940s, then, the CNR had become firmly allied with the institutions, agenda, and personnel of the postwar French state, which elevated the role of technical and economic experts in national politics and policy-making. Although the CNR has never officially been part of the government's bureaucracy, the agency's close financial, institutional, professional, and epistemological ties to the state, especially after World War II, effectively brought the agency under the state's oversight.[90] As Alexandre Giandou has argued, by the late 1940s, the CNR had essentially become a regional arm of the French state.[91] In the process, reconstructing the Rhône was rendered integral to the state's postwar development agenda for France.

This long history of river management, and especially the contentious debates over the river's remaking that took place between 1871 and 1930, affected how the Rhône was transformed after World War II. Naturally the war and its legacy also shaped this process. Growing anxiety over energy supplies, the CNR's evolving financial and institutional relationship with state agencies including the EDF, and a widespread fervor for technological and scientific prowess together contributed to a hydroelectric paradigm within the CNR that dominated the agency until the early 1960s. This model defined the CNR's envirotechnical regime for the Rhône during the postwar era and consequently molded the specific goals and designs of the CNR's first postwar project, Donzère-Mondragon, as France's political and technical elites sought to reconstruct the war-torn nation through the transformation of nature.

2

IMAGINING THE NATION'S RIVER

On October 25, 1952, French president Vincent Auriol inaugurated the CNR's first postwar project, Donzère-Mondragon. Earlier that fall, Auriol's son, serving as his father's attaché, had written to the CNR's president, Emile Bollaert, explaining that "the President of the Republic wants to generate the most publicity possible—within France as well as abroad—about the trip that he will take to Donzère-Mondragon in October and through this make known France's successful efforts to increase its energy potential."[1] To prepare the country for the inauguration, Radiodiffusion Française, France's national radio network, broadcast stories about Donzère-Mondragon throughout the month of October, and articles appeared in newspapers and popular magazines like *Paris match* and *Science et vie*.[2] Bollaert hoped to imprint Donzère-Mondragon permanently on the nation's memory. Five days after the project's inauguration, he recommended that the Ministry of the Post Office, Telegraph, and Telephone Services issue a stamp commemorating Donzère-Mondragon's "global significance." In October 1956, the Ministry released just such a stamp as part of its "Grandes réalisations françaises" series.[3]

The inauguration of Donzère-Mondragon was a grand, extravagant affair that cost over one hundred million francs and was intended to impress not only the nation but also the world. Fifty-three delegates from forty-five countries attended the festivities. Mayors, prefects, and representatives from every major government agency brought the number of dignitaries present to over 250. After taking a train from Paris to the village of Viviers the night before the event, these dignitaries accompanied President Auriol on a morning boat ride up the Rhône to Donzère-Mondragon. As their boat, the *Frédéric Mistral*, named for the Nobel Prize–winning Provençal poet, passed through the project's locks, a symbol of the nineteenth century became literally overshadowed by a new symbol of the twentieth. At high noon, Auriol entered the alternator room of the André Blondel hydroelectric plant. Surrounded by politicians and journalists in buttoned suits, with camera bulbs flashing, he started the plant (Figure 2.1). In reality it was a symbolic act, since several of the plant's six turbines had been operational for months, but the next day, newspapers across France ran photographs of Auriol hitting the switch on their front pages.

At the lavish banquet that followed the president's ceremonial gesture, Auriol, Bollaert, Jean-Marie Louvel (minister of commerce and industry), and Édouard Herriot, then president of the National Assembly, each spoke in turn. In his patriotic speech, Auriol declared that Donzère-Mondragon demonstrated France's rise from the ruins of World War II and the country's return to its former greatness. He proclaimed that "the names of Donzère-Mondragon not only evoke a prestigious achievement of work and of French science and technology, but also mark a date in the history of our national recovery." Donzère-Mondragon proved that

Figure 2.1. *Inaugurating Donzère-Mondragon*. This photograph shows the CNR's general director, Pierre Delattre, Prime Minister Antoine Pinay, and President Vincent Auriol deliberating before Auriol hit the switch during Donzère-Mondragon's inauguration on October 25, 1952. Versions of this photograph were published in numerous newspapers and magazines across France. (Courtesy of Compagnie Nationale du Rhône.)

France was a modern Prometheus. After the banquet, the dignitaries were driven south along the Rhône on Route Nationale 7 to Avignon, where they dined in the Palais des Papes. Here the country's modern achievements again met the nation's great past. Late that night the dignitaries returned by train to Paris.[4] With the aid of the spectacle that had enacted national redemption, renewal, and regeneration, the regional Rhône of Herriot and Perrier had become the nation's river.

The extravagant ceremony and nationalistic speeches at Donzère-Mondragon's inauguration explicitly compared the state of the nation in October 1952 with that of 1944, the year of France's liberation from German occupation. In particular, the inauguration enrolled the country and its citizenry in a patriotic celebration of France's national recovery. The spectacle celebrated the economic, political, and moral reconstruction of the country less than a decade after its welcome yet opprobrious liberation by the Allies. Donzère-Mondragon's enormous concrete structures and sophisticated technological system confirmed France's return to greatness. Its stark, modernist presence on the shores of the Rhône soon came to serve as a literal representation of the country's refound *grandeur*.

Invocations of "greatness" *(grandeur)* and "radiance" *(rayonnement)*, especially after World War II, tapped into a rich discourse on the glorious civilization of France that extended back to the reign of Louis XIV, the intellectual and artistic achievements of the late medieval era, and even Charlemagne. Recent scholars have shown how science and technology played an increasingly important role in the definition of *grandeur* after 1945, an association epitomized by Charles de Gaulle and his commitment to the development of an independent nuclear power program. They have also demonstrated how debates over national identity were not simply imposed on technological projects such as France's nuclear power program, the Train à Grande Vitesse, or the Concorde. Rather, the development of "technological" projects like those of the CNR and notions of national identity worked in tandem. The postwar reconstruction of the Rhône was thus twofold: material and technological reconstruction was enacted simultaneously with cultural reconstruction.[5] While these studies have enriched the substantial literature on national identity in modern France, few scholars have considered the centrality of ideas about nature and the interrelated definitions of nature and technology in the construction of national identity—in France or elsewhere.[6]

The representations of the Rhône in the press, politicians' speeches, and technical publications were much more than images and ideas. They provided a powerful language of legitimation for a series of large-scale, state-sponsored projects of modernization. Because the CNR underwent financial, political, and technical review for each project, the agency needed both to maintain support for its general program and to generate backing for each individual project. As subsequent chapters show, the Rhône thus became interwoven into a fabric of ongoing discussion over the proper foundation and direction of the French nation throughout the second half of the twentieth century.

One important thread of these multifaceted and often contentious debates involved the relationship between the region and the nation in post-1945 France, and specifically the relative location of power within the state. As with any concepts that are defined as much through juxtaposition with one another as on their own terms, the nation did not entirely supersede the region in the postwar era. Likewise, the construction of Europe as a unified economic and political entity beginning in the 1950s did not completely shift this balance from the nation toward a transnational Europe. Nonetheless, while Herriot and Perrier had hoped to develop the Rhône by and for Rhodanians during the early twentieth century, the experience of World War II and its legacy, the onset of the Cold War, and the perceived hegemony of the United States helped reframe the meaning of the river's development after 1945. Together these factors tipped the spatial scales from regional to national after World War II. Accordingly, the political, economic, and cultural contexts of the postwar era remade the Rhône into the nation's river, a shift perfectly encapsulated in the maxim emblazoned on the side of the CNR's dam at Seyssel, completed in 1951: *"Le Rhône, au service de la nation."*

Since World War II, various groups connected representations of nature and technology to debates about national identity and the process of state building in France. A distinct divide in cultural understandings of the river emerged during the second half of the twentieth century (Chapter 6 addresses the latter period). Between 1945 and the late 1960s, politicians, engineers, and writers shared a modernist conception of the river, a perspective not unique to France. In *Seeing Like a State,* James C. Scott explored the ways a "high-modernist ideology," which exudes confidence in "scientific and technical progress, the expansion of production, the growing satisfaction of human needs, the mastery of nature (including human nature), and, above all, the rational design of social order commensurate with the scientific understanding of natural laws" underpinned the ideology, politics, and projects of many states during the twentieth century.[7] In post–World War II France, intellectual, political, and technical elites celebrated the conquest of the Rhône's obstinate nature through science and technology by male technical experts in the name of the nation, legitimating the CNR's objectives as well as the projects that embodied these goals through several contradictory strategies of naturalization and historicization. During the postwar era, a variety of groups thus framed the Rhône as the nation's river, shifting the contested spatialization of the river from a reconciliation of regional and national interests seen in the early twentieth century to the domain of the nation-state.

Taming the "Furious Bull"

As noted, the 1935 proposal of the CNR's technical committee had emerged, in part, from centuries of experience with the Rhône's floods. These floods were a mixed blessing for the river's human neighbors. Their rich alluvial deposits promised future agricultural productivity, but they also wrought destruction. Seasonal and annual floods caused less damage than thousand-year floods like those of 1840 and 1856, but they did so more frequently. Individuals, towns, and government agencies began to dredge the Rhône and build dikes and other infrastructure in an attempt to keep the river's waters within its ever-changing banks.[8] These efforts met with varying degrees of success. Each failure inspired hope of finding a better scheme to master what the nineteenth-century historian Jules Michelet memorably called a "furious bull that leaps from the Alps to the [Mediterranean] Sea."[9]

Ambivalence about the "bullish" Rhône characterized many commentaries from the 1940s through 1960s, a period in which the CNR undertook its largest projects. Writers and artists viewed the river's ecological features—including its flow, descent, unpredictability, and tendency to flood—as powerful and impressive, but also dangerous and threatening. In a 1948 article in *France Illustration,* Henri le Masson described how the Rhône "often brutally plunges" through its course south of Lyon, and another writer characterized the river's waters as "fiery and turbulent."[10]

Many authors referred to the Rhône as a "torrent," uncontrollable by implication, while others called it "undisciplined," "capricious," even "impetuous."[11] In the 1953 acclaimed short film *Crin-Blanc: Cheval sauvage* (White mane), award-winning filmmaker Albert Lamorisse depicted both the sheer power and idealized beauty of the Rhône. In the film's final scenes, the narrator warns that the Rhône threatens to wash out to sea a young boy and his newly tamed Camargue horse. Yet the film concludes with the two companions fleeing ranchers who hope to capture "Crin-Blanc" by plunging into the Rhône's powerful flow, supposedly to be carried to an island where children and horses can remain friends forever.[12] Gilbert Tournier, then general director of the CNR, also captured this ambivalence about the Rhône in his 1952 book *Rhône, dieu conquis:* "for at least a century, engineers have had all kinds of problems when they tried to impose a little order on the [river's] morphology—a morphology that the capricious God [the Rhône] often changes profoundly in the space of just a few hours."[13] The Rhône's impressive power thus implied a dark underside: humans could not always predict and manage the river, as floods periodically demonstrated. Coupled with images of the Rhône's raw power, this language of an unruly river justified attempts to control it.

During the postwar era, politicians, technical experts, and the press portrayed the CNR's development of the Rhône as an epic "battle against nature." Unsurprisingly, they frequently employed metaphors of domestication and conquest.[14] In so doing, they reflected the country's historical *mission civilisatrice* and Western claims of domination over the natural world, not to mention the recent memory of World War II. Perhaps more important, the ways they used these metaphors suggested that France's engineers would prove triumphant. Press coverage repeatedly dubbed Donzère-Mondragon a "victory."[15] Amid reconstruction, the portrayal of river development as a combat—and one France would ultimately win—might redeem a nation haunted by fresh memories of defeat. Winning the war on the banks of the Rhône could be a crucial first step to restoring France and regaining the nation's international standing.[16]

Metaphors of domestication also indicated how gender ideology mediated descriptions of the river and human interactions with it. Notably, the aspiration to master a "capricious" "torrent" targeted a major river that was gendered masculine.[17] Dating back at least to the mid-seventeenth century, the Rhône was personified as a Greek god complete with rippling muscles, flowing beard, and laurel wreath. In Lyon's *Hôtel de ville,* Guillaume Coustou captured this iconography in stone. His sculpture of the Rhône and Saône rivers, which meet in the southern part of the city, are quintessential classical figures. Leaning on a lion, the Rhône confidently extends his hand to the Saône, who reclines: a voluptuous, bare-breasted woman looking up passively at the god-like Rhône.

At least one writer asserted that nature itself explained why the river had been gendered masculine. In the early 1950s, Jean-François Virenque wrote that "man decided to attack the [Rhône] himself and to domesticate its power because the river is power itself. It is young and has the force of youth." Assuming a bit of poetic license, Virenque added, "Moreover, isn't it the only of our rivers that we left with a masculine name?"[18] Associating the river's features, especially its flow and strength, with masculine qualities, he argued that these ecological characteristics differentiated the masculine Rhône from other, presumably feminine, rivers. Through his gendered construction of the Rhône, Virenque essentially reproduced an ideology of masculinity centered on, as he put it, "power" and "force."

As Virenque's interpretation suggests, these metaphors of battle and domination highlighted the role of men in the supposed conquest of nature, thereby forging even stronger ties between masculinity and industrial technology. In the preface to the 1953 book *Donzère-Mondragon,* which celebrated the project's recent completion, Herriot rejoiced that "once again, man had tamed the most impetuous forces of nature." Other writers emphasized that the CNR and its supporters hoped to master the Rhône's entire course within France, making the river "the servant of man."[19] Numerous newspaper and magazine articles, along with the prize-winning 1950 documentary film *L'or du Rhône* (The Rhône's gold)—an overt allusion to *Das Rheingold,* the prologue to Richard Wagner's epic opera cycle "The Ring"—all made visual links between man and conquest.[20] These included dozens of photographs of men laboring at the site and the different machines they used to carry out their work. For instance, in an issue of *Bibliothèque de travail,* a series of educational publications for schoolchildren, an imaginary boy, Jean, leads the reader through the construction site. The images in the pamphlet not only reinforced close associations between masculinity, industrial work, and technology, a connection with roots in the industrial revolution, but also suggested fundamental ties between national identity, gender, and the transformation of nature.[21]

Yet not all contemporaries expressed confidence in the mastery of nature through large-scale industrial technology. In 1952, the journalist Jacques Sabran explained that during the Mistral, the strong, biting wind that flows down the Rhône valley from the Alps to the Mediterranean, "the old river, now domesticated, wakes up and furiously battles [Donzère-Mondragon's] dam."[22] Men may have "battled" the Rhône, but Sabran indicated that the river too might resist, challenging whether it had been "domesticated" at all. In fact, during the 1950s, several episodes diminished the CNR's supposed conquest. Floods caused setbacks when they interrupted construction in 1950 and again in 1951.[23] Then an unusually cold winter during 1955–1956 froze parts of the river, threatening the integrity of the recently completed project. Despite the CNR's aspirations, even Donzère-Mondragon could not always perfectly regulate the Rhône. Such problems already began to undermine the coupling of nature and nation (a theme I will explore further in the next three chapters). Overall, however, the French press as well as the CNR and its advocates exuded confidence that the agency's projects could control a river that had caused centuries of anxiety for those along its banks. These groups continued to hope that the Rhône might be conquered once and for all.

The challenges were many. Industrial machinery threatened men and their masculinity because it had largely replaced human (and specifically male) labor. Enormous pieces of industrial equipment carried out much of the construction of Donzère-Mondragon, especially when compared with the building of the Suez Canal less than a century earlier.[24] However, such concerns went beyond questions of job security. Several journalists, invoking Frankensteinian imagery, portrayed the new project as a traumatic disturbance that turned these machines into masters of nature and man alike. Journalist Claudius Deriol lamented that "on the work site, one only meets mechanical things, attentive and blind, like the lunar monsters that [H. G.] Wells described. Where are the men?" Other reporters emphasized that these were not just "mechanical things" but "new monsters [that] have names: bulldozers, scrapers." Moreover, they "devour the fields with terrifying bites." The world of technology had become monstrous and apocalyptic, emasculating and even dehumanizing. The writer J. P. Aymon described Donzère-Mondragon's construction zone as a "concrete palace" in which "we were lost in this cave of still damp concrete that resounded with the terrifying grinds of the newly domesticated powers. Lilliputians lost in the world of robots, our heads filled with these inhuman noises, our vision blurred." Indeed, contemporary images conveyed how the work site's immense scale literally overshadowed human presence there. Seven thousand laborers may have worked at the height of the project's construction, but they resembled ants in photographs and film footage, evoking Emile Zola's description of miners in his nineteenth-century novel *Germinal*.[25] Even worse, Aymon bemoaned, in these conditions humans themselves had become machine-like, "walk[ing] like robots."[26] Despite this discourse of technological apocalypse, critics of the industrial age voiced limited opposition to the river's development. In general, representations of the Rhône as a difficult river and of technology as a means to conquer nature bolstered the case for its taming.

Tensions of Modernism: Naturalizing Technology, Invoking History

Although man and machine may have battled against the Rhône, the "bullish" river did have one asset that both inspired and justified development. In a 1949, J. Labadie wrote of "the colossal productive power of the lower Rhône," while another journalist described the river as a "marvelous mine of energy" just waiting to be tapped. Some even saw the Rhône as "by far, the most beautiful mine in France."[27] By depicting the Rhône in terms of its potential energy capacity, these writers translated the river's ecological features into a language of energy and financial production. This rhetorical move reduced the river to a mere hydraulic object, thereby blurring the boundaries between nature and technology.

It is not surprising that CNR officials employed this strategy of naturalization. Invoking the "nature" of technology helped depoliticize it while justifying the agency's projects (and presumably profits). The CNR's General Director Tournier stressed the supposed symbiosis of the Rhône and the agency's projects in his book *Le Rhône, fleuve dieu, vous parle* (The Rhône, God-River, speaks to you). Assuming the river's voice, Tournier wrote, "I find the palaces where your engineers capture my power worthy of me."[28] Journalists echoed Tournier's assertion that CNR engineers and their technological "palaces" honored, rather than destroyed, the "God-River." A journalist from *Paris match* argued that there was an organic fit between Donzère-Mondragon and this reach of the Rhône valley: "here, for once, nature and men seem to be allied [together] to attempt a new adventure."[29] Other writers too alluded to this new collaboration: "I was daydreaming there, sitting along the banks of the Donzère[-Mondragon] canal, facing a spectacular sunset, looking not without the hostility of a peasant at this disruption of the earth. Nonetheless, the *grandeur* of the project imposed itself on me. When measured with the Rhône, [Donzère-Mondragon] does not dishonor [the river]. These works of pure lines and this flow of water are harmonious. They do not betray the spirit of the river."[30] Even Tournier, who had framed the transformation of the Rhône as a conquest, maintained that the river was a "conquered God, and not a beaten God; because one conquers by love."[31] Put another way, engineers knew and appreciated nature through their work and their mastery of that nature.[32] By invoking the naturalness of Donzère-Mondragon, this rhetorical move naturalized dramatic ecological change while helping create a sense of inevitability.

The CNR and its supporters even suggested that it was nature itself who authorized the agency's development program. As writer Pierre de Latil asserted, "nature designated [the Rhône] . . . to give France the greatest hydroelectric powers." Other advocates, without granting nature quite so much agency, argued that the Rhône's features inherently led to civilization. One writer professed that the Rhine and Rhône were twin rivers with "the Rhône finding, in its axial orientation, its peaceful vocation: to be one of the great arteries of Latin civilization" through which it would unify Europe. Similarly, in a 1959 article on the history of the Rhône in the children's series *Bibliothèque de travail,* Raoul Faure stated that "the Rhône valley is a route of natural penetration for civilization." The CNR's administration repeated this sexually charged assertion in an internal bulletin distributed to employees.[33] Such remarkable metaphors echoed earlier representations of the Rhône and its supposed relationship to French civilization. For instance, Gabriel Hanotaux (1853–1944), statesman and historian, went as far as to declare that "by the Rhône, France became France. France participates in its great Greek and Italian heritage because she has the Rhône. Without the Rhône, she would be turned uniquely toward the cold mists of the North. It is the Rhône that insinuates in her light, heat, the joy of being under the great living sun."[34] According to these views, French civilization sprang from the land or, in this case, the Rhône's flow. Moreover, this postwar emphasis on French civilization contrasted Gallic *grandeur* with German savagery. By this logic, not only was the river the source of civilization and the basis of France's special status but also it offered the ultimate sanction for development.

Historicization offered another powerful tool of legitimation.[35] The CNR and its supporters repeatedly cited the river's crucial role in the history of France and even the world as justification for their current efforts. The CNR's President Bollaert credited Herriot and Perrier with "intervening in the development of hydroelectric power," thereby creating "a means by which to give back to the Rhône its historic importance."[36] Other commentators stressed the point. A 1960 article in the technical journal *Water Power* claimed boldly that "since time immemorial, the Rhône has been one of the main arteries of communication in France," while Université de Lyon professor André Allix claimed that "the Rhône has always played in the history of the world a role out of proportion with its physical importance."[37] These observers all suggested that the CNR's projects were only the most recent step in a long lineage of French greatness.

Several writers invoked the Rhône's destiny, sustaining the idea that the CNR was simply fulfilling the river's fate. In *Rhône, dieu conquis*, Tournier titled his second chapter "In the Course of Destiny." The cover of Virenque's popular book of the early 1950s, *New Destinies of the Rhône*, even depicted the Rhône at the center of a wheel showing the six spokes of its "new destinies": navigation, electricity, irrigation, technical aid, energy, and financial benefits.[38] Fostering a sense of historical inevitability further naturalized the CNR's program for the Rhône.

Supporters employed another rhetorical strategy of historicization when they envisaged the future celebration of the Rhône's remaking. In other words, they implied that contemporaries were the fortunate witnesses to history. At Donzère-Mondragon's inauguration, headlines proclaimed that the celebration "marked a day in the history of France."[39] Others predicted that the names Donzère and Mondragon would soon be put alongside Suez, Panama, Tennessee, and Dnieprostroi, alluding to other famed projects around the globe. One journalist declared that the great-grandchildren of workers who had built Donzère-Mondragon "will learn the history of the great achievements of French genius and work." He even thought that the project might become the eighth wonder of the world.[40]

This construction of historical meaning for contemporary decisions came into tension, however, with claims about the project's relationship to the past. Although proponents of the Rhône's development repeatedly linked projects like Donzère-Mondragon to history, they simultaneously emphasized their newness, presenting them as ruptures with that past. As one newspaper headline exclaimed in 1952, "the new Rhône is born."[41] Other press coverage described the CNR's projects in ways that underscored their political and ecological genesis. In 1951, the journalist Marcel Carrière recounted the history of the Rhône's development, dating the idea to 1901, when "the idea of a giant project that would give to this corner of France a new physiognomy was born." As Professor Allix elaborated, "one can adopt the expression and speak here of constructing a river." Yet "the landscape will be less denatured than enriched," he claimed. Rather, the modern "face of France" would be, as of June 1952, "the new version of nature, desired and achieved by men seeking a better state of being."[42] While history might be the parent of river development, the CNR's projects also broke bonds with this heritage in order to redefine a nation that was being created, in part, on the Rhône. The experience and legacy of World War II helps account for the clash between these views of the projects as both manifestations of tradition and icons of modernity. While the projects might be justified as extensions of the past, perhaps the notion of "the new Rhône" helped break ties with certain aspects of that past: defeat, occupation, and collaboration.

This friction within a legitimating language of historicism illustrates one of the inherent contradictions of modernism. David Harvey has explained that an attempt to create the new while avoiding destruction of the old is one of the fundamental tensions within a modernist worldview.[43] One way to allay the view of the CNR's projects as disruptive was to portray them as achieving a balance between modernity and tradition. As an article in *Paris match* put it, Donzère-Mondragon enabled the "return to old sources [of energy] in order to forge the future," and other journalists emphasized that the CNR's "development plan" would "safeguard Rhodanian patrimony" for "future generations" and continue to allow traditional activities such as fishing.[44] The disruption that dramatic change brought could be partially masked, then, by forging strong continuities with the past. While these rhetorical strategies of naturalization and historicization served as powerful tools for advocates of the CNR's projects, they also highlighted modernism's intrinsic contradictions.

"The Symbol of French Radiance"

The historicization of the CNR's projects smoothed the past, present, and future of the Rhône and France into an unproblematic yet profoundly political narrative. It also revealed the insecurities of a country recovering from war. Postwar anxiety, coupled with constructions of a great past, contributed to the search for new symbols for the nation. "The new Rhône" seemed to answer that call. In 1952, Tournier waxed poetic in *Rhône, dieu conquis:* "promised glory, achieved not without pain! The sun, father of rivers, waits for the Rhône at the most beautiful part of its course and, in our descent of history, it there becomes the symbol of French radiance."[45] It was not by chance that Tournier made such sweeping claims in the same year of Donzère-Mondragon's inauguration.

Reconsiderations and reconstructions of national identity after World War II became closely intertwined with the CNR's transformation of the Rhône, which was epitomized by Donzère-Mondragon.[46] In his inauguration speech, President Auriol asserted that with this project "we have the right to mark the place of France."[47] Herriot also boasted that "more than proof of its renewal, with Donzère-Mondragon France offers to the universe a radiant demonstration of its vitality."[48] According to Herriot, it was not simply France, Europe, or even the world that would be honored by Donzère-Mondragon; the project was truly cosmic in scale. Journalists echoed these pronouncements in ample press coverage, especially during the weeks surrounding the inauguration. One reporter asked rhetorically on the day of the inauguration, "isn't [Donzère-Mondragon] a magnificent record of achievement, a splendid crown to place on the nation's head?"[49] For a country that had faced the humiliation of defeat and occupation just a few years earlier, this project could be held up as "a new victory on the French flag," one that eliminated any lingering doubts about the "greatness of our nation."[50] Politicians and journalists made Donzère-Mondragon a symbol of the new nation and especially its newfound *grandeur.*

What exactly was so great about Donzère-Mondragon? On one level, politicians, engineers, and writers highlighted the project's sheer scale and size, the literal meaning of the word *grandeur.* Contemporary reports frequently referred to Donzère-Mondragon as "a gigantic undertaking," "a giant work," "a colossal development," and "this undertaking of Titans."[51] These phrases often made their way into headlines and peppered patriotic exclamations in newsreels.[52] Journalists also emphasized Donzère-Mondragon's relative scale within France. They proudly noted that it was the largest construction site and the biggest hydroelectric plant in the country at the time.[53]

This strategy of comparison also worked on the symbolic level. Writers and politicians paired Donzère-Mondragon with familiar symbols of greatness. Like other new technologies, including the nuclear power program, Donzère-Mondragon became a modern-day temple.[54] In describing Auriol's visit to the project, a journalist reported that Donzère-Mondragon's hydroelectric plant was "in the middle of the plain like Chartres Cathedral," while de Latil noted how "the amount of concrete, which made up the dam, plant, and locks [of Donzère-Mondragon] reached three times the volume of Notre Dame of Paris."[55] In a glossy coffee-table book about Donzère-Mondragon, an illustration represented how much dirt had been moved in order to build the project, an amount equal to the area of the Place de la Concorde filled to the height of the Eiffel Tower.[56] These writers thus constructed mental maps for readers in terms of famous landmarks, not to mention some of France's most historic icons.

Of course, Donzère-Mondragon's aesthetics made comparison with these religious and architectural symbols problematic. Its massive, stark concrete structures hardly resembled the heaven-bound architecture of a Gothic cathedral or the Eiffel Tower, that late-nineteenth-century masterpiece of steel modernity.[57] But as Figure 2.2 shows, contemporaries were quick to seize on any similarities between these architectural icons. Furthermore, one author argued that Donzère-Mondragon had its own different, decidedly modern, but still attractive aestheticism. Virenque declared that his contemporaries should look on Donzère-Mondragon with more accepting eyes. "In several years," he correctly predicted, "other giant work sites will open up and continue to transform this valley and its cities. Without a doubt, some people will cry vandalism." But he decried those who condemned these future projects because they might "disrupt the admirable landscapes, which have been here for centuries and centuries."

Virenque countered that these projects in fact fit within a logic of historical greatness that originated on French soil and at the hand of man. Tapping into the sexualized discourse on the Rhône and its role in French civilization, Virenque asked rhetorically, "Isn't it in the destiny of this land to be marked by civilization forever? Isn't the greatness of these landscapes composed of what the touch of man has discovered there?" Although "cement, metal, and concrete" might appear antithetical to French *grandeur,* Virenque suggested otherwise. After all, these materials "[do] not prevent the arc de triomphe at Orange or the ramparts

Figure 2.2. *Interior of André Blondel hydroelectric plant.* This photograph, dated November 10, 1952, shows part of the generator room of the Blondel hydroelectric plant at Donzère-Mondragon soon after the project's inauguration. It was reproduced in several CNR brochures, as well as newspaper and magazine articles. The windows and lighting suggest the interior of a church, a convenient parallel for those who framed postwar hydroelectric projects as the nation's new cathedrals. (Courtesy of Compagnie Nationale du Rhône.)

of Avignon from challenging the centuries. Moreover, these new walls of concrete, these giant locks also have their own beauty. A beauty that one must learn to appreciate, without a doubt, because one isn't used to it, [it is] a beauty of a new age: the age of dams and pioneers, since the age of cathedrals is over."[58] Virenque maintained that France could—and should—be a modern, technological nation, and that Donzère-Mondragon helped the nation take another step toward its new future and new identity.[59]

The *grandeur* of Donzère-Mondragon and the CNR's other projects was also measured relative to public works projects around the world. As a newsreel from 1950 put it, "one of the biggest work sites in the world at the present is in France."⁶⁰ These international comparisons revealed the extent of the country's postwar insecurities. As shown in Figure 2.3, the CNR and press compared Donzère-Mondragon most frequently to the Suez Canal and, though less often, the Panama Canal. A headline in *Science et vie* declared simply, "bigger than Suez and Panama."⁶¹ Others quantified and ranked the projects in terms of units of dirt, labor, and machines.⁶² Repeated references to the world records that Donzère-Mondragon had set further reinforced the display of its international greatness.⁶³ This competitive spirit even entered into children's books. One showed how the project's excavations were twenty-five times the volume of that of a single Egyptian pyramid.⁶⁴

Amid these international comparisons, the United States earned special attention. More than any other country, it threatened French *grandeur*.⁶⁵ Labadie attempted to assuage these insecurities when he claimed that the "gigantic project concedes nothing to those in America."⁶⁶ Other journalists argued that "with its fifty million cubic meters of excavations, the work site at Donzère-Mondragon ... belongs with the most audacious American realizations."⁶⁷ In fact, "the biggest building site in the world," they bragged, "is under way not in America but in France."⁶⁸ Apparently, the nation could hold its own. One writer even dismissed American achievements when he claimed that the CNR's projects on the Rhône posed greater technical challenges than recent projects that had been completed on the Colorado and Columbia rivers.⁶⁹

Yet silences were also telling. Discussions of Donzère-Mondragon usually ignored the fact that American dollars channeled through the Marshall Plan had subsidized over 90 percent of the project's costs.⁷⁰ The CNR received almost 2 percent of all Marshall Plan funds distributed in France between 1948 and 1951, the vast majority of which were directed to building Donzère-Mondragon.⁷¹ In a twenty-page, photograph-filled story on the project, the reporter Albert Plécy noted that "all material

Figure 2.3. *Side view of Donzère-Mondragon.* This CNR-produced image, which was included in a promotional brochure about Donzère-Mondragon in the early 1950s, offers not only a side view of the project but also striking comparisons with the Kembs and Suez canals. Most of the maps, charts, and other data about Donzère-Mondragon in the brochure were highly technical, and the profile shown here is certainly no exception. However, the juxtaposition of these canals emphasizes Donzère-Mondragon's impressive standing among such projects and indicates how conceptions of *grandeur* were literally inscribed into even the most technical documents. (Courtesy of Compagnie Nationale du Rhône.)

PROFILS EN TRAVERS DE LA DÉRIVATION ET COMPARAISON AVEC D'AUTRES CANAUX

Echelle: 1/2000

CANAL D'AMENÉE DE DONZÈRE-MONDRAGON

CANAL DE KEMBS

CANAL DE FUITE DE DONZÈRE-MONDRAGON

CANAL DE SUEZ

used at Donzère-Mondragon is new," neglecting to mention that the Americans had paid for this equipment.[72] These convenient omissions distanced the project from its financial dependency on the United States and consequently reinforced its supposed "Frenchness." Publicizing American financing might undermine the project's national, not to mention nationalist, prestige.

The obsession of journalists, politicians, and engineers with the "Frenchness" of Donzère-Mondragon illustrates the specific links they forged between national identity on the one hand and science and technology on the other. Popular accounts highlighted the important roles that "French technology," "French technologists," and "French genius" played in the project's successful completion.[73] The inauguration of Donzère-Mondragon particularly stressed the project's "Frenchness." As one headline blazed: "Donzère-Mondragon: French victory."[74] However, the nationality of its experts and even construction of the project on French soil made neither the technology nor the project inherently "French." Rather, that politicians and the press explicitly invoked French nationality and associated it with Donzère-Mondragon shows how they mutually constructed national identity and technology.[75] Nature and its transformation played, however, a crucial role in this formulation.

Even those who criticized Donzère-Mondragon for its working conditions and social organization celebrated "French" technical genius. According to the leftist *Allobroges,* a Grenoble-based newspaper for the Alps and Rhône valley, Donzère-Mondragon confirmed "a victory of technology and French genius."[76] Journalists who envisioned a different France than the one the CNR helped build may have taken issue with the labor relations at the site or with the relationship between the CNR and the state, but not with the agency's overarching mission for the Rhône. From a wide range of political positions, then, the "French" technology and river's remaking embodied at Donzère-Mondragon verified the nation's restoration.

Despite the many references to France's refound *grandeur* in political, popular, and technical documents, Aymon bemoaned how few French citizens celebrated Donzère-Mondragon, criticizing "the absence of French pride" in "this colossal undertaking." He complained that foreigners were more excited about the project than many of his fellow citizens. Yet Aymon also wondered when tourists would quit thinking of France as a quaint country composed of peasants. He scoffed at those who traveled to Montélimar to sample the town's famous nougat when only a few miles away, Donzère-Mondragon was then producing 20 percent of France's electricity.[77] Like Virenque, Aymon welcomed a new, technological nation, one that promised to restore French *grandeur* amid the polarization of international relations during the Cold War and the fading power of France on the international stage. The new Rhône embodied and would help achieve this goal.

River of Promise

From 1945 through the 1960s, those who promoted the development of the Rhône invoked Donzère-Mondragon and other CNR projects not only as symbols of the country's political and economic reconstruction but also as the means to achieve it. In a 1960 book, *En descendant le cours du Rhône,* Tournier traced the river's journey from the Swiss Alps to the Mediterranean as he mapped out France's history over the previous two millennia. He concluded with an extensive discussion of how the CNR's projects would enable France's modernization.[78] Through these projects, a transformed Rhône promised to rebuild the nation, materially and culturally. Donzère-Mondragon would help ensure that the country had ample electricity, but the project's importance extended into the symbolic realm as well. A journalist from *Le progrès* held up Donzère-Mondragon as "the centerpiece of our national economy," and de Latil proclaimed that the CNR's program "changes [France's] future."[79] These narrative frameworks, both textual and visual, presented the transformation of the Rhône as a way for France to achieve wider economic and industrial progress.

The process of rebuilding after the war, which was exemplified by projects like Donzère-Mondragon, entailed extensive expropriation, social dislocation, and environmental transformation. Most politicians and writers represented the disruption wrought by the CNR's projects as positive, focusing on the projects' numbers—from the hectares of land appropriated to the volume of concrete needed and the quantity of electricity it promised to produce. However, these figures quantified, objectified, and thus abstracted dramatic changes that were at once social, economic, and environmental.[80] For instance, in depicting scenes of construction, the film *L'or du Rhône* portrayed the march of progress through its juxtaposition of imagery and music. Regional musical themes accompanied scenes of a quaint, provincial countryside before construction, triumphant, brass-dominated melodies accompanied images of razed homes seen just seconds before.[81]

The documentary therefore conveyed how the development of the Rhône aimed to bring France into the modern industrial world. In a 1950s *Paris match* article, one journalist claimed that the broader mission to develop the Rhône valley, including the CNR's projects, "will bring industrial France into the Europe of the twenty-first century." It was here, along the central Rhône, that "France already takes its face of tomorrow." He credited mayors, bureaucrats, and CNR leaders with "giving birth to the twenty-first century along the banks of the Rhône." Accompanying photographs showed the suited visionaries huddled around a large conference table, deep in conversation, with large maps of the river valley on the wall behind them.[82] Since all of these leaders were male, this was an unusual birth indeed.

Furthermore, proponents of the Rhône's development declared that projects like Donzère-Mondragon could achieve economic and political renewal, and at the same time social unification. These comments hinted at the extent of social and political instability during the postwar era. France suffered from massive strikes after World War II, and the Communist Party was expelled from the tripartite government in 1947. The state attempted to transcend deep-seated hierarchies through cooperation between trade unions and capitalists while avoiding the centralization of a command economy.[83] Amid this social strife, the transformation of the Rhône aimed to reconstruct not only the material infrastructure and political economy of the country but also the social fabric of a troubled nation.

Media coverage of Donzère-Mondragon's inauguration emphasized the project's ability to reach across social and economic divisions. In his speech, President Auriol somewhat blurred class and profession when he referred to the engineers, the artisans, and the seven thousand workers who had together "contributed to the conception or the execution of the great common undertaking."[84] Auriol asserted that large-scale projects of national reconstruction could unify a populace fractured by war. "Collective labor accomplishes miracles," he proclaimed, adding that "the undertaking that we celebrate today, the product of a team of workers from all orders and all ranks, is an example for the union of French citizens."[85] The CNR included Auriol's speech in the first issue of its internal company bulletin after Donzère-Mondragon's inauguration, sending a message to workers who had in fact organized dozens of strikes during the project's five-year construction and even shut down the work site for over an entire month.[86] Several years later, Tournier echoed Auriol's collaborative message while emphasizing that this "common" work could be undertaken without violence or revolution: "here in the valley blessed with sunshine, public collectivities have parity with industrialists in a cooperative effort to make the Rhône, navigable waterway of a great past, work by ensuring a source of energy for a great future and contributing to the revitalization of the Rhône."[87] The Rhône therefore served as a powerful laboratory for reconstructing not only the river but also the whole of French society.

Standing in the way of progress, however, were peasants and locals—at least according to the project's advocates, who might have argued that the Third Republic's efforts to turn "peasants into Frenchmen" was far from complete, even in the early 1950s.[88] Several journalists juxtaposed peasants' protests over the loss of "their land" with the completion of a modern, technological marvel. Sabran called Donzère-Mondragon "one of the most efficient triumphs of modern science" and noted that an impressive modern landscape now stood where almond trees, symbolizing the cultivated land of the peasant, had bloomed just six years before. Sabran believed, perhaps hopefully, that locals would eventually be persuaded; patriotism would ultimately override the loss of land and livelihood. Another journalist told of "plows . . . being replaced by machines," concluding that "progress was stronger everywhere." By portraying local communities as ignorant and traditional, the project's proponents suggested that the nation would ultimately move forward with or without them.[89]

A broad coalition, therefore, backed the CNR's program for the Rhône, but some residents in the surrounding region joined the apparently reticent peasants in harboring misgivings about Donzère-Mondragon. Few left their mark on official records, but numerous letters written to the CNR about the project's effects on groundwater stand as testament to their story (the focus of Chapter 4). Although some locals demanded repairs or reparations, most sought ways to live with the project. No one fundamentally challenged Donzère-Mondragon; most simply complained about the impact on their land and lives.[90] These critics were not environmentalists, and their arguments were not grounded in support of nonhuman nature. Rather, they inveighed against the socioenvironmental costs imposed on those living in the shadow of the project.[91] Locals' apparent ambivalence may be partially explained by the fact that many did profit from it; the CNR's projects subsidized rural electrification and irrigation. While those who lived near Donzère-Mondragon experienced both the costs and benefits of development, only a handful of journalists tracked the environmental impact of the project or regretted the "heavy tribute" that locals had paid for "the great undertaking" during the postwar era.[92]

The most vocal critics of the Rhône's postwar development objected to its social, rather than environmental, consequences. Labor unions and left-leaning political parties did not oppose the CNR's projects per se. Rather, they lobbied against the organization of the projects and their relationship to the state, sharing a modernist confidence in large-scale technological systems but envisioning a different organization and purpose.[93] These criticisms did not have a significant impact on the Rhône's transformation after World War II, but (as Chapter 6 will show), they became central to debates beginning in the late 1960s.

Many of the ways politicians, engineers, and journalists articulated their understandings of the Rhône and its remaking during the postwar era reflected not a specifically French sensibility but the intrinsic contradictions of modernism. Advocates justified the river's transformation by creating a smooth lineage between the nation's past and the CNR's projects, while at the same time claiming that these projects marked a rupture with that very past. They naturalized the CNR's projects by asserting that they fulfilled the Rhône's inherent qualities yet also characterized these projects in terms of dramatic change, disruption, and ultimately progress.

Although this discourse and the tensions embedded within it may not be uniquely French, its articulation and deployment were intimately tied to the specific context of post-1945 France. Here World War II was crucial. The war and its legacy played central roles in mediating how France's political and technical elites developed the Rhône. On one level, the particular context of the postwar period shaped the literal remaking of the river. The nation's energy crisis, among other factors, pushed the CNR to make hydroelectric generation its top priority even though the agency was constitutionally bound to build multipurpose projects.

The postwar context also shaped the discursive reconstruction of the Rhône: the representations and meanings ascribed to the CNR's projects and the larger process of industrialization and modernization of which they were a part. Amid the ruins of war, France's leaders and citizens debated the future contours of the nation's political, economic, and social orders. They were also faced with the ways the Cold War began to reconfigure the international landscape. In this context of political and social flux, politicians, engineers, and writers held up the CNR's project, Donzère-Mondragon, as an icon of national *grandeur*. According to this formulation, the country was once great and could be great again. These groups sought to restore their country's former glory and appropriated technological artifacts like Donzère-Mondragon in order to realize these claims. This project was not the only technological system celebrated in this way. Nor was France the only nation that created such narratives around large-scale technology during the twentieth century. Nonetheless, the postwar transformation of the Rhône exemplifies the mutual construction of technology and culture as well as the centrality of nature and technology to conceptions of national identity and the nation-state.

In fact, the seeming restoration of French *grandeur* at Donzère-Mondragon actually masked deep divisions that threatened the nation. Continued political instability and intensified conflict between owners and workers marked the early years of the Fourth Republic. Despite—or more likely because of—this instability, the coupling of nature and nation initially held firm. Though their visions of reconstruction and national identity differed, diverse groups supported the Rhône's development and shared a belief in its tangible demonstration of French radiance. During this period, the CNR's administration, engineers, and workers began to translate an ideal vision into literally concrete "technological" artifacts and systems. The nationalization of the Rhône after World War II thus set the stage for the materialization of new enviro-technical systems that remade the river and, in the process, France itself.

3

POSTWAR TRANSFORMATIONS

Between the end of World War II and the late 1960s, various groups and institutions enrolled the Rhône in their agendas for postwar France. In addition to the CNR, several state agencies, including the Commissariat à l'Energie Atomique (CEA), EDF, Génie Rural, and Service de la Pêche (Fishing Service) each undertook projects. Members of local communities and organizations also involved themselves in the river's development as advocates, critics, observers, and participants. The projects' diverse sponsorship suggests how the postwar Rhône served an array of human users with disparate interests.

While these groups' distinct priorities led to many disagreements, all regarded the river as a technology—an artifact or instrument with which to do something, a common means to quite different ends. For CNR officials, the Rhône promised to produce vast quantities of electricity while satisfying the agency's requirements to promote navigation and agriculture. For those associated with the CEA and EDF, the river appeared to ensure that France could develop nuclear capabilities, both in terms of energy production and weapons. The Génie Rural's agricultural engineers saw the Rhône's potential in terms of modernized farming. This chapter examines how these groups eventually completed numerous projects to redesign the Rhône in the form of new envirotechnical systems in order to achieve the goals of their respective envirotechnical regimes.

In tracing the history of the Rhône's postwar transformations, this chapter explores how these envirotechnical systems could be at once complementary and contradictory. For instance, the nuclear development of the Rhône depended on the CNR's management of the river. Yet the CEA's and EDF's atomic ambitions for the Rhône eventually reshaped the CNR's management of it. Similarly, completion of the CNR's envirotechnical system increased the amount of available irrigation water, expanded irrigation infrastructure, subsidized rural electrification, and reorganized farmland, thereby increasing agricultural productivity in the central and lower Rhône valley. Yet the agency's projects, planted in the middle of fields and vineyards, also created problems for the valley's farmers. But conflicts *among* different agencies, and their respective envirotechnical systems and regimes, were not the only ones to emerge. Even the CNR's technical elites faced challenges designing the agency's own system because of its multipurpose mandate. While these projects together transformed the Rhône in the years following World War II, ultimately their sponsors had to work with and adapt to the river's hydrology and ecology—even if their knowledge of these systems was partial (as Chapter 4 will show). In fact, the Rhône's "natural" attributes not only exposed but also heightened conflicts among different institutions' envirotechnical systems.

This chapter thus describes the contested process by which multiple constituencies literally reconstructed the Rhône during the postwar era. Drawing on methods from the history of technology, I focus on the design and development of these projects, emphasizing the negotiations over their objectives and "technical" features. After examining the CNR and its efforts, which were dominated by hydroelectricity, I turn to the CEA's and EDF's nuclearization of the river before concluding with a discussion of the CNR's and Génie Rural's agricultural reforms in the central and lower Rhône valley. These three dimensions of the Rhône's remaking demonstrate how diverse groups' attempts to transform the river into new envirotechnical systems were repeatedly complicated by other systems and ecological processes.

Deciding the "Rhône Formula"

By 1935, only a few months after the state granted the CNR its concession, the agency's technical committee, a body composed of political and technical elites from the CNR and several key state agencies, had already generated a blueprint for the probable locations of the CNR's future projects along the river's course through France. The CNR's technical committee had not yet identified, however, the model for these projects, what would become known as the "Rhône formula" *(la formule du Rhône)*. This "formula" was critical for at least two reasons. Not only would it guide the CNR's design of each individual project but also it would illuminate how the agency intended to transform the Rhône to realize its overarching mission. In particular, the CNR's "formula" needed to balance the demands of hydroelectricity, navigation, and agriculture to fulfill the agency's multipurpose mandate.

During the first few months of the CNR's existence, its engineers and administrators considered four options as models for the Rhône's development. First, they studied a high-chute project, which would entail construction of a high dam in the middle of the river and result in a long reservoir. Second, they debated building a series of low-chute dams and locks that would essentially turn the river into a channelized cascade of small dams. Third, they considered a long lateral canal paralleling the Rhône south of Lyon with a series of hydroelectric dams and locks; this option envisioned construction of a second "river" roughly paralleling the first. The fourth design combined elements from the second and third; it involved building a series of shorter diversion canals. Each canal would send the Rhône's waters into a separate channel on which a low-chute dam and lock would be located; the canal would then rejoin the river before being diverted once again downstream.[1] Although the distinctions among these options may appear wholly technical at first glance, historians of technology have demonstrated how "technological" choices inevitably involve social, political, economic, and cultural considerations. They have rarely, however, fleshed out the way these considerations both embody and shape interactions between humans and the natural world.

The CNR's technical committee quickly turned to the fourth option, the diversion approach (see Map 3.1), as its model for the Rhône's development. At the committee's first meeting in December 1934, it recommended that "long diversions" be studied further so that it could formulate an overall recommendation. At its next meeting, however, engineers from the prestigious Conseil Général des Ponts et Chaussées questioned the diversion approach because it appeared "more costly" than other alternatives. They preferred the second option, the channelization of the Rhône. Disagreement between the two agencies proved short-lived, however. Although engineers from Ponts et Chaussées expressed reservations in early 1935, their opposition appears to have dissipated by 1941.[2]

Despite the fact that France was soon embroiled in war, facing battles on the home front and a paralyzed economy, the CNR forged ahead with its redesign of the Rhône. In many ways, it is astonishing that the

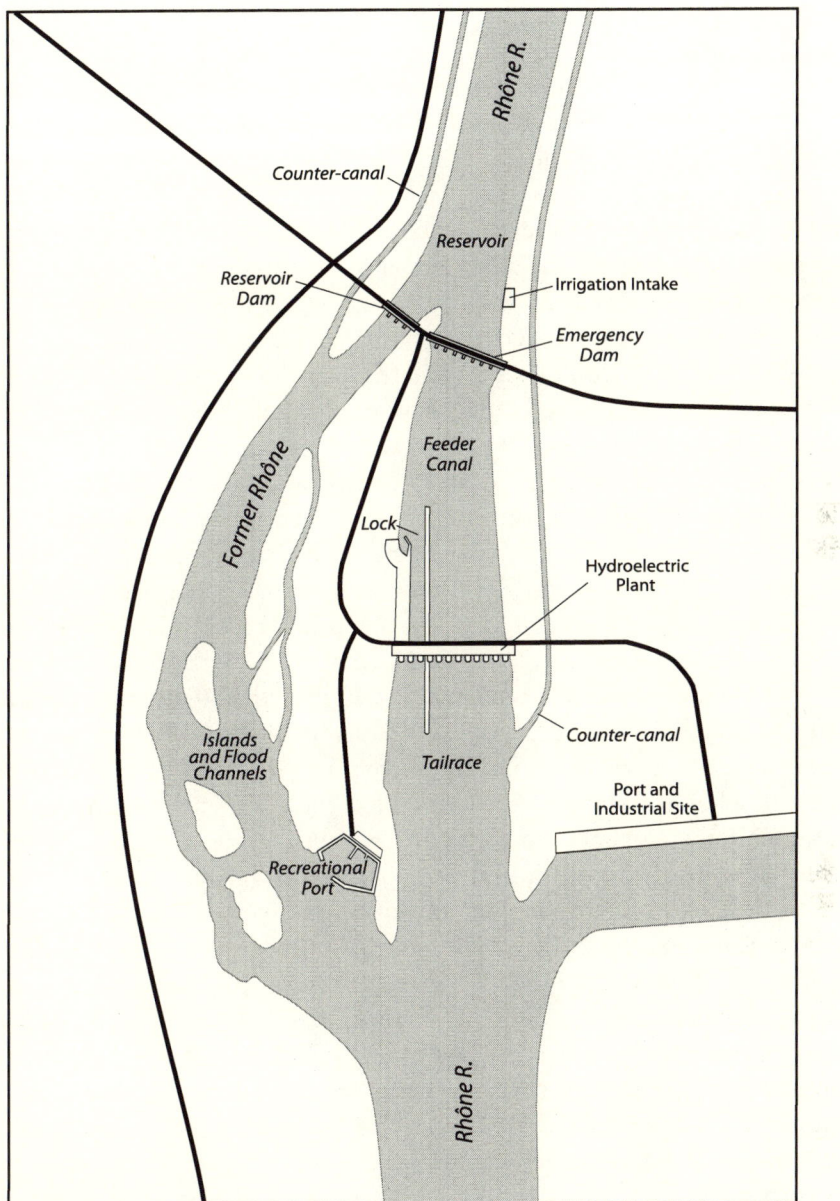

Map 3.1. *The Rhône formula.* The CNR's "Rhône formula" aimed to foster energy generation, navigation, and agriculture simultaneously through what was called the "diversion approach." This drawing shows the formula's major features. The diversion canal (including feeder canal and tailrace), former Rhône, hydroelectric plant and lock, reservoir dam, and emergency dam are particularly notable. (Map by Joseph W. Stoll, Syracuse University Cartographic Laboratory.)

agency made any progress, given the country's dire wartime conditions, but while construction may have slowed, planning proceeded apace. The CNR's technical committee continued its work while France was actively at war and then under German occupation. At a meeting in 1941, the committee outlined the proposed features of Donzère-Mondragon, what would become the agency's first postwar project.[3] Located in the central Rhône valley between Valence and Avignon and named after the towns at the northern and southern ends of the canal, the project included a diversion canal 25.8 kilometers (approximately sixteen miles) long. This long canal was a central feature of the project, to be sure, but by 1946 it had also come to serve as the principle guiding the agency's future development of the French Rhône. As the CNR's President Jean Aubert explained in 1943, "the system that is currently envisioned, and I hope will soon be in the process of construction, consists of channelizing this central third [of the Rhône] and associating low-chute dams with very long diversions." According to this design, each project's diversion canal, hydroelectric plant, and reservoir complex would serially reroute most of the river's flow. Notably, the exception to the "Rhône formula" was the CNR's first project, Génissiat, undertaken in 1937 but only completed after World War II. Located in the steep gorges of the upper Rhône, Génissiat did not have a diversion canal but rather, like the Hoover Dam on the Colorado River, a high-chute dam that completely blocked the river before a dramatic sixty-nine-meter drop.[4]

The CNR's technical committee and administration cited a combination of economic, political, and environmental factors to explain why they had selected the diversion approach as their "formula" for the Rhône's remaking. First, the CNR's administrators had to deal with serious financial constraints, especially given the economic crises France faced during and immediately following World War II. The CNR's President Aubert stated that his agency had rejected the lateral canal, the technical committee's third option, because of the "enormity of expenses" associated with its construction. He added that CNR officials had selected a single diversion canal rather than double diversion with separate canals for the hydroelectric plant and locks also because of cost.[5]

In addition to these budgetary considerations, CNR leaders and their supporters maintained that the diversion approach offered an effective way to comply with the agency's multipurpose mandate. As a government publication explained in 1948, it had been "conceived to accomplish simultaneously all three [of the agency's] missions." Apparently, however, this declaration did not convince all interest groups in the Rhône valley. Even after construction of Donzère-Mondragon began in 1947, navigation constituencies continued to question the compatibility of energy generation and navigation. Perhaps not surprisingly, CNR officials defended their agency's "formula." In 1946, J. Bonnier, a CNR engineer, explained that his agency's projects would "capture the maximum power of the river, but equally bring to navigation more certainty and speed." He asserted that the projects even promoted "the idea of progressively transforming the Rhône into a great artery for modern navigation." Government officials came to the CNR's defense, avowing that the diversion approach effectively met the demands of multipurpose river development.[6]

The CNR's administrators also argued that the natural features of the Rhône valley played an important role in the agency's decision. As the Rhône flowed south of Lyon, the river alternately bumped up against mountain ranges on its eastern and western banks. At times, these mountains were close together, making repeated short stretches of the valley narrow. Consequently, the central Rhône valley's topography made the construction of a lateral canal extremely difficult because there was literally little room for it. In addition, a number of rivers and streams fed into the Rhône, forcing the agency to address how it would preserve runoff patterns. The CNR's general director, Pierre Delattre, concluded that the "habitat and richness" of the Rhône valley "forbade vast submersions," which would result from the construction of a high-chute dam. It was less nature per se than the probable environmental and political repercussions of such a design that made high-chute dams unfeasible. After all, if built, these projects would inundate thousands of acres, including cities, villages, and transportation networks. Furthermore, such flooding would submerge vast tracts of farmland, thereby undermining the CNR's mandate to promote agriculture. As Bonnier wrote in 1946, "having considered the topography of places [along the central Rhône], [the CNR] can no longer even think of creating large reservoirs in this section of the river, but rather must conceive of development in the form of diversion." Because of the valley's topography and long human presence, "the CNR had, therefore, to design the development [of the Rhône] as a continuous chain of stations exploiting low heads with distances of about fifteen to thirty kilometers between each [project]." Top CNR administrators added that "new countries" like the United States could build high-chute projects, but France could not.[7] These concerns solidified the agency's selection of the diversion approach as its envirotechnical "formula" for the reconstruction of the Rhône.

Although the CNR's elites might have interpreted the topography, hydrology, and ecology of the Rhône valley as impediments to achieving their ambitious mission, they tried to use these features to their advantage. In 1952, Delattre characterized the alternating mountainous topography of the central Rhône as "propitious to the construction of large diversions" because his agency could build a diversion canal in each widening of the valley. The CNR's officials hoped to improve navigation by taking advantage of this alternating topography in two ways. First, the CNR's diversion canal offered a shortcut to the river's long curve as it flowed through wider parts of the Rhône valley. For example, Donzère-Mondragon's diversion canal shortened the curve of the Rhône as it descended to the west from the town of Donzère back to the east toward the village of Mondragon. Second, this canal eliminated hazardous rapids, which had endangered the lives of mariners and limited the Rhône's navigability for centuries, and offered a safer alternative to the original riverbed.[8]

The CNR's officials also hoped to use the river's floodplain to their agency's benefit. Preserving an active floodplain might better protect the CNR's projects while decreasing the severity of floods by sending excess water into the surrounding plain and thereby delaying the flow of this water downstream. Alternatives to the diversion approach could not make such a promise. A high-chute dam and large reservoir would have already inundated thousands of acres upstream of the dam. A series of low-chute dams in the middle of the riverbed would keep too much water in the Rhône during floods rather than allowing excess water to spill into the surrounding plain, intensifying flooding downstream and undoubtedly inciting local complaints. The diversion approach appeared, then, to strike an effective compromise among upstream and downstream users as well as with the CNR's multipurpose mandate. Although the CNR's engineers sought to rationalize and simplify the river, on some level they also aimed to work with the ecological features of the Rhône valley.

While a diversion canal may have avoided certain problems, it too proved difficult during times of flood. If, as predicted, the diversion canal raised the level of the river's flow several meters above low-water levels, then large dikes would be needed to brace the canal. Aubert stated that locals would probably oppose such infrastructure. Furthermore, trying to channel all of the Rhône's waters, including extremely high flows during thousand-year floods, might only worsen flooding downstream, since huge dikes would not allow water to be released from the diversion canal into the surrounding floodplain. During such extreme floods, the canal might even increase the likelihood of dike breakage and consequently severe localized flooding.[9] The CNR's engineers concluded, therefore, that Donzère-Mondragon's diversion canal needed to be large enough to channel sufficient water for its hydroelectric plant—the exact quantity of which proved a controversial decision (as we will see)—but also needed to be able to send excess water into the river's floodplain. This confluence of financial, political, and environmental constraints together help to explain why the CNR's administrative and technical elites ultimately selected the diversion approach as the Rhône formula. Over the next five decades, this "formula" guided the design and construction of the vast majority of the CNR's projects along the Rhône, including Donzère-Mondragon (see Map 3.2).

Rerouting the Rhône

Consensus in favor of the diversion approach did not determine, however, *where* Donzère-Mondragon's canal would go. The precise route of the diversion canal was important for at least two reasons: the path determined how much electricity the project could produce and which communities it would affect. France's postwar energy crisis and reconstruction efforts pushed the CNR's leaders to seek, and ultimately select, a route that favored high energy production. After all, the farther the river's flow dropped, the greater the project's energy potential and thus the CNR's financial yields. While the agency's engineers hoped to take advantage of the river's steep course through the central Rhône, other factors complicated their selection of the diversion canal's path. They had to integrate Donzère-Mondragon into existing social, natural, and built environments without exceeding the CNR's limited budget.[10] Politics, economics, and the environment all challenged the CNR's aim of maximizing the Rhône's hydroelectric potential.

Population centers and productive farmlands seemed to limit the CNR's choices for the diversion canal's route. Expropriation and social dislocation potentially entailed high costs, both financial and political. Nonetheless, although the CNR's administration often maintained the

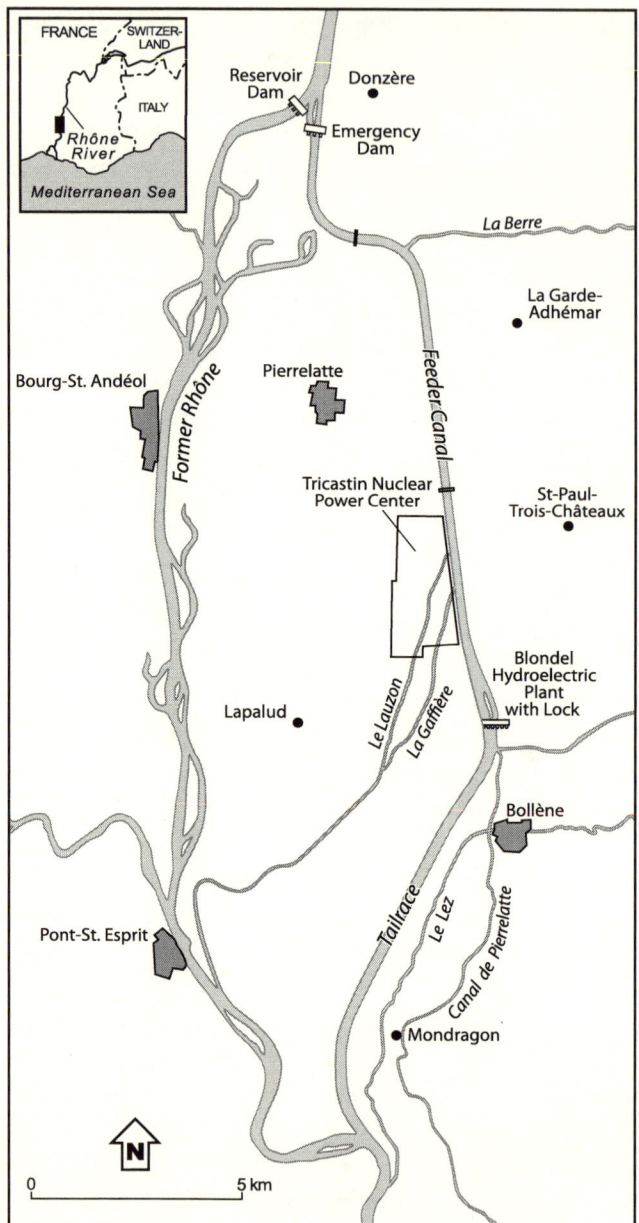

Map 3.2. *Donzère-Mondragon*. The Rhône formula at work at Donzère-Mondragon. Comparing Maps 3.1 and 3.2 suggests how the political, economic, and environmental specificities of this region mediated the particular realization of the Rhône formula in this part of the central Rhône valley. (Map by Joseph W. Stoll, Syracuse University Cartographic Laboratory.)

impossibility of flooding towns and farms, the agency proved largely insensitive to the projected impact of Donzère-Mondragon on locals. In June 1947, one CNR engineer acknowledged that the CNR's preliminary route for the canal had incited ardent protest, before adding cynically that whatever route the agency selected, there would always be incensed farmers.[11]

The CNR also had to contend with the dynamic hydrology of the Rhône's watershed. Rivers and streams that fed into the Rhône presented the CNR's engineers with several challenges. Alpine streams are more predictable and therefore less dangerous than the tributaries along the western bank of the Rhône, whose sudden and severe floods during the rainy season are volatile and difficult to control.[12] Accordingly, the CNR's engineers decided to place Donzère-Mondragon's diversion canal to the east of the Rhône (see Figure 3.1).

In addition, the CNR's technical experts needed to ensure the integrity of the project, especially the locks and the Blondel hydroelectric plant, through their selection of an appropriate geologic site. In fact, several engineers thought the agency's highest priority should be to locate the project on a solid geologic foundation. In July 1946, the CNR's technical committee reported that geologists had found "an absolutely negligible difference of risk" between the two routes the agency was considering at the time. Despite these assurances, CNR officials continued to express concerns about the project's geologic stability. Moreover, members of the CNR's technical committee acknowledged that location had significant financial implications—for better and for worse. By selecting a site with geologic properties that better fit the project's ostensibly technical features, the CNR hoped to take advantage of the valley's characteristics and thereby to minimize dredging and the amount of infrastructure it would need to build.[13]

These issues help explain why the CNR's engineers revised the preliminary route proposed by the agency's technical committee. In its initial proposal, the committee had planned a series of short locks between *point kilométrique* (PK) 17 and 22 (the reach of the diversion canal seventeen to twenty-two kilometers after the point of diversion), with the Blondel plant located at PK 21.8. However, in 1947, several private companies recommended that both the locks and hydroelectric plant be located on solid bedrock and identified a promising site near PK 17. They urged the CNR to replace the series of short locks with a single lock twenty-six meters high, and to move both the locks and hydroelectric

Figure 3.1. *Building Donzère-Mondragon: the tailrace.* This CNR photograph of the tailrace of the diversion canal was taken in September 1949. It captures some of the extensive excavations undertaken to complete the project. (Courtesy of Compagnie Nationale du Rhône.)

plant to PK 17. They emphasized that these design modifications had significant benefits for the CNR, from improving the project's geologic security and saving the CNR money to boosting Donzère-Mondragon's hydroelectric potential by making the water drop further before passing through the plant's turbines.[14] By July 1947, the CNR's *conseil d'administration* had approved these changes.[15] Construction of the Blondel plant soon followed (see Figure 3.2).

As these modifications suggest, the CNR's leaders had to reconcile the project's multiple goals, the agency's wider political and economic context, and the particularities of the local environment. The CNR's decisions often came at the expense of local human communities. In 1952, Delattre recounted that his agency had altered the original route of the diversion canal to take advantage of the region's geology. He admitted that the "inconveniences" of the new route were "far from negligible," since the diversion canal now traversed the most inhabited region of the river's left bank, including the towns of Donzère, La Garde-Adhémar, Bollène, and Mondragon. He even noted that numerous properties had

Figure 3.2. *Building Donzère-Mondragon: Blondel hydroelectric plant.* This CNR photograph taken in November 1949 shows the plant at an early stage of construction. The vast excavations, which increased the project's hydroelectric potential, are most significant here. The lone man walking uphill (right) gives a sense of scale. (Courtesy of Compagnie Nationale du Rhône.)

been expropriated and the CNR's lengthy diversion canal bifurcated many others. Yet Delattre asserted that the drawbacks of this new route were far outweighed by its advantages, from preserving the river's floodplain to reducing costs by eliminating the need for high dikes.[16] While these may have been benefits for the CNR, they were costly to many locals. But when faced with either burdening local people or contending with the local environment through its selection of a route for Donzère-Mondragon's diversion canal, the CNR chose not to risk the wrath of nature.

Regulating the Rhône

After the CNR's administrative and technical leaders had identified *la formule du Rhône* and a specific route for Donzère-Mondragon's diversion canal, they still needed to decide how to manage the river itself: they needed to determine *how much* of the Rhône would flow *where* and *when*. Controlling the Rhône's waters was by no means achieved by a single technological artifact. In fact, regulating the river's flow required the CNR's entire large-scale, envirotechnical system.

Several factors informed how CNR leaders attempted to redirect the Rhône's flow. The agency's management practices after World War II especially reflected the strength of the hydroelectric paradigm. Yet the river's hydrology also constrained the agency. The CNR's engineers could not design—or rather, redesign—the Rhône into an envirotechnical system of their choosing. In fact, because the river's envirotechnical possibilities were finite, the CNR's leaders were forced to adapt their ideal to their understanding of the river's hydrology. In order to determine the design and operational procedures for their projects, CNR engineers needed substantial information about the Rhône, such as its flow during average and extreme periods, so that they could assess the full range of the river's possible behavior. In other words, they needed to construct a "normal" and "abnormal" river.

To do so, CNR engineers culled hydrologic data from 1921 through 1938.[17] No explanation remains as to why they selected this temporal window, particularly one so limited in scope. It seems likely that the CNR's engineers embarked on detailed studies in the mid-1930s as they began to plan the Rhône's transformation and included data back to 1921, the year the state authorized the river's development. This short sample is especially perplexing, however, given the flood of research, so to speak, on the Rhône's hydrology that was sparked by the floods of 1840 and 1856. Nonetheless, from this limited seventeen-year sample, engineers calculated "average" seasonal and annual flows, including designations of ten-, hundred-, and thousand-year floods. They concluded that the Rhône's annual floods averaged 2,600 cubic meters per second (m^3/s), and biannual floods reached 5,500 m^3/s.[18] These calculations had vast implications. The engineers' understanding and classification of the river's flows established the most basic assumptions by which the CNR operated. These data guided the design and operational procedures of not only Donzère-Mondragon but all of the CNR's projects over the next fifty years.[19]

Armed with these figures, the CNR's engineers needed to determine three related design features in order to regulate the flow of the Rhône: the height of the reservoir, the quantity of water allotted to the diversion canal, and the amount of water designated for the original riverbed. Because the CNR's administrators hoped to maximize Donzère-Mondragon's energy potential, they aimed to send as much water as possible through the diversion canal on which the hydroelectric plant was located.

Several factors constrained the agency's regulation of the Rhône's flow, however. For starters, floods posed serious threats to the CNR's projects. Agency officials needed, therefore, to manage the Rhône's waters in ways that would ensure the safety and integrity of Donzère-Mondragon. If all of the Rhône's potential floodwaters were to pass through the diversion canal, the CNR would have to build enormous dikes far exceeding catastrophic flood levels to stabilize the canal. Initial studies estimated that dikes four meters above thousand-year flood stage were necessary to guarantee the project's integrity; the probable costs of this infrastructure were astronomical and therefore prohibitive. Consequently, CNR officials ruled out this option and instead built an emergency dam at the head of the diversion canal so they could divert waters at or above a certain flood stage into the original riverbed.[20]

The CNR's technical experts also needed to consider the long-term impact of the height of Donzère-Mondragon's reservoir on the rest of its envirotechnical system. This project remade the Rhône along its sixteen-mile stretch between the towns of Donzère and Mondragon, but recall that the CNR intended to build others. The technical committee's 1935 blueprint envisioned twenty projects along the three hundred miles of the French Rhône. The CNR needed to bear in mind, then, the relationship of Donzère-Mondragon to its other proposed projects, especially those upstream. For example, just a few miles north of Donzère-Mondragon, the CNR had already begun planning Montélimar, the second project the agency eventually completed after World War II. If the CNR's engineers raised the level of the reservoir behind Donzère-Mondragon's dam, they would reduce the drop of the river's water at Montélimar. In other words, increasing Donzère-Mondragon's energy production might decrease Montélimar's yield. Ponts et Chaussées engineers warned that the productive potential of Montélimar should not be "mutilated" by a poorly selected reservoir height for Donzère-Mondragon. Even Delattre argued that the "harmony" between the two projects had to be maintained. These comments served as an important reminder that Donzère-Mondragon was only one component of the agency's larger envirotechnical system for the entire French Rhône.[21]

Finally, the CNR's engineers had to ensure that the project fit with existing envirotechnical systems. The agency may have wanted to create a new system, ideally one to maximize hydroelectric generation, but others preceded and therefore constrained it. For instance, the agency's projects both facilitated and threatened existing transportation systems. Raising the average level of the river upstream of Donzère-Mondragon through the gorges of Donzère would improve navigation in one of the "most difficult" reaches of the Rhône by eliminating dangerous rapids.[22] And yet, if the Rhône's flow was too high, boats and barges might not pass safely under bridges that had been built decades, if not centuries, earlier. Somehow, the CNR's technical experts needed to identify a reservoir height that met all of these demands.

Although engineers from Ponts et Chaussées recommended a reservoir height of 58 Nivellement général de la France (NGF), France's national standard for altitude, at the entrance of the diversion canal to account for the bridge near the town of Donzère, CNR engineers initially preferred a lower reservoir height of 56.6 NGF.[23] At this height, the reservoir would increase the level of the Rhône upstream of the dam approximately 4.25 meters above the river's low-water level.[24] As a point of comparison, the greatest known flood, that of 1856, reached an average height of 60 NGF there.[25] According to the CNR's preliminary recommendation, then, Donzère-Mondragon's reservoir would elevate the river to a level about halfway between low-water and thousand-year-flood stage, thus increasing the project's hydroelectric yields while somewhat limiting the amount of land flooded by its reservoir and preserving navigation. This recommendation seemed to strike a compromise between the CNR's hydroelectric paradigm and existing systems.

Within a year, however, the CNR's technical elites revised their initial proposal and increased the height of the reservoir to 58 NGF, the original recommendation of the Conseil Général des Ponts et Chaussées. Now it was almost one and a half meters higher than the agency's preliminary recommendation and only two meters below the level of a thousand-year flood. If adopted, this height would create a reservoir approximately ten to twelve kilometers long behind Donzère-Mondragon's dam and require large dikes along its diversion canal. Evidence suggests that the agency's goal to increase electricity generation spurred the design modification, since elevating the level of the reservoir would increase the water's drop at the hydroelectric plant.[26] By October 1945, 58 NGF had become the CNR's standard for the project, guiding construction of the reservoir dam, among other features (see Figure 3.3).[27]

In addition to determining the height of Donzère-Mondragon's reser-

Figure 3.3. *Building Donzère-Mondragon: reservoir dam.* This CNR photograph taken in August 1949 of the dam's construction once again suggests the transformative effects of the project's construction on the surrounding landscape. (Courtesy of Compagnie Nationale du Rhône.)

voir, the CNR's hydraulic experts needed to decide the relative proportion of water that would flow into the diversion canal and the original riverbed, what CNR engineers now called the "former Rhône" *(l'ancien Rhône),* or even the "dead Rhône" *(le Rhône mort).* Needless to say, these terms were loaded and conveyed how CNR officials viewed the river, their projects, and the agency's mission to develop the Rhône.[28]

Historically, these two phrases, the "former Rhône" and "dead Rhône," were interchangeable terms that referred to the river's erstwhile channels. Countless "former" and "dead" Rhônes had existed before the CNR's projects, since the river had meandered for millennia. Earlier envirotechnical systems, including dikes, canals, mills, and irrigation networks, had modified the river's hydrology, at times changing and even creating "former" Rhônes. In many ways, then, the CNR's projects were only the most recent chapter in an extremely long history of river management—though one that dramatically intensified the scale and scope of the river's development. Yet it is significant that the CNR's elites used the same nomenclature to render all the "former" Rhônes equivalent, thereby naturalizing the dramatic environmental consequences of their projects and, by extension, their envirotechnical system and regime.[29]

Calling the original riverbed the "former Rhône" implied that the diversion canal was the current and active, rather than historic, river. The expression "dead Rhône" obviously made this point even more explicit. Yet contrary to its discursive implications, the river had not simply died. The CNR's scientific and technical experts had *created* a "dead Rhône" by building a second river, redirecting most of the river's flow through the diversion canal, and dramatically reducing the amount of water flowing through the original riverbed. By the early 1950s, the CNR's experts had begun to alter radically the hydrology and ecology of the original river in the process of trying to create a new one in the image they desired.[30]

CNR President Aubert admitted that establishing flows for the diversion canal and the "former Rhône" was a "delicate issue."[31] For instance, deciding how much water would flow through the canal consequently shaped how much electricity the Blondel plant could produce. Flows less than this amount would reduce production, a consequence CNR officials hoped to avoid. For one, France's reconstruction efforts depended on plenty of energy. Failure to meet electricity goals would have serious repercussions for national reconstruction, both materially and symbolically. On a more pragmatic level, it also meant the agency would receive less money from the sale of its electricity to the EDF, ultimately jeopardizing its financial state. These motives therefore pushed the CNR's political and technical leaders to enhance Donzère-Mondragon's hydroelectric potential.

However, the agency's engineers could not design Donzère-Mondragon for just any hydroelectric capacity. The Rhône's hydrologic regime constrained the project's parameters, as did CNR experts' knowledge of that regime.[32] The CNR's officials had to take into account the river's average flow and its seasonal and annual variations. The CNR's hydrologists concluded that the river's "average flow" *(débit moyen)*, which they defined as the flow that was met or surpassed at least 150 days annually, fell between 1,600 and 1,625 m^3/s. But this average, problematically, did not provide a sense of the river's ranges during the rest of the year, let alone its variations from year to year. In fact, the Rhône's flow could vary dramatically: from as little as 560 m^3/s during periods of low water, which occurred ten days per year, to as much as 9,500 m^3/s, the CNR's marker of a thousand-year flood. In other words, the Rhône could channel from barely one-third to nearly six times its "average" flow, a wide range that had important implications for the CNR's hydroelectric paradigm. During times of low water, only one or two of the Blondel plant's six turbines might have enough water to operate. Engineers determined that for five of the turbines to run, the diversion canal's flow needed to be at least 1,160 m³/s.[33] The Rhône's seasonal, annual, and extreme hydrologic variations thus threatened to undermine the CNR's objective to produce as much electricity as consistently as possible.

At first, the CNR's engineers appear to have adapted their goal to the river's variable flows. Between 1943 and 1945, agency officials debated several possible flows in the diversion canal, which ranged from 1,100 to 1,530 m³/s. Most reports, including the state's mandatory "Results of Public Inquiry and Hearings" from October 1945, indicated that the CNR's engineers initially recommended a flow between 1,060 and 1,320 m³/s, in which case the maximum capacity of Donzère-Mondragon's diversion canal (and thus of the Blondel hydroelectric plant) would fall below the river's average flow of 1,600 m³/s.[34] In other words, the Rhône would be able to meet the canal's maximum capacity for a larger share of the year than if it were set at higher levels. As a result, more of the river's water would be left to the "dead Rhône." These preliminary recommendations suggest that the CNR's engineers initially designated a more conservative flow for the diversion canal that adapted the agency's hydroelectric paradigm to the river's average flows. However, this compromise did not last for long. Less than eighteen months later, CNR officials had boosted the diversion canal's capacity, thereby augmenting the project's potential energy production and the CNR's coffers. By February 1947, as the CNR began to shift from planning to building Donzère-Mondragon, its recommendation for the diversion canal's flow had jumped to 1,530 m³/s. This figure eventually served as both the diversion canal's capacity and the CNR's target flow to maximize energy generation.[35]

While Donzère-Mondragon now promised to produce more energy, at least during certain times of the year, the project had also become more susceptible to the river's variations, particularly in times of low water. While a lower capacity could be met more frequently, the CNR's new standard could be achieved for less than half of the year. For the rest of the year, CNR officials would have to be satisfied with less than maximum capacity in the diversion canal and thus less than maximum hydroelectric generation. Raising the canal's capacity, then, intensified pressure on agency officials to fulfill it as much as possible. Even in the late 1990s, CNR engineers admitted that during low-water periods, technicians increased the flow in the diversion canal by sending less than the mandated volume of water to the former Rhône, essentially sacrificing the river's already vastly reduced flow to boost power generation.[36]

The amount of water apportioned to the "former" Rhône further revealed how the CNR had turned the diversion canal into the river's primary channel, even though it could not serve the same ecological function as the original river. Between 1941 and 1953, the CNR, Administration des Eaux et Forêts (the state's water and forest agency), and locals deliberated over how much water should be allotted to the dead Rhône, or what CNR officials called "reserved flow" *(débit réservé)*. Like "former" and "dead Rhône," this term was loaded, implying that the water was being set aside or "reserved" from the river's primary flow, which now passed through the diversion canal. Furthermore, while the CNR raised its recommendation for the canal's capacity, agency officials quickly homed in on 60 m^3/s as the volume of water that should be rationed to the former Rhône.[37] In short, the agency deemed that less than 4 percent of the river's average flow was appropriate for a so-called dead Rhône.

Several locals and representatives from state agencies disagreed. Even in 1943, M. de Pampelonne, a representative of the Vichy government's Ministry of Agriculture and Revitalization, warned that the "considerable reduction in the flow of the [former] Rhône will modify the current equilibrium between the flow of the river and that of the Rhône's tributaries in the region." He also presented the CNR with a long list of issues that the four departments bordering Donzère-Mondragon (Vaucluse, Ardèche, Drôme, and Gard) would face if the agency implemented a reserved flow of 60 m^3/s. The prefect of the Ardèche echoed de Pampelonne's worries two years later when he stated that a substantial decrease in the former Rhône's flow would damage many homes and farms in his department.[38] The expected ecological consequences of the CNR's new envirotechnical system foretold that the agency's project would conflict with other uses and users of the river, impacting especially severely those groups living along the former Rhône. These groups included nonhuman species.

Several state officials expressed concerns about the effect of the CNR's management of the former Rhône on fish populations even during the project's planning stage. In May 1945, nearly two years before construction of Donzère-Mondragon had begun, specialists from the state's

Service de la Pêche conducted studies of the project's probable impact on fish habitats and populations. They predicted that the reservoir dam at the head of the former Rhône would alter the river's average height and speed upstream of the dam and consequently submerge some branches of the river. They also projected that the river's flow would be dramatically reduced downstream of the dam, drying up vegetation and some of the river's secondary channels, thereby decreasing the Rhône's biological capacity. Service de la Pêche officials estimated that the former Rhône would suffer a 25 percent decrease in biological productivity and a 75 percent decrease in its capacity for fish reproduction. Moreover, they expected that the "cutting" *(coupure)* of the river's primary channel by the reservoir dam would block fish migration. They admitted that some fish might be able to migrate through the diversion canal, but they generally remained skeptical as to whether fish would identify this alternative route. Service de la Pêche experts estimated that after Donzère-Mondragon's completion, only 10 percent of the former amount of fish would continue their usual migratory patterns.[39] Overall, these experts expressed a deep pessimism about the compatibility of Donzère-Mondragon and the river's fish. Their concerns exemplified another divergence within the state over the reconstruction of the Rhône, including challenges to the CNR's envirotechnical system.[40]

Despite these concerns, CNR officials remained wedded to a reserved flow of 60 m³/s. Initially, they seemed willing to guarantee that *at least* that amount of water would be sent down the former Rhône. In a January 1944 study of Donzère-Mondragon's potential energy yield, one CNR official "admit[ted] that the reserved flow downstream of the intake [for the diversion canal] should *not be less than* 60 cubic meters per second." But the agency soon sought to reduce the already dramatically diminished flow of the former Rhône. After 1944, CNR officials aimed to reserve *exactly* 60 m³/s for the former Rhône. When officials from the Génie Rural demanded, during the period of public inquiry in October 1945, that the CNR be required to ensure "*at least* 60 m³/s at all times of the year everywhere," CNR representatives responded that "the reserved flow is clearly fixed *at* 60 m³/s by the *cahier des charges* [the legal statutes governing the CNR's rights and responsibilities for the project] established during the public inquiry."[41] If agency officials were successful in implementing this figure as their standard, the CNR would bolster the hydroelectric returns of the Rhône by redirecting any additional water to the diversion canal and therefore the turbines of the

Blondel hydroelectric plant. The CNR's leaders appeared entirely unsympathetic to concerns about the ecological and social consequences of this policy, even when expressed by representatives of state agencies.

In fact, the CNR's administration sought even more water for the turbines of Donzère-Mondragon. Already decreased from "at least" to exactly 60 m^3/s, CNR officials hoped to reduce further the reserved flow. In 1951, Marc Henry, CNR's director of research and construction, noted that the former Rhône would become a series of small lakes whether it was allotted 60 or 30 m^3/s. Because the environmental outcome was ultimately the same, Henry hinted, then a reduction to 30 m^3/s made sense. Although Henry mentioned widespread protests against Donzère-Mondragon, locals' strong reactions evidently did not affect his recommendation to lower the reserved flow.[42]

Even as construction of Donzère-Mondragon neared completion, the CNR's technical elites still had not finalized their redesign of the river's flow—or, more accurately, the rivers' flows—and continued to press for a decreased reserved flow. In February 1952, CNR President Emile Bollaert thanked the government's Commission on Sites and Historic Monuments for its recommendation to reduce the reserved flow to 30 m^3/s, since it "would result, as you indicate, in a notable increase in [energy] production." One week later, Delattre told the director of EDF that "the very important question about the flow left in the Rhône between the [diversion canal's] intake and restitution remains. It would certainly be interesting to be able to reduce this flow from 60 to 30 m^3/s as the Commission on Sites envisions."[43] That the amount of water CNR officials hoped to allocate to the former Rhône steadily decreased illustrates just how deeply the hydroelectric paradigm permeated the agency and thus how it guided the agency's attempt to redesign the Rhône.

Ultimately, despite their ambitions, the CNR's administrators and engineers were obliged to ensure that a minimum of 60 m^3/s of water flowed through the former Rhône. In December 1953, fifteen months after the project's inauguration, the state issued its final *cahier des charges*. It is unclear why the state approved these by-laws after the project's completion, not before its construction. Nonetheless, the state ordered "a superficial, continuous flow of *at least* 60 m^3/s in the bed of the [former] Rhône." The project's by-laws aptly depicted the post-Donzère-Mondragon ecology of the new "former Rhône": although its flow might be "continuous," it was "superficial" indeed.[44]

The CNR's administration failed to cap the reserved flow at exactly 60 m³/s, but even if they had succeeded, the reserved flow still would have varied during floods. During these periods of high water, this variation benefited the agency. Once the diversion canal reached its maximum capacity, the CNR's engineers used the dead Rhône as a safety valve for its envirotechnical system, since the dikes of the diversion canal had not been built to support more water. In these conditions, the CNR's technicians would close the emergency dam at the head of the diversion canal, completely open the reservoir dam at the head of the former Rhône, and send any additional water down the original riverbed. The greater the flood, the more water descended the former Rhône and spilled into the river's floodplain. During a thousand-year flood, the former Rhône would channel more than six times the volume of water in the diversion canal.[45] Apparently, then, the "dead Rhône" could again come to life. Yet the general attitude of CNR administrators and technical elites toward the reserved flow indicates that the agency sought to transform the Rhône into an envirotechnical system best suited to hydroelectric generation. To do this, the diversion canal had to become the new Rhône, while the ecology of the "former Rhône" had to become what its name implied.

The CNR's allocation of water to Donzère-Mondragon's diversion canal and the former Rhône altered the historic blending of ecological and technological systems in unprecedented ways. The CNR created a new river channel and rerouted 96 percent of the Rhône's average flow through the linear, concrete-lined diversion canal. In the process, the agency changed, even reversed, the Rhône's "natural" and "technological" dimensions, challenging these categories and their implications. The former Rhône was certainly no longer "natural." For half of the year, less than 4 percent of the river's average flow passed through it. The CNR's management practices had significant repercussions on surrounding hydrologic, biological, and ecological systems, radically remaking the ecology of the entire floodplain. Yet the diversion canal was not entirely "unnatural" either. After all, most of the river's flow now passed through it. The Service de la Pêche even hoped that the Rhône's fish would treat the diversion canal as the river's primary channel; their survival undoubtedly depended on it. At the same time, the linear channel of the diversion canal essentially isolated this reach of the river from the surrounding floodplain, preventing water that flowed through the canal from being connected to wider hydrologic and ecological processes. In some ways, then, the CNR's diversion canal became more of a river than the former Rhône, but it was a rationalized, industrialized, and simplified Rhône. Indeed, this is precisely what CNR officials intended. Their particular remaking of the river sought to serve the hydroelectric priorities of the CNR's envirotechnical regime. It did not necessarily meet, however, the needs or demands of local populations, human and otherwise. By the early 1950s, the CNR's engineers had remade the central Rhône into two rivers, both part of the agency's larger envirotechnical system, whose design was strongly shaped by its hydroelectric paradigm. While the CNR and hydroelectricity transformed the Rhône after World War II, this agency and its envirotechnical regime were not the only ones to reshape the river during the postwar era.

Nature Fit for the Nuclear?

The central government's commitment to developing an independent nuclear power program after World War II also played a major role in the reconstruction of the Rhône. The complete history and management of nuclear installations along the central Rhône is complicated and beyond the scope of this book. I focus here on the initial siting, design, and implications of three facilities built along Donzère-Mondragon's diversion canal between the late 1950s and the 1970s: the CEA's Centre Nucléaire de Pierrelatte (Pierrelatte), the EDF's Centrale Nucléaire de Tricastin (Tricastin), and Usine Eurodif, which was managed by the European Gaseous Diffusion Uranium Enrichment Consortium (Eurodif). All of these facilities eventually became part of the enormous nuclear complex known as the Site Nucléaire du Tricastin (Tricastin Nuclear Power Center). (See Map 3.2).

In many ways, it was the Rhône itself and the features of the surrounding valley that first drew the attention of the country's politicians, engineers, and other proponents of the atomic age. Nuclear reactors need an abundant and steady stream of water within a certain temperature range for their cooling processes. The Rhône, France's most powerful river, was particularly attractive for these purposes. As CNR General Director Delattre stated in October 1958, "if the Commissariat à l'Energie Atomique has set itself up along the length of our [diversion] canal [at Donzère-Mondragon], it is probably because [its new project] will require a significant supply of water." The river's voluminous flow also offered a convenient way to reduce potential pollution. Even in the mid-1970s, once the ecological sciences and environmental movements had challenged the idea that "the solution to pollution is dilution," Claude Gemaehling, then the CNR's general director, bragged that the Rhône's flow allowed for "the necessary dilution of radioactive effluent." A third advantage was the Mistral, whose periodic strong winds could reduce hazardous contamination by quickly dispersing noxious elements.[46] Advocates of the atomic age therefore saw some of the features of the Rhône and the surrounding valley as both advantageous to the facilities' daily operational procedures and crucial for the prevention of potentially catastrophic conditions.

If the "nature" of the Rhône valley attracted France's atomic agencies, how exactly did their plants use the river? Several processes appropriated the Rhône's water. In part due to security concerns, little description is to be found in archives, but the limited evidence available describes how the Rhône's water diluted several kinds of contaminated water produced by the nuclear plants. This industrial water, now called "treated effluent," then passed through settlement basins filled with river water. Only once the river's captured flow further diluted the industrial water and chemical effluent were they returned to Donzère-Mondragon's diversion canal.[47]

The discharge of water from nuclear facilities to the diversion canal serves as an important reminder that the CNR had already spent nearly two decades remaking the Rhône by the time France's two primary state agencies involved in nuclear activities, the CEA and later the EDF, targeted the river for its atomic potential. Construction of the first nuclear reactor at Marcoule, located along the Rhône farther south near Orange, had already begun by the early 1950s. Then, in 1958, the CEA selected the central Rhône as the site of its new uranium enrichment plant at Pierrelatte; it was only the first of several nuclear installations built along this reach of the river.[48] As Delattre had noted, the CEA located the plant near Donzère-Mondragon's diversion canal, which the CNR had completed six years earlier. The CEA's siting of Pierrelatte was tied closely to the CNR's recent construction efforts. In fact, although the "natural" river offered certain advantages, the CNR's transformation of the Rhône into a new envirotechnical system offered an "improved" "nature" that better suited nuclear technologies. The massive volume and powerful flow of the Rhône, which made it an appealing location for nuclear facilities, simultaneously posed threats to those projects because of the river's unpredictability; but the CNR's management of the Rhône seemed to diminish those threats. The agency's construction of diversion canals and regulation of the river's flow appeared to yield a steadier supply of water. The CNR's projects thus facilitated later development, including nuclear projects, which would rely on a more constant, regulated, and simplified river. For France's nuclear engineers, then, the CNR's new envirotechnical system provided a critical foundation for their subsequent transformation of the Rhône.

Conflicts soon emerged, however, among the three hydrologic regimes along the Rhône's central reach: the waterway's "natural" hydrology, the CNR-modified river, and the CEA- and EDF-managed Rhône.[49] Each agency's envirotechnical system reshaped the river's hydrologic processes and thus created its own hydrologic regime. Yet the regimes were less distinct than the categorization suggests. Indeed, the intersection and interaction of these envirotechnical systems resulted in many conflicts. One example is the attempt to create a new, nuclear-oriented hydrologic order that built on "second," not "first," nature.[50] The CNR-modified Rhône facilitated the subsequent nuclearization of the river, but how the CEA and EDF managed it did not necessarily serve the CNR's objectives or comply with its existing management practices. Meanwhile, although the river's ample flow seemed to make the Rhône an ideal river for nuclear purposes, it simultaneously challenged the CNR's hydrologic order and even the hydrologic orders of the CEA and EDF as well. In the end, internal state conflicts, extreme natural events such as floods and periods of low water, and the emergence of an antinuclear movement together threatened to undermine de Gaulle's ambition to make France a global nuclear power, a status central to postwar conceptions of *grandeur*.

"Nuclear Reactor Hydrology"

The construction of nuclear reactors in the Rhône valley beginning in the 1950s both depended on the CNR-managed river and at once transformed it. Contemporaries acknowledged that they were blending complex technological systems and equally complicated ecosystems in new ways. In November 1959, CEA representative R. Galley wrote to the chief engineer of the Génie Rural, indicating that he wanted to discuss the "hydrology of the CEA's new nuclear facility at Pierrelatte." In particular, procedures governing the intake and discharge of the Rhône's water at the Pierrelatte plant concerned Galley. For the next five years, technical elites from the CNR, CEA, and Génie Rural debated "nuclear reactor hydrology," then the phrase suddenly and inexplicably disappeared from official correspondence.[51]

The idea of "nuclear reactor hydrology" offered a powerful tool for technical and political leaders involved in the development of France's nuclear capabilities. The concept helped them not only to understand but also to justify the nuclearization of the Rhône. Hydrology is the scientific field that studies the properties, distribution, movement, and quality of water on the earth's surface, in the soil and geologic formations, and in the atmosphere.[52] By discussing the "hydrology" of nuclear reactors, these elites made analogies between the cycle of water through the air, soil, and groundwater and the cycle of water through France's new nuclear facilities, two of which were located on the Rhône, with more soon following. The phrase thus helped naturalize nuclear technologies by invoking a vocabulary drawn from descriptions of the natural world. This strategy of naturalization resembled the one the CNR's engineers employed in their construction of the "former" and "dead Rhône." These descriptions performed strategic work by blurring the boundaries between the natural and the technological.

Although this conceptual tool conveniently legitimated atomic development, these technical and political elites were correct to portray nuclear plants as having their own hydrologic orders. Some of the Rhône's waters flowed through these plants, and France's atomic agencies returned most, but not all, of this water to the CNR's envirotechnical system. Furthermore, officials from the CNR, the CEA, and later the EDF discussed explicitly how they should manage the flow of water through each nuclear plant. They debated where to take river water, how much, and where to discharge it. By making such choices they consciously *designed* the hydrology of nuclear reactors; but there were two complications here. First, this hydrology remained connected to the Rhône, and second, the river was now managed by the CNR. When planning the hydrology of their agencies' nuclear facilities, or these atomic envirotechnical systems, CEA and EDF officials relied on the CNR's own system. Although they appropriated it for their own purposes, they often failed to recognize the implications of doing so: enrolling the CNR's Rhône into their own envirotechnical system exacerbated conflicts among the agencies, because the CNR's management practices did not always serve the CEA's and EDF's demands.

By the late 1950s, nuclear facilities had become part of the landscape of the Rhône valley. The nuclearization of the Rhône expanded over the next thirty years as France's growing atomic ambitions continued to reshape the river, as they did other landscapes in France and in its colonies, its soon-to-be former empire, and even beyond.[53] In the process, the CNR-managed Rhône became an essential component of the nuclear plants' own "technological" systems. At the same time, cooling towers, effluent, and radioactive isotopes became parts of the Rhône's ecosystem. "Nuclear reactor hydrology" aptly described, then, the new relationship between the Rhône and the atomic age in postwar France, encapsulating how the CEA and EDF blended, both metaphorically and materially, the ecological and technological through their construction of atomic facilities. Moreover, it became increasingly difficult to separate the Rhône's hydrologic processes from "nuclear reactor hydrology," and vice versa. The boundaries of what was part of the Rhône and what was part of the reactor became more and more indefinite on the ground—or rather, in the river—as these systems intersected in increasingly complex ways. In many respects, the term "nuclear reactor hydrology" epitomized the ways the Rhône had become several envirotechnical systems. But by diminishing distinctions among the river's nuclear, CNR-managed, and "natural" hydrologies, state elites justified the nuclearization of the Rhône and paved the way for France's new status as a global nuclear power.

Enrolling the Rhône in the Atomic Age

In October 1958, Delattre predicted correctly that Pierrelatte, the first nuclear project built at Donzère-Mondragon, would require a substantial supply of water to enrich uranium. He did not foresee, however, that the CEA's demands for water to ensure the plant's safe operation would increase over the next four years. In November 1959, the CEA calculated that it would need 8,000 cubic meters per hour (m^3/h) of water, but eight months later, CEA agent Galley dramatically reduced this estimate to just 1,000 m^3/h when he informed the CNR of his agency's actual requirements. While Galley may have hoped to downplay his agency's demands on the CNR, he did stress the importance of this water when he stated that the "safe functioning and security" of the conduit linking the Rhône's waters to the Pierrelatte plant was "essential."[54] Galley's comment indicates the way the river's water and the infrastructure necessary for its transmission played critical roles in the CEA's envirotechnical system by reducing some of the grave potential risks associated with the project.

Yet the CEA soon increased its water demands on the CNR, not once but twice. Galley wrote to the CNR in June 1961 explaining that his agency had reevaluated its water needs: they were 2,000 m³/h "at the present." Eight months later, the CEA's request for the increase had been approved. Then in 1964, a new agreement granted the CEA 3,000 m³/h of water after its revised demands had been deemed "insufficient today."[55] Although the CEA's intake was still less than half what Galley had predicted in November 1959, its demands for Rhône water had tripled in four years. No evidence remains to suggest whether locals protested these increases.

By the 1970s, France's latest nuclear facilities required even more water than Pierrelatte's growing thirst. Two installations were built along Donzère-Mondragon's diversion canal that decade: Tricastin, an EDF-run electricity-producing facility that was built after the central government issued its "all nuclear" *(tout nucléaire)* energy policy in 1974, following the oil crisis the previous year, and Usine Eurodif, another uranium enrichment plant, overseen by a European consortium. Early studies predicted that Usine Eurodif would need 6,000–7,500 m³/h of water. Though CNR officials complained about the project's demands, the official 1979 agreement allocated 6,000 m³/h.[56] But it was not just the volume of water that mattered. In their discussions of Tricastin's needs, several EDF engineers stressed that their project required a *constant* supply of water when they hinted at the risks of noncompliance.[57] Not only were Usine Eurodif's water requirements twice that of Pierrelatte but also both the CEA and EDF framed the continous supply of river water as essential to the safe operation of their respective nuclear facilities. France's status as a nuclear power and its growing reliance on nuclear energy brought the CNR-managed Rhône into an atomic envirotechnical system that, in turn, made these two developments possible.

Beginning in the 1950s, then, the Rhône began to provide water for nuclear facilities along its banks. But *which* Rhône? In the context of the post-Donzère-Mondragon landscape, these facilities could tap the Rhône from the diversion canal, the former Rhône, or drainage networks that locals had built. They could also tap the CNR's counter-canal network, which had been constructed to deal with the perceived impacts of Donzère-Mondragon on surrounding hydrology (I examine this story in detail in the next chapter). The CNR's administration worried about the risks associated with certain intake sites for the nuclear facilities. Depending on where the CEA took water, the CNR's flood management and groundwater recharge programs, as well as locals' use of established drainage and irrigation networks, could all be disrupted. They might even be made ineffective. Debates over the proper source of water for nuclear facilities thus revealed conflicts among the Rhône's hydrologic regimes and its competing envirotechnical systems.

The CNR and CEA's disputes over the source of water for Pierrelatte's cooling processes centered on Donzère-Mondragon's counter-canal system. Although the CEA concluded that it was the cheapest and most convenient source of water, CNR engineers expressed reservations about how intake from these canals might affect both the quantity and quality of water the CNR pumped into the water table. In 1960, CNR official G. Audebrand wrote to his superiors that the loss of 1,000 m^3/h from the counter-canals to Pierrelatte was comparable to the loss of four or five groundwater recharge pumps. Because the CNR was eventually held responsible for fixing damages that Donzère-Mondragon had caused to the region's groundwater, even a modest deficit could be significant. Alternatively, Audebrand proposed that the CEA tap the CNR's feeder canal (the reach of the diversion canal upstream of the Blondel hydroelectric plant) and use the CNR's counter-canals only for emergencies.[58] Moreover, the CEA's intake increased to 3,000 m^3/h by 1964, presumably tripling its potential impact on the CNR's groundwater recharge program. Because CNR officials worried about the consequences of the CEA using the agency's counter-canal network on the recharge program and thus local communities, they advocated for a different solution to the CEA's intake needs.

The CNR successfully opposed the CEA's recommendation. As Audebrand had proposed, the agency persuaded the CEA to build intake infrastructure off the diversion canal upstream of the hydroelectric plant. Two intake pipes with a capacity of 1,000 m^3/h each linked the feeder canal to Pierrelatte. Later the CEA added a third pipe to meet the project's growing demands. Only during emergencies could the CEA tap water from the CNR's counter-canals.[59] This resolution attempted to meet the needs of both agencies' envirotechnical systems.

Thirteen years later, the CNR and EDF avoided replicating this intake debate. Early documents outlined Tricastin's proposed intake and discharge policies. The EDF planned to send water directly from the CNR's diversion canal to Tricastin.[60] It made no mention of the CNR's counter-canals. The CNR's successful defense of its counter-canal network thus became inscribed into the landscape of the central Rhône valley, as intake pipes linked Pierrelatte and later Tricastin to Donzère-Mondragon's diversion canal. The three agencies had successfully negotiated one potential conflict between the CNR- and nuclear-oriented envirotechnical systems.

Where this water went after it passed through Pierrelatte, Tricastin, or Usine Eurodif was, however, another issue. Debates over the discharge of water at Pierrelatte reveal deep conflicts between the CNR's existing commitment to flood management and groundwater recharge and the CEA's goal of a convenient and economical solution to its water evacuation problem. Moreover, the CEA needed to discharge two kinds of water: surface water from rain and from drainage networks on the site of the Pierrelatte facility, and effluent produced by it. The fact that these were different kinds of water from different sources mattered. The source and type of the water determined what path it would (and could) take once outside Pierrelatte.

Controversies over the discharge of first kind of water—surface water—revolved around its destination. At a meeting in September 1959, Génie Rural agent J. Fioravante proposed that the Gaffière and Lauzon, old drainage canals, built by local communities, could together accommodate up to 6 m^3/s of surface water. However, their capacity would prove inadequate during floods.[61] Consequently, Fioravante suggested, discharge into Donzère-Mondragon's tailrace (the reach of the diversion canal downstream of the Blondel plant) "seems to be the better solution." But he acknowledged that the CNR's counter-canals would be easier for the CEA and asked the CNR for its opinion. Evidently he did not confer with locals.[62]

Just as the CNR attempted to keep its counter-canals out of the CEA's intake system, the CNR also hesitated to open them to the CEA for discharge. Delattre responded indirectly to Fioravante's query by explaining how the CEA might use the Gaffière, but he also voiced concerns over the efficacy of this supposed solution. As Fioravante had noted, a constant flow of water from Pierrelatte through the Gaffière and other drainage canals could diminish their ability to diffuse the Rhône's floods. Moreover, said Delattre, these floods could be so great as to actually reverse the flow of some "drainage" canals, thereby threatening the safe evacuation of water from Pierrelatte. Although the CNR's administration preferred that the CEA discharge water to the Gaffière and Lauzon canals rather than to the counter-canal network, the relationship of these canals to the Rhône weakened their effectiveness. Neither the established drainage canals nor the CNR's counter-canal system was an ideal solution. The CEA's officials recommended a hybrid solution (evacuating rainwater into both the Gaffière and the counter-canals). The final deciding of this question is not recorded in the archives, but a 1961 study by the Génie Rural recounts that Pierrelatte's runoff had to be sent to the Gaffière alone. Apparently, the CNR's concerns about its counter-canal network proved persuasive after all.[63]

Meanwhile, the CEA also had to deal with the second kind of water—the discharge of Pierrelatte's industrial effluent. Here debates over possible destinations centered on the quality of the water. At first, CEA officials sought to discharge industrial water into existing drainage networks or the CNR's counter-canal system. The CNR questioned both proposals because of their potential consequences: discharge into drainage canals might affect established hydrologic patterns, while evacuation into the counter-canals risked contaminating groundwater. The CNR instead proposed that the CEA use Donzère-Mondragon's diversion canal, although the CNR's technical experts did not explain why they believed it offered a safer alternative to the CNR's counter-canal system. Since some of the water from the diversion canal ended up in the water table because the counter-canals diverted it there, perhaps the CNR engineers believed that dilution in the diversion canal would eliminate, or at least reduce, the risks of contamination that concentrated, direct dumping in the counter-canals could not. Although not always explicitly, CNR officials seem to have perceived environmental hazards associated with industrial water discharge.

In fact, the CEA echoed the CNR's concerns about the potential effects of the discharge of industrial effluent on groundwater quality. Galley explained that "water, more or less polluted as it exited the various machines and processes [of Pierrelatte], would be evacuated in the canals traversing the plant site [only] after having been correctly treated in order that it presents no noxious character." In 1961, CEA officials admitted that "it would be interesting to return this [industrial] water via pumping back into the [CNR's] feeder canal. It is necessary, however, to know the noxiousness of these waters at the point at which they exit the [Pierrelatte] plant before giving an opinion on the most appropriate outlet." Even the CEA acknowledged, then, that the quality of effluent might affect where it could go. For these reasons, CEA officials ultimately decided to discharge Pierrelatte's industrial water into Donzère-

Mondragon's diversion canal (specifically the feeder canal), presumably because of its ability to dilute the industrial effluent.[64] Thus CNR leaders had succeeded in persuading the CEA to adopt the CNR's definition of risk, which focused on the potential hazards caused by the direct discharge of industrial water into the region's groundwater. Discharge pipes linking Pierrelatte's cooling processes to Donzère-Mondragon's feeder canal embodied this decision.

Fifteen years later, with both Tricastin and Usine Eurodif in the works, CNR officials continued to oppose discharge into the agency's counter-canal network because of the role it played in groundwater recharge. As Gemaehling stated in 1974, "no dumping, even accidental, is admitted to groundwater and thus to the [CNR's] counter-canals." Other CNR officials warned sternly, "we do not accept the dumping of these [industrial] waters, even treated, in our counter-canals." They informed the CEA that it could instead send industrial water to Donzère-Mondragon's feeder canal or tailrace. A representative from the 6ème Circonscription Electrique, the state agency to which the CNR reported, backed the CNR's position.[65] During the early and mid-1970s, the CNR continued to defend its groundwater recharge program and thus the area's water quality.

Less than one year after its stern warning, however, the CNR had apparently given in. In 1975, it agreed to let Eurodif dump limited amounts of surface water, "hot industrial waters 500 l/day, [and] waters from the [plant's] cooling process [totaling] 5 m^3/hour," into its counter-canals.[66] Although the quantities were small, the CNR allowed surface and industrial water, both previously banned from its counter-canals, into its recharge network. After fifteen years of defending the region's groundwater from potential contamination, the CNR succumbed to pressure from France's atomic agencies. It was not the last concession the CNR would make to the CEA, EDF, or Eurodif.

Guaranteeing Water

Conflicts over the volume, source, and destination of the Rhône's flow as nuclear plants were to make use of it worsened when the CEA and EDF attempted to wrest guarantees of water from the CNR for their facilities. Periods of both low and high water made such guarantees difficult. Periods of low water reduced the flow in Donzère-Mondragon's diversion canal because the river was simply channeling less water; in addition, recall that the former Rhône was ensured a reserved flow of at least 60 m^3/s.

Paradoxically, floods also resulted in reduced flows in the diversion canal because during floods, the CNR decreased how much water passed through the canal to safeguard its project. While this policy protected the CNR's infrastructure, it threatened the CEA's and EDF's nuclear facilities. Since the plants' intake pipes were located on the diversion canal (because the CNR had defended its groundwater recharge network), how the CNR managed the diversion canal's flow had serious implications for the nuclear facilities. As a result, the CEA and EDF attempted to modify the CNR's management practices to prevent one risk associated with nuclear reactors: the drastic reduction or complete loss of a constant stream of water for the plants' cooling processes. If the CEA and EDF could successfully lobby to receive water guarantees, then the CNR would have to change its existing policies. Periods of low water and floods, times when the river's "natural" hydrology was most apparent, thus particularly exposed tensions among the Rhône's competing envirotechnical systems.

Major floods in 1954 and 1955 showed that maximum flows in the diversion canal threatened its dikes. Between the mid-1950s and mid-1970s, then, the CNR's technical elites remained committed to the flood management policy they had developed after the completion of Donzère-Mondragon. As CNR Director of Research and Construction Henry explained, once floods reached 5,000 m³/s, the CNR's technicians decreased the diversion canal's flow from its usual capacity of 1,530 m³/s to 500 m³/s. At that point, the "dead Rhône" channeled not only any floodwaters above 1,530 m³/s but also the difference between the diversion canal's normal and flood-induced reduced flow, or an additional 1,030 m³/s.

The CNR's flood management policy concerned CEA and later EDF officials for two interrelated reasons: lowering the diversion canal's flow both reduced the amount of water available for the reactors' cooling processes and decreased electricity generation. Both were important issues for France's atomic agencies. In fact, Marcoule and Pierrelatte consumed a large share of Donzère-Mondragon's electricity. Henry admitted that major floods "happen frequently enough (almost annually) that the [Blondel] plant cannot furnish the 180 M[ega]W[atts] desired [by the CEA]." He explained that increasing the diversion canal's flow to more than 500 m³/s during floods would reduce the frequency of this energy deficit, but it would demand modifications to Donzère-Mondragon's infrastructure, since the project had not been built to meet these requirements.[67] Because the CNR's and the CEA's envirotechnical systems literally intersected, two risks thus came to a head: the CNR's flood concerns conflicted with the CEA's demands for electricity and cooling water, which in turn required minimum flows in the CNR's diversion canal.

CNR officials concluded that they could amend their agency's practices during low flows so that the CEA would receive a minimum amount of energy, but not during floods. M. Boudrant, EDF's director of production and transport, stated that in periods of low water, the CNR could guarantee production of at least 120–140 megawatts of electricity by backing up water into Donzère-Mondragon's reservoir for short periods of ten to sixty minutes. Then the CNR could send intermittently a larger stream of water through the Blondel plant's turbines. Although Boudrant implied that "natural," CNR-managed, and nuclear hydrologies could be reconciled during periods of low water, he argued that no alternative was feasible during floods, because the flow of water in the diversion canal had to be reduced drastically to protect the surrounding dikes.[68]

Despite Boudrant's verdict, CEA leaders continued to pressure the CNR to change its policies during floods. At first, CNR officials maintained their opposition to the CEA's demands. In 1962, Henry stated succinctly that although the CNR could implement Boudrant's new regulations during periods of low water, it was "totally impossible to approve such an agreement for floods." It was "imperative," he added, that the diversion canal's flow not exceed 500 m^3/s during floods greater than 5,500 m^3/s, even for short durations. Henry attempted to appease the CEA, however, when he reminded the agency that the CNR's latest project at Châteauneuf could probably help meet its energy needs.[69]

Although Henry's proclamation was resolute, the CNR's defense of its flood management policy gradually weakened, as officials at France's atomic agencies, apparently, ramped up their pressure on the CNR and the Cold War only strengthened their claims. By the early 1970s, the demands of the CEA's and EDF's envirotechnical systems had begun to supersede the needs of the CNR's. In 1973, EDF engineer P. de Gaujac explained that "since the installation of Pierrelatte, the [CNR's] Blondel plant is obliged to guarantee a minimum supply of energy to the CEA, [thereby] necessitating a flow of 900 m^3/s through its turbines [and thus the diversion canal]." Even he admitted that the change represented a "nonnegligible risk."[70]

The stakes over water guarantees intensified when the French government passed its "all nuclear" energy policy in 1974. Perhaps not surprisingly, EDF officials stepped up their pressure on the CNR to modify its flood management policy during 1974 and 1975. At first, the CNR's experts agreed only to minor changes. The CNR engineer P. Bayard indicated that his agency would be able to meet the EDF's new demands for guaranteed flows as long as they were modest. By 1976, however, other CNR officials were willing to go far beyond Bayard's minor modifications and allow significant increases in the diversion canal's flow during floods.[71] If the CNR adopted these changes, the agency would be reversing its long-standing flood management policy.

This is exactly what happened in September 1976, when CNR officials boosted the diversion canal's flow during floods to either 1,000 m^3/s or 1,500 m^3/s, depending on the severity of the flood. Later that month, the CNR again increased the amount of water it estimated the canal could handle.[72] As a result, by the mid-1970s, the CNR's technical experts accepted flows in the diversion canal during floods far surpassing what they had recommended over the previous two decades. These higher flows appeased the EDF's worries, but they appeared to flout the CNR's historic concerns over the risks associated with its own system.

These changes illustrate a broader shift in the Rhône's management from the 1950s to the 1970s. Between 1945 and the late 1950s, the CNR played a major role in the river's management. In the late 1950s, it began to share these responsibilities with France's nuclear agencies, because the central government pursued the development of nuclear power and the Cold War further justified such ambitions. Consequently, technical and political leaders from the CNR, CEA, and EDF had to juggle their management of the Rhône. This complicated, contested process of negotiation and eventual compromise resulted in a growing emphasis on nuclear power. This objective was certainly not exclusive: hydroelectricity, navigation, and agriculture did not disappear from the landscape of the Rhône valley. However, the Cold War and the risks associated with nuclear power offered particularly compelling reasons for the CNR to modify and ultimately weaken its envirotechnical regime.

Despite the political weight of such arguments, extreme conditions like periods of low water and floods not only exposed the limits of these envirotechnical systems but also heightened tensions among the envirotechnical regimes of the CNR, CEA, and EDF. In trying to forge a better fit between "natural" and "nuclear" hydrology, the CEA and EDF came into conflict with the CNR's Rhône. Although CNR officials did not believe initially that they could respond to the demands of both natural and nuclear hydrologic regimes, increasingly they adapted their agency's practices to suit the atomic age, as all three agencies attempted to defend their envirotechnical systems. In the process, the Rhône became an atomic river. This trend would only accelerate after the 1970s oil crises. (See Map 3.3).

Machines in the Rhône's Garden

As noted, the enormous industrial facilities built in the Rhône valley after World War II enrolled the river in the industrialization and modernization of France while simultaneously transforming the Rhône. The relationship between these envirotechnical systems and Rhône valley agriculture was ambiguous, because the projects created both opportunities and difficulties for the valley's farms. On the one hand, agricultural interest groups attempted to harness the river's newly managed flow to expand irrigation and rural electrification. The CNR and Génie Rural oversaw a number of these projects in the central Rhône valley during the postwar era. On the other hand, the CNR, CEA, and EDF had built dams and reactors in the middle of productive farmland, displacing families, dividing properties, and modifying the area's environment in ways that did not always sustain existing agricultural practices.

The CNR's constitution required it to develop the agricultural potential of the Rhône valley in addition to fostering energy and navigation. Donzère-Mondragon illustrated the difficulties in achieving all three goals simultaneously. Augmenting hydroelectric generation by building a higher dam and bigger reservoir, for instance, might come at the expense of farmers by inundating more farmland. The design and construction of Donzère-Mondragon thus revealed potential incompatibilities *within* the CNR's own envirotechnical system, let alone *among* the systems of the CNR, CEA, EDF, and other institutions. Donzère-Mondragon marked the first time the CNR had to face the implications of its multi-purpose mandate. Because the steep gorges of the upper Rhône encased Génissiat, that project's impact on farming had been limited.[73] In contrast, Donzère-Mondragon was located halfway between Lyon and the Mediterranean, in the middle of the rich plain of the central Rhône valley. Donzère-Mondragon seemed to be, therefore, a big machine in the provincial garden of France.[74]

Contemporaries predicted that the construction and subsequent operation of Donzère-Mondragon were going to affect its rural hinterland. During preliminary hearings regarding the proposed project, Génie Rural representatives enumerated some of Donzère-Mondragon's probable impacts by department. In the Drôme, they predicted that Donzère-Mondragon's dam and reservoir "will perceptibly modify" the hydrology

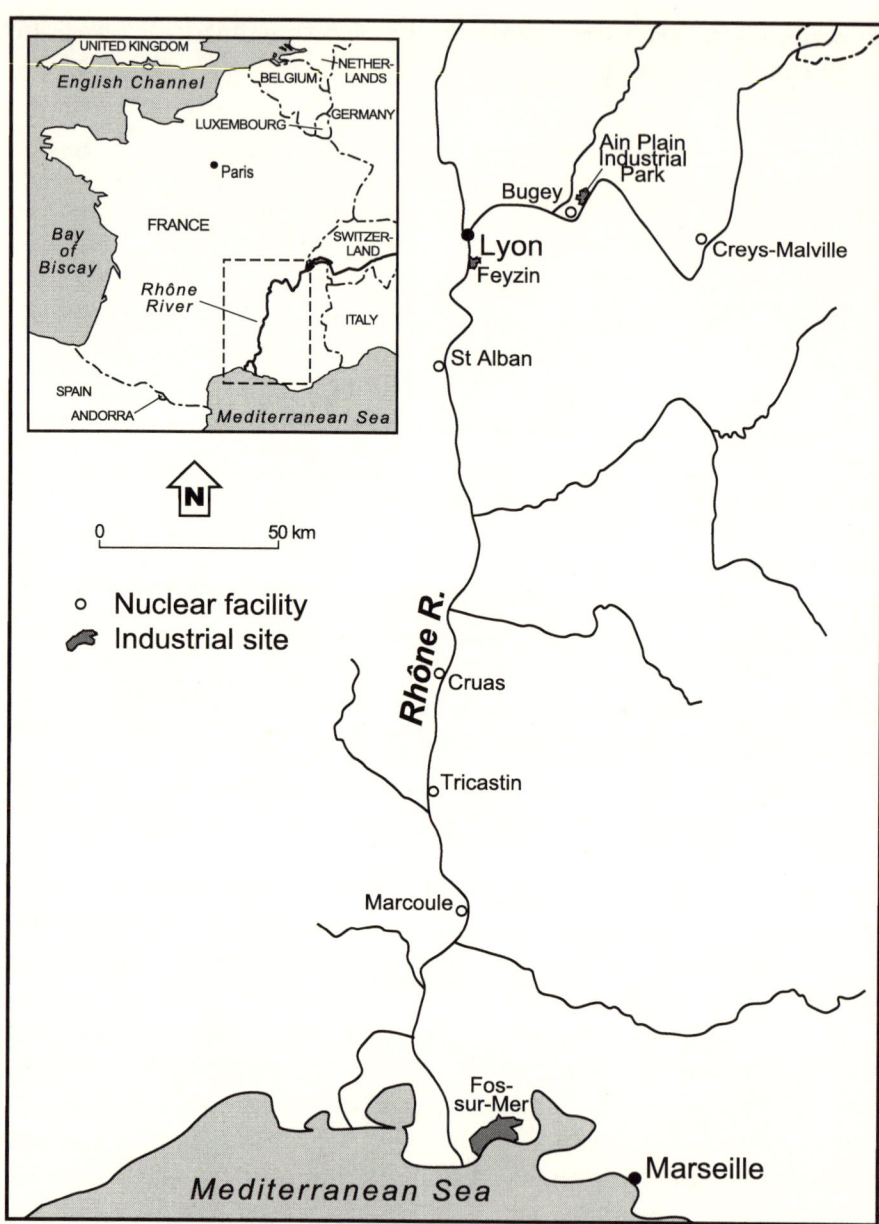

Map 3.3. *Nuclear and industrial development in the Rhône valley.* The CNR's projects were not the only large-scale development efforts in the Rhône valley after World War II. Others included several nuclear installations (Creys-Malville, Bugey, St. Alban, Cruas, Tricastin, and Marcoule), large industrial zones such as Ain and Feyzin, and an oil refinery at Fos-sur-Mer. (Map by Joseph W. Stoll, Syracuse University Cartographic Laboratory.)

of the Drôme River, located upstream of the dam. They also expected that the project would alter drainage patterns downstream of the dam because the CNR's diversion canal intersected with two streams that helped drain surrounding farmland. Génie Rural officials concluded that the entire region surrounding the former Rhône would be "completely transformed." Others worried specifically about the project's "important modifications to the [area's] agricultural regime." For instance, L. Nourrit anticipated that "certain properties will be completely destroyed, others will be made smaller." He also predicted that the CNR's project would make parts of established irrigation systems nonfunctional and ruin key transportation and commercial networks.[75]

Not all departments were expected, however, to be affected in the same way or to the same extent. For example, in contrast to the Drôme, rural engineers calculated that Donzère-Mondragon "improves more than aggravates" agriculture in the department of the Vaucluse. Although they expressed reservations about the project's likely impacts, agricultural experts ultimately concluded that the CNR would be able to ensure the continued flow of water *(écoulement des eaux)* for agricultural purposes. Deciding that the CNR's envirotechnical system would help realize their agency's own goals, Génie Rural officials therefore decided to give the project their blessing.[76]

Once the CNR had completed Donzère-Mondragon and initiated normal operating procedures, communities along the Rhône began to perceive considerable changes around them. In July 1952, the mayor of Aubenas wrote to the CNR's General Director Tournier informing him of several problems: in addition to issues with the town's potable water and sewer systems, his community was experiencing problems associated with shifts in the water table. He also criticized the haphazard way the CNR was dealing with agricultural issues: although officials from the CNR and Génie Rural were working together to develop an irrigation program for the Pierrelatte plain, they had no plan for the worrisome situation in the town of Bourg-Saint-Andéol. The mayor warned that these irrigation problems had to be solved.[77] Although Donzère-Mondragon was supposed to improve farming, locals and some experts from state agencies expressed concerns that it was actually having a detrimental effect on agriculture.

The CNR had superimposed its project on an extensive system of irrigation and drainage canals that had historically facilitated agriculture. This network had regulated stream flows, diverted water, drained wetlands, and modified groundwater patterns, thereby creating a hydrologic order more conducive to farming. Some canals, like the Canal de Pierrelatte, which served as the primary irrigation canal for expansive secondary and tertiary irrigation networks that traversed the Pierrelatte plain, dated to the seventeenth century.[78] Now the Donzère-Mondragon project literally cut through these systems, often leaving them ineffective, if not obsolete. Not only, then, did the CNR's project alter the established hydrologic order; for some locals it also created a problematic new one. Furthermore, consultation between locals and CNR and state experts was apparently limited.

The construction of Donzère-Mondragon, then, resulted in a hybrid scheme of agricultural water management that sought to combine old systems with the project's new possibilities. Yet rarely was the right amount of water in the right place at the right time, and realizing Donzère-Mondragon's agricultural potential in some places meant its simultaneous downfall in others. State engineers rushed to fix these problems, but their solutions frequently proved only temporary, often with mediocre results. Such episodes exposed the complexity of the post-Donzère-Mondragon landscape, which scientific and technical experts had a difficult time understanding and modeling, let alone controlling (see Chaper 4).[79]

Given these changes, some state agricultural officials continued to express concerns about Donzère-Mondragon, but others were more ambivalent about the project and its long-term implications for the valley's agriculture. Donzère-Mondragon had created serious problems for certain farmers, but it also promised others rich possibilities. For instance, while the Canal de Pierrelatte had historically supplied farmers with 5 m^3/s of irrigation water, Donzère-Mondragon now distributed 25 m^3/s. Several state agricultural officials seized on this opportunity. In 1946, A. David, chief engineer of the water division of the Génie Rural in the department of the Bouches-du-Rhône, declared that it was absolutely necessary to modify some aspects of Donzère-Mondragon to reduce agricultural "damages." He admitted that these modifications would be substantial but hoped that they might ultimately improve the area's agriculture.[80] When addressing the relationship of Donzère-Mondragon to its rural hinterland, David raised two key issues that would receive considerable attention over the next fifteen years: agricultural reconstitution and improvement.

"Reconstituting" Agriculture

The construction and operation of Donzère-Mondragon had a range of consequences for the valley's agriculture, and various groups viewed these changes differently. But what, if anything, was the CNR required to do about them? Article 13 of the CNR's *cahier des charges* for Donzère-Mondragon obliged the agency to "reconstitute" agriculture after the project had been completed. "Reconstitution" essentially meant fixing damages that state-sponsored projects like Donzère-Mondragon had caused. Because the term's definition and applicability both had significant financial consequences for the CNR, it is not surprising that its leaders quibbled over the term's meaning and interpretation. These early controversies also indicate that at least some state officials had much wider ambitions for the future of agriculture in postwar France.

Early drafts of article 13 describe how the CNR "will be required to contribute to the reconstitution of agricultural production, which had been reduced because of its projects." Then they outline the measures the CNR would have to undertake: first, *remembrement,* or structural land reform that reorganized agricultural land through the "rational" reallocation of property into larger units, and second, the reestablishment of water flows *(reécoulement des eaux)* so that farmers could carry out irrigation "in the future"—that is, once Donzère-Mondragon was finished—"as easily" as they had in the past. Initially, the CNR planned to devote 30 m^3/s of the Rhône's flow to irrigation. Subsequent negotiations reduced it to 25 m^3/s.[81]

The CNR's leaders hoped that the state would not require their agency to carry out full reconstitution, however. They attempted to reduce the CNR's obligations, both in terms of expenditures and spatial scope, but were challenged by Génie Rural representatives, who not only sought a strict enforcement of the CNR's commitments but even advocated a broad interpretation of reconstitution that would expand the CNR's responsibilities and boost investment in agriculture. The Génie Rural's inclusive definition worried CNR officials because of its financial implications. In addition, the CNR's administration questioned the fine line between agricultural "reconstitution" and "improvement," suggesting that it was difficult to identify where repairs ended and improvements began.[82]

State ministries and the CNR also disagreed over whether locals who had suffered damages should receive financial indemnities (cash payments for expropriated property) or indemnities *en nature* (a different but supposedly equally valued piece of land). The CNR did not want to give locals a choice, but the Ministry of Agriculture did. The state's agricultural bureaucrats worried that cash indemnities might encourage farmers to leave the countryside for the city, thereby exacerbating France's postwar rural exodus. Issuing compensation *en nature* seemed more effective at keeping farmers tied to the land, even if not the same piece of land. The Conseil d'Etat eventually concluded that if locals demanded indemnities, they had to be financial, because the state wanted to preserve uniform national indemnity laws, but the CNR's administration could always try to encourage farmers to accept exchanges of property instead of cash.[83] Few records indicate what locals thought about these policies.

Agricultural reconstitution originally responded, then, to the economic, social, and ecological dislocation caused by projects like Donzère-Mondragon. Put simply, reconstitution was supposed to fix damages such projects had caused. Yet one of its central methods, *remembrement*, suggested that the boundary between agricultural "reconstitution" and "improvement" was indeed unclear. Furthermore, the fact that *remembrement* was one of two primary policies of "reconstitution" indicated that state experts and technical elites had much broader aims for agricultural reform in the Rhône valley, even in the early 1950s.[84]

Cultivating Agricultural Improvement

State elites used the multifaceted disorder that Donzère-Mondragon and other projects generated as an opportunity to extend the state's program of modernization into rural France and to help transform its agriculture. Policies such as *remembrement* initiated this process in places like the central Rhône valley, where state projects were built during the late 1940s and early 1950s, but the practice soon expanded well beyond the immediate area of project sites. Agricultural engineers began to redraw property boundaries and reallocate land so as to make what looked like a haphazard, chaotic landscape more orderly. The goals of increasing agricultural efficiency and productivity motivated their "rationalization" of farmland. This policy did much more, then, than simply repair damages caused by Donzère-Mondragon. The legal obligation of reconstitution offered both a convenient and powerful opening for greater state intervention in rural France, couched in the tempered language of mere restoration. The CNR's projects and its obligation of reconstitution thus created an opportunity for enacting major agricultural reform in the Rhône valley beginning in the late 1940s.

By the early 1960s, agricultural engineers and state elites basically defined reconstitution *as* improvement. This interpretation was precisely what David had recommended in 1946 and what CNR leaders had feared. Although the CNR's administration had worried that broadening the definition of reconstitution would increase their agency's financial responsibilities, this ideology of agricultural improvement could actually help, rather than hurt, the agency. If a proposed project's potential benefits included agricultural improvements along with increased kilowatts and commerce, it might be easier for the CNR to justify. The promise of greater agricultural value might even make a financially marginal project tenable. At the same time, the growing importance of agriculture during the 1950s and 1960s amplified tensions within the CNR's multipurpose mandate: if agriculture could no longer be minimized, how were the CNR's elites supposed to balance it along with energy and navigation? In effect, the growing importance of agriculture during these decades exacerbated certain problems while mitigating others, complicating the CNR's multipurpose mandate but also suggesting that the CNR's agricultural critics might not have to view its machine as a threat to the Rhône's garden. Furthermore, while large-scale technology might improve farming, agricultural benefits might in turn help to justify the CNR's mission. Perhaps, then, the industrial machine and the garden could actually be symbiotic rather than at odds.[85]

The development of an ideology of agricultural improvement therefore recast the relationship of the CNR and its projects to the Rhône valley's agriculture. As Pierre Salenc, head of the Génie Rural in the department of the Ardèche, noted in a 1963 report, no longer was the CNR "strictly repairing the loss of agricultural potential due to its work, but rather it was carrying out the development of specific programs for *remembrement,* the improvement of irrigation (or drainage), or the commercialization of agricultural products in surrounding regions in every possible way." By 1970, Salenc even stated that "it is not adequate to reestablish the prior situation by strictly repairing the losses of agricultural potential. . . . Instead, [the goal is] to promote and achieve all kinds of work that will enable a veritable reorganization of agricultural conditions that are instead oriented toward the development of its potential."[86] In short, agricultural reconstitution was no longer enough. Instead, the CNR's projects offered valuable opportunities for rural reform and the means to do so.

While many state agricultural engineers were initially ambivalent, even antagonistic, toward the impact of CNR programs on the Rhône valley's farms, many came to view the agency's projects as a valuable way to ensure France's agricultural future. For example, two Génie Rural engineers lamented the "underutilization" of water in Rhône valley agriculture and hoped to expand production there. Similarly, Fioravante, the Génie Rural expert, recommended that his agency improve *(améliorer)* irrigation and establish a "modern network" in the Pierrelatte plain. Meanwhile, the prefect of the Drôme expressed excitement in 1971 about the potential of the CNR's agricultural plans for his department.[87] This ideology of agricultural improvement therefore appeared to defuse conflicts between the CNR and farming interest groups. Furthermore, by espousing this ideology, agricultural engineers shared with hydraulic engineers and other technical elites a high modernist view of the Rhône and its potential.

By the early 1960s, then, the regulation and use of the Rhône's water promised, at least in principle, to improve farming while producing hydroelectricity and nuclear power. Advocates of the agricultural transformation of the Rhône argued that water, and specifically water from the river, should play a vital role in the agricultural redevelopment of southeastern France. In 1962, Salenc wrote that although insufficient and poorly managed water had limited agricultural production in the Rhône valley in the past, "agricultural hydraulics" *(l'hydraulique agricole)* could now serve as the foundation of a new *hydroagricole:* a blending of agricultural and hydrologic systems, or an entirely new, cutting-edge envirotechnical system that maximized hydraulic resources for the benefit of farming.

Furthermore, the management of the Rhône for agriculture aimed to do more than simply "improve" farming. It was intended to serve as the technical means toward greater socioeconomic ends: modernizing agriculture would aid local communities and strengthen the economy of the Rhône valley. Making broader links between water management and the area's socioeconomic status, Salenc asserted that "the lack of water is one of the essential causes of underdevelopment." Accordingly, he argued that it was "essential" to develop land that could use existing water supplies. Salenc and other agricultural engineers aspired, however, to go beyond the hydrologic status quo; they wanted to increase farmers' aquatic resources. After condemning "archaic" irrigation and celebrating structural land reform, Salenc advocated the "augmentation of resources," including *remembrement,* improved production methods, and, "most of all, the water factor." Fioravante echoed Salenc, asserting that irrigation increased the competitiveness of agriculture and opened up vast possibilities for the area's farmers. The repercussions of expanded water supplies and agricultural advancements were therefore far greater than higher yields. "It has been largely proven," Salenc declared, "that improvements have social characteristics." Expanding electrification, increasing potable water supplies, and improving rural housing, among other efforts, "act as an efficacious brake on the exodus from the countryside." Using water effectively and efficiently offered the key to solving pressing, interrelated problems in rural France, from antiquated agriculture to a lagging rural economy. Managing the Rhône's water resources was thus the cornerstone of provincial redevelopment. Water was, then, not just a natural object; it was a powerful means to social and political ends. As one study memorably put it, "it must not be forgotten that the [Rhône] waterway is the mother-way [*voie-mère*] of regional planning."[88] Once again, from the Rhône flowed forth the promise of progress, but this time it was rural provincial progress (see Chapter 5).

Legally, a *convention agricole*—a formal agreement between the CNR and the Ministry of Agriculture that outlined a general program for agricultural development in areas where the CNR's projects were located—described the agricultural dimensions of the Rhône's transformation. A project's preliminary technical dossier included drafts of the *convention agricole*, while the final agreement reflected compromises that representatives from the CNR, Génie Rural, and professional agricultural organizations had reached. The *conventions* had two main objectives: first, "to develop the use of water for agricultural needs," and second, "to maintain a sufficient harmony between the development of agricultural and other activities (industrial, etc.)." By outlining a rural development program, these *conventions* identified priority zones, based on relative impact of and proximity to the project. The 1961 *convention agricole* for CNR's project at Beauchastel marked the first time that agriculture carried considerable weight in the agency's decision-making and design processes. It had taken almost three decades since the CNR's founding for agriculture to receive its new, elevated status in the Rhône valley.[89]

The CNR's *mission agricole* thus reflected the agency's latest interpretation of "development," as well as that of the French state, between 1945 and the early 1960s. Agricultural experts had transformed "reconstitution" from remediating Donzère-Mondragon's impacts into a much broader program of agricultural modernization that aimed to revitalize rural France, an ambitious vision that would underlie the agricultural laws of 1960 and 1962. As the language and objectives of rural reform changed, so too did the place of agriculture in the CNR's program. With Donzère-Mondragon, the CNR had simply carried out "the basic repair of agricultural damages resulting from its work." Now the CNR's development of the Rhône integrated *"l'aménagement agricole"* into its mission.[90] Such pronouncements contrasted sharply with the CNR's earlier tendency to emphasize its "hydroelectric development" of the Rhône. While the CNR's administration had initially feared this shift, it proved to have considerable benefits for the agency. In fact, preliminary studies for the state's Fifth Plan (1966–1970) suggested that agricultural expansion in the Rhône valley had the potential to be as profitable as hydroelectricity.[91] The CNR may have concentrated on "developing the energy resources of France as much as possible" after World War II, but now it began to focus on the role of the Rhône valley in the "economic activity of the country."[92] The CNR's agricultural efforts could help "increase the value" of, even "get maximum value" from, the Rhône valley.[93] By the early 1960s, the goals for the transformation of the Rhône had once again been recast. The new priority given to agricultural modernization reflected the state's shifting political economic objectives and especially the transition from postwar reconstruction to a broader strategy of economic expansion and redevelopment, particularly for France's supposedly "backward" regions.

The Emergence of *la Politique Agricole*

Several factors help explain why the reinterpretation of agricultural "reconstitution" as modernization became persuasive in France during the 1950s and early 1960s. First, agricultural engineers perceived the environment of southeastern France as a mixed blessing. The region's climate and geology presented both "natural" benefits and limits. Soil degradation and erosion threatened the future of farming. Especially important was water, or the lack of it: in the Rhône valley, Salenc declared, "water is the limiting factor for agricultural production everywhere." On the other hand, through the same "miracle of water," adequate irrigation and drainage systems might compensate for these "natural" limitations.[94]

In addition to environmental variables, state agricultural officials began to perceive the crops farmers chose to grow as problematic. They felt that too many farmers relied almost exclusively on vineyards, producing as a result too much cheap wine, especially in the departments of the Gard and Hérault. State agricultural experts recommended instead a program of agricultural reconversion to what they believed were high-quality crops. They also maintained that many Rhône valley farmers were trapped in a "crisis" of overproduction, insufficient diversification, high production costs, and poor-quality crops, all of which made their participation in the emergent agricultural market less profitable.[95] The creation of the European Economic Community (EEC), or Common Market, in 1957 undoubtedly heightened state officials' concerns about the future of French agriculture and encouraged them to foster the modernization of the nation's farms.

Although the most famous postwar agricultural problem, *l'exode rural,* did affect some areas of the Rhône valley, it was not nearly as severe as in other parts of France. In fact, the southeast grew dramatically after World War II. During the 1960s, the population increases in the three regions of Provence-Côte d'Azur, Rhône-Alpes, and Languedoc-Roussillon ranged from 9.8 to 17 percent: rates that actually rivaled Paris. High immigration rates, especially by repatriates from North Africa after the Algerian War, contributed to this growth.[96] Nevertheless, national public discourse about "the rural exodus" offered a powerful framework for fostering agricultural reform in the southeast.

Finally, as hydroelectric, nuclear, and industrial projects transformed the Rhône during the postwar era, some state elites and intellectuals wondered if the country had focused on industrialization at the expense of agriculture. In 1969, Fioravante authored a thick study titled "The Hydro-Agricultural Development of the Mid-Rhône," in which he declared: "The Rhône will thus be able to become an important source of agricultural riches. If not, we must fear that the Rhône valley offers this distressing spectacle of an industrial beehive stretched between two new French deserts that has taken away men and dried up all other activity. To the contrary, with irrigation along a perimeter as vast as possible, we will witness the the blooming of a prosperous agricultural sector."[97]

The CNR, CEA, EDF, and other institutions had built Fioravante's "industrial beehive" after World War II. These agencies had constructed enormous projects, the industrial sectors of Lyon and Marseille were booming, and other large ventures were already in the works. Fioravante voiced some contemporaries' fears that France had industrialized too much and too quickly after the war. Agriculture, albeit "reconstituted" agriculture, offered a necessary antidote to industrialization. According to this view, farming composed an "indispensable part of the balanced development program of the Rhône valley." Moreover, it was "impossible to have a harmonious equilibrium between urban and rural space without the appropriate development of agriculture." In short, industrial expansion without sufficient agricultural investment risked exacerbating economic, geographic, and demographic instability. "It is therefore essential," proclaimed one Ministry of Agriculture official, "that agricultural development accompany industrial development."[98]

As these arguments suggest, postwar intellectuals and state elites expressed growing concern over the vast economic disparities across France, a sentiment captured by Jean-François Gravier's 1947 book *Paris et le désert français*. They worried that postwar industrialization had only worsened France's existing inequalities. These concerns pushed the state's economic policy toward regional development, which attempted to foster economic growth in diverse sectors of France's provincial economies to reduce their "drain" on Paris. Agricultural renewal was a key component of this policy. *La politique agricole* thus emerged by the early 1960s not only to fulfill the "natural" potential of the Rhône valley but also to counter the earlier industrial transformation of the river, to maintain rural France, and to revive the lagging provincial economy. Now the fruits of the river would include not only kilowatts and commerce but also fruits in the literal sense. An interpretation of agricultural reconstitution as improvement simultaneously addressed the related problems of the so-called rural crisis and the abysmal state of the provinces.

Changes in the Farmland

These factors together renewed a commitment to agriculture in the Rhône valley. But this was not the same agriculture that most farmers' parents and grandparents had practiced. Postwar agriculture actually had a great deal in common with the industrial development it was intended to offset. As agricultural reconstitution turned into improvement, farming in the Rhône valley became rationalized, capitalized, and technologically intensive; in a word, modernized. It was perhaps ironic, then, that the industrialization of agriculture aimed to balance other forms of industrialization.

Rationality permeated state officials' discourse and dominated their framework for remaking agriculture. In 1958, government agricultural experts asserted that the Rhône valley's agriculture needed to be managed rigorously with rational methods and viewed new irrigation technologies as the ideal means to inscribe rationality into the landscape. This agenda shaped the practices, policies, and artifacts they devised and eventually implemented. In 1963, Salenc recommended that area farmers pursue "rational irrigation," adding that the "rational use of water, especially by the best means, that is, sprinkler irrigation systems," yielded significant benefits for farmers. State officials also noted that these irrigation technologies worked most effectively when they followed structural land reform. Salenc even warned that irrigation without *remembrement* was "useless." Agricultural engineers tended to view new irrigation technology, and all of the changes associated with implementing this technological system, as a silver bullet. It required comprehensive changes in agricultural practices but also promised to yield great rewards. In one sweeping statement, Ministry of Agriculture officials declared that implementing these new irrigation systems would expand agricultural production, improve quality, increase yields, lower prices, offer more flexibility so that farmers could adapt to changing market demands, and promote overall the rationalization and modernization of agriculture.[99] Apparently, there was little these new technologies could not achieve.

One means of rationalizing the Rhône valley's agriculture was institutional. New organizations and committees were formed, both within and among state agencies, professional groups, and local communities, to assess and improve the area's agricultural possibilities. These groups illustrated the renewed commitment to farming but also reflected ongoing tensions between development objectives. One such committee emerged from the Plan. Its leaders created a working group to study the development of the Rhône that met for the first time in March 1969. Financial issues were of particular concern, and the group hoped to compel the Ministry of Agriculture to pay for a share of the Plan's projects because agricultural objectives were now receiving greater priority than in the past. Officials from the Ministry of Agriculture complained, though, that the CNR's plans and the financing of its projects tended to favor navigation and *l'aménagement du territoire* (literally "territorial development" but usually translated as "regional planning"). They did not see agriculture as one of the CNR's real priorities. Consequently, the Ministry of Agriculture developed its own vision of the Rhône's future and in April 1969 founded its own committee, Le Groupe d'Etude des

Perspectives Agricoles et Rurales Rhodaniennes (GEPAR), whose mandate was to study the past, present, and future of agriculture and rural life in the Rhône valley. After meeting several times between 1969 and 1972, it published a massive tome in May 1972.[100]

These institutional methods of modernizing agriculture in the Rhône valley engaged a variety of sociotechnical means. In 1975, Salenc itemized agricultural reforms that had been carried out directly by the CNR or indirectly through financial contributions to state agricultural agencies. They included structural land reform, irrigation systems, drainage networks, flood control, the commercialization of agricultural products, and energy development; but Salenc noted that *remembrement* and irrigation were the two most important elements of the CNR's multifaceted program.[101]

Indeed, *remembrement* was one of the two cornerstones of agricultural reform and, according to advocates, the necessary first step. Even in 1950, agricultural engineer J. Arrighi de Casanova complained about the valley's "structural disorganization" and lamented its multifarious consequences. Other studies declared that a host of problems resulted from the division of agricultural land into many small parts *(morcellement)*, not to mention the partitioning of a single property owner's land into numerous distinct, often widely scattered parcels *(parcellement)*. According to agricultural experts, "structural disorganization" was inefficient and diminished agricultural productivity. "In order to assure the rational development of the region," one reform advocate argued, "it is therefore indispensable to reduce the number of parcels as much as possible and have them in a convenient organization. Structural reorganization allows for this regrouping of parcels." Moreover, such land reform would reduce the cost of irrigation by creating large swathes of contiguous land that could be irrigated by a single system and would thus "assure the most rapid development of the region." Agricultural experts believed that *remembrement* and irrigation went hand in hand, but land reform had to come first.[102]

If land reform was the first step to agricultural modernization, ensuring that sufficient water was in the right place at the right time was the second. Numerous factors made achieving this seemingly simple objective difficult. Historically, some parts of the Rhône valley had too much water, so locals had formed drainage syndicates to share the burden of water management. Because Donzère-Mondragon added a new variable to the area's hydrologic equation, the CNR had to deal with drainage issues, expanding drainage networks in areas that now had a higher water table and granting financial compensation to affected syndicates.[103]

Agricultural engineers, reformers, and locals focused particularly, however, on areas that had a shortage of water. For one thing, the CNR and state agricultural services had to deal with old irrigation systems that became ineffective after the completion of Donzère-Mondragon, such as the Canal de Pierrelatte; Donzère-Mondragon's diversion canal bisected it, thereby affecting its distribution of water to surrounding farms. State engineers attempted to modify the Canal so that it would continue to work in the post-Donzère-Mondragon landscape, eventually sending water from the diversion canal through the Canal downstream from the point of their intersection. Upstream, however, locals who had depended on the Canal were stuck. Of course, the CNR's *cahier des charges* required that the agency fix these problems. The CNR's solutions varied according to the terrain, the system that had been in operation historically, and the proximity of the diversion canal. Wherever possible, CNR officials attempted to maintain existing systems, rather than building new intakes off Donzère-Mondragon's diversion canal. By 1965, the Canal de Pierrelatte channeled 12 m^3/s of water, a significant increase over its historic flow of 5 m^3/s. To accommodate the Canal's increased volume, engineers added irrigation pumps and appropriated seasonal streambeds to serve as supplementary irrigation canals.[104]

Such efforts reflected the shift in the meaning of agricultural reconstitution from remediation to improvement between the 1940s and early 1960s. In short, agricultural improvement involved more water applied to larger, contiguous properties through more "modern" technology. The CNR increased flows in existing irrigation canals like the Canal de Pierrelatte. The Génie Rural also sought funds to modernize older canals like the Canal de Bourne. Other agricultural officials replaced individual irrigation systems with large-scale, collectively owned networks.[105] Studies assessed the possibility of building new irrigation canals off Donzère-Mondragon's diversion canal, although CNR officials were hesitant to do so.[106] In their efforts to improve the Rhône valley's agriculture, rural engineers targeted especially the modernization of irrigation technology. In their view, this objective meant encouraging farmers to adopt sprinkler irrigation. Salenc maintained that this approach was the best way to maximize the "economy of water."[107] In total, the agricultural development of the Rhône valley south of Lyon eventually reserved 175 m^3/s of the river's flow for 350,000 hectares of farmland. The

CNR's projects between Lyon and the Mediterranean sent a total of 100 m³/s into the countryside, while a new agency diverted an additional 75 m³/s near Beaucaire for use in the departments of the Gard and Hérault along the Rhône's western bank.[108]

In addition to land reform and irrigation, electrification transformed the Rhône valley's agriculture. While state elites often pushed reform in this area, some local communities also initiated change, organizing cooperatives in order to adopt new electrified irrigation systems. The Cooperative des Irrigants de la Basse Vallée de l'Arc, for instance, intended to electrify its members' irrigation pumps, in the hope that all farmers in the area might eventually join the cooperative. Initial studies determined, however, that adopting the new irrigation system was "almost entirely impossible" because it involved "extremely onerous" costs. Consequently the cooperative, arguing that electrified irrigation networks acted as a "heavy handicap" on farmers who could not expand gravity-fed irrigation systems, sought subsidies from the government for new irrigation technologies, or at least the institution of peak and nonpeak rates, which might reward farmers who came up with creative ways to use more water during times of lower demand.[109] Such proposals indicated that some rural consumers wanted more energy made available to them. The state did not always force modernization onto provincial subjects; at least some locals desired and actively sought change.[110]

Growing energy demands by agricultural consumers may have been used by the CNR's political and technical elites to justify their agency's projects, but reserving electricity for farmers and subsidizing its cost did truly concern some officials. Article 18 of the CNR's *cahier des charges* specified that the Ministry of Agriculture could "requisition some energy coming from the CNR's plants within the limit of available power." The idea of "available power" had been outlined in a 1948 accord between the CNR and EDF for the CNR's first two projects, Génissiat and Donzère-Mondragon. Before the nationalization of France's electricity agencies in 1946, the CNR had signed a contract with the Société du Sud-Electrique to supply the lower Rhône valley with three thousand kilowatts of electricity annually for agricultural purposes. In its 1948 accord with the CNR, the EDF agreed to uphold this contract but renegotiated its rates. The agreement also included provisions for scheduled future increases as additional CNR projects came on line. By 1953, the CNR would supply agricultural groups with seventy-five hundred kilowatts. The CNR negotiated a total of 18,500 kilowatts for agricultural purposes in the entire Rhône valley. By 1971, 10 percent of the electricity generated at each CNR project was reserved for use and distribution as the Ministry of Agriculture saw fit. However, the CNR's administration sold this electricity to the Ministry grudgingly, characterizing it as a "sacrifice." Even worse, the CNR was forced to sell this electricity at subsidized rates.[111]

In addition to *remembrement,* irrigation, and electrification, the CNR supported a number of measures under the general category of "rural development." The increase in available irrigation water enabled the adoption of new water-intensive crops such as maize. Other changes in crop selection reflected attempts to make the valley's agricultural production more amenable to the market. In the Vaucluse, agricultural engineers aimed to replace traditional polyculture with more intensive crop cultivation that better suited an export market. In the Gard and Hérault, the goal was the same, but the method was reversed: here state experts promoted a program of agricultural reconversion based on the transition from monoculture, primarily low-quality wine, to high-quality polyculture. Other state projects focused on the infrastructure necessary to take agricultural products to market. For instance, the Ministry of Agriculture highlighted investments it had made in fruit refrigeration and processing stations. Agricultural engineers also strongly encouraged mechanization. In 1960, Salenc gave a speech at "the day of mechanization" in the village of Privas. Although he admitted that mechanization could either save or ruin agriculture, ultimately he recommended the adoption of tractors and other means of mechanized production. The moment had come, Salenc argued, to mechanize rationally and to determine the ideal configuration of property size, tractor, and labor hours (both human and mechanical) in order to maximize agricultural productivity and profit. Criticizing the persistence of small properties, Salenc advocated widespread *remembrement* to facilitate mechanization. By the early 1970s, the CNR's agricultural investments included a variety of projects that fostered agricultural production, hydraulic infrastructure, tourism, and rural development.[112]

Other institutions had begun to send the Rhône's waters even farther afield for these purposes. The state created the Société d'Etudes des Canaux de la Rive Droite du Rhône in 1951 to study an ambitious proposal to irrigate the plains of the lower Rhône's right bank (see Chapter 5). The organization ran into immediate conflict with locals, but the proposal moved ahead. In 1955, the state established the Compagnie

Nationale d'Aménagement de la Région du Bas-Rhône et du Languedoc (BRL) to oversee the development of the lower Rhône and Languedoc.[113] The CNR and BRL collaborated to divert 75 m³/s of the Rhône's flow to promote agricultural reconversion, rural development, and tourism in the departments of the Gard and Hérault.

By the early 1960s, the Rhône's waters cultivated farmland throughout the central and lower river valley and even beyond. The CNR's postwar remaking of the Rhône facilitated, in turn, the industrialization of agriculture. While the river's water could serve reactors or its hydroelectricity fuel industry, the CNR's projects could also cultivate farming: they regulated the river's flow, generated electricity, and expanded irrigation infrastructure. Moreover, the projects' siting and construction had created dislocation, which the state used to extend its program of rural modernization. Between 1945 and the 1960s, state projects increased production of export-oriented crops by applying an industrial model to farming. In effect, this new Rhône facilitated the remaking of agriculture and of rural France.

In less than two decades, the Rhône came to little resemble the river of 1945. Additional changes extended across the valley into the surrounding countryside, as new forms of development and the extent of river management expanded considerably during the postwar era, until the Rhône's flow eventually powered hydroelectric turbines, cooled nuclear reactors, and irrigated thousands of acres of farmland. Without a doubt, these technologies dramatically reworked and altered the Rhône—from where and how the river flowed to its temperature. At the same time, these technologies still fundamentally depended on the river. Thus, as competing agencies together oversaw the comprehensive, large-scale, state-sponsored, multipurpose development of the Rhône from the Swiss border to the Mediterranean Sea, the boundaries and relationships between supposedly distinct ecological and technological systems shifted in dramatic new ways. In the process, the Rhône came to be multiple envirotechnical systems.

Although the groups and institutions involved in the Rhône's transformation after World War II expressed different visions and priorities for the river, they generally shared a high modernist view of it. They continually expanded their ambitions and increased their expectations of the river's political, economic, and cultural potential. But realizing the river's development gave rise to many conflicts within the state, between the state and local communities, and between institutions' projects and the valley's physical constraints. One of the most heated was over Donzère-Mondragon and its relationship to the groundwater and the people in the surrounding area. To that story we now turn.

4

LOCAL RESPONSES

In the late 1940s, residents along the Rhône south of Montélimar and north of Orange began reporting disturbing changes in their communities. In some areas, fields became saturated with water that exceeded the capacity of existing drainage networks. Farm animals slogged through sodden pastures, and crops dependent on drier soil began to rot. In other places, water had become difficult to access. When wells ran dry, farmers found they could no longer pump groundwater to irrigate crops and supply their homes. Large crevasses even scarred the landscape in certain locales.

In 1947, the CNR had begun building Donzère-Mondragon. Was there a causal link between construction of the project and the swampy fields and dry wells? Answering this question proved extremely contentious, in part because it was freighted with legal, financial, and political implications. Opinions varied according to author, audience, and historical moment.

For instance, in March 1949, CNR General Director Pierre Delattre wrote to M. Kirchner, chief engineer of the 6ème Circonscription Electrique, "We do not think that the diversion of the Rhône River [for Donzère-Mondragon] will have a perceptible effect on groundwater levels." A few years later, Charles Varennes, who lived near Mondragon, disagreed with Delattre. In December 1952, Varennes "inform[ed]" the CNR "with the greatest urgency" that "my house, located in the area known as Paty, located on the road to Mondragon, is *extremely cracked* because work on the Donzère-Mondragon canal has provoked a drying of the soil. I ask that you send me an expert in order to judge the situation. The land has begun to gape open. A man could fall into certain crevasses."[1] In December 1956, Jacques Vollant, a resident of the town of Pierrelatte, agreed with Varennes and wrote the CNR, asserting that the complexity of the local environment supported residents' claims: "I am sure you will admit that it is almost impossible to obtain certainty scientifically that the CNR's work has not modified the regime of subterranean water: too many uncontrollable phenomena are involved."[2] The CNR's political and technical elites proved unable to uphold Delattre's easy dismissal. The agency soon found itself in a quagmire of environmental change, competing knowledge claims, and the problem of uncertainty.[3]

Modernizing groundwater was not a goal of either the CNR or the French state, but the attempt to maximize hydroelectric generation immediately following World War II appeared to have resulted in significant changes to the central Rhône valley's hydrology. At least, locals believed this was the case. Historians usually frame such environmental changes as unintended consequences, but they were not unexpected by locals and even some CNR officials. Indeed, French engineers had a long tradition of building large water engineering works, including the seventeenth-century Canal du Midi, which had altered surrounding groundwater. In some ways, then, the history of groundwater debates in the Rhône valley after 1945 is a typical account of conflict between locals and experts, small communities and the centralizing impulse of the French state. As such, it exemplifies some of the limits of the high modernist state.[4]

If we dig a little deeper, however, a half century of conflict over the Rhône valley's hydrology yields a richer, far more complicated story about the politics of water, knowledge, and envirotechnical systems. These conflicts reveal some of the epistemological and political complexities of expertise, causality, and uncertainty in mid-twentieth-century France. Drawing on methods from STS, especially so-called controversy studies, which take seriously competing knowledge claims, this chapter juxtaposes the various positions of groups involved in the groundwater debates.[5] The CNR's officials and locals disagreed about the cause—or causes—of hydrologic change along the central Rhône, as well as the extent to which human or environmental factors were to blame. It took two decades for a consensus about the existence and general character of hydrologic change in the region to emerge. Moreover, comparing CNR leaders' rhetoric, in both official statements and internal discussions, with their actions complicates the agency's account of the new hydrology of the central Rhône. Despite the apparent limits to the CNR's knowledge, the agency's models guided its formulation of solutions that continued to transform the region. However, these "solutions" repeatedly failed to control not only the landscape of the Rhône valley but also local communities. In the end, tensions within the CNR's envirotechnical system and local responses to that system threatened to undermine the ambitious project of national reconstruction of which Donzère-Mondragon was supposed to be an icon.

Locals' Perspectives

Locals worried about the potential effects of Donzère-Mondragon on groundwater as early as January 1945, during the project's planning stage. In public hearings, the mayor of Bourg-Saint-Andéol, a town on the Rhône's western bank, and administrators from the department of the Ardèche voiced their concerns. The mayor predicted the project would "fatally" decrease the level of water in the Rhône itself and in the surrounding water table. Because his town depended on these water supplies, the mayor could not accept the project "as is."[6] In the coming months, the CNR and supervisory state agencies essentially ignored his objections.

When Donzère-Mondragon appeared to validate their fears, farmers, residents, and local politicians did not remain silent. The CNR's archives offer unusually rich insights into the perspectives of locals on these developments.[7] The agency received letters from about one hundred individuals—some mayors or business leaders, a few property owners from Paris and Marseille who leased their land to tenant farmers, but mostly small farmers, their widows, and their descendants. The vast majority of their letters are short and direct, usually starting with a quick identification of the author's profession and home: say, a farmer with so many acres of land from a particular commune. Then they briefly describe the specific problems the writer and his family are facing: a well has been dry for two weeks, groundwater is no longer potable, cracks have begun to appear in the walls of their house and outbuildings. The letters always conclude with the formal closing salutation expected in all official correspondence in France.

In their letters, locals employed several rhetorical tactics to forge what they saw as an unquestionable link between hydrologic change and the CNR's project, and accordingly to request technical or financial assistance. Not surprisingly, those writing to the CNR to file complaints or ask for assistance held the agency responsible. However, the volume of letters sent, including a petition with over six hundred signatures for a region of approximately fifteen thousand people, suggests just how widely communities attributed hydrologic change to Donzère-Mondragon. Some even held the agency responsible into the 1980s.[8] Thus, the most striking feature of these letters is their unanimity in blaming the CNR.

In dozens of letters, locals made unambiguous connections between Donzère-Mondragon and hydrologic change.[9] How exactly locals constructed these narratives of causality can be categorized in three ways.[10] First, about half of the authors stated that *"les travaux de la CNR"* (the work of the CNR) had caused their water woes. It was the most general phrase used. For example, Nouguier, president of the "Federation of Those Expropriated by the Compagnie Nationale du Rhône in Bollène," reported, "several inhabitants of the commune of Bollène have had their wells dried up by the work undertaken by the Co[mpanie] Nationale du Rhône." Second, most other letter writers targeted *"le canal"* of Donzère-Mondragon, by which they meant the long diversion canal. As Henri Bastet put it, "The canal of the Rhône, having dried up the water, has created fissures in my farm. I ask you to come and fix them."[11] Third, and less frequently, locals referred to the tailrace *(le canal de fuite)*, the stretch of the diversion canal downstream of the Blondel hydroelectric plant. In one such case, Fournier expressed his unequivocal opinion to the CNR's general director: "There is indisputably a clear relationship between the dredging of the tailrace after the Blondel plant and the drying of my well, which, by completely depriving my farm of potable water, takes away from me the possibility of renting or selling it."[12] The overwhelming use of the two most general terms, *les travaux de la CNR* and *le canal,* suggests that although locals clearly associated hydrologic change with the CNR's project, most did not know or try to identify what specifically was at fault. Locals' use of a vague terminology of causality indicates they may not have had the knowledge to understand and describe the new landscape that seemed to be causing them so many problems. However, CNR and state experts also appeared to have limited knowledge. Locals' accounts of causality therefore reflected a larger shift in control over environmental management in the central Rhône valley from their communities to the CNR and its envirotechnical system.

Some locals questioned the CNR's explanations of hydrologic change in their letters to the agency. As I show below, the CNR's officials did not entirely deny responsibility, but the agency's elites argued that several other factors played significant roles. In response, only one letter, a petition from residents of La Garde-Adhémar, accused CNR officials of using drought as an excuse to cover up their own culpability.[13] Instead, most writers presented evidence in an effort to undermine the agency's assertions.

A letter written by three farmers from the town of Donzère in December 1949 exemplifies this rhetorical strategy:

> The undersigned, Joseph Roussin, Simon Gilles, and André Larmand, would like to call the attention of the Chief Engineer of the CNR in [the town of] Donzère to the state of their wells, which have been dry for a considerable period of time.
>
> At first, they [Roussin, Gilles, and Larmand] attributed the state [of their wells] to last summer's terrible drought, though in similar situations in the past, their wells had never dried up. However, since abundant rains have fallen, the level of our wells have not stopped dropping, and thus, we think that this [problem] must be the effect of the pumping required for the construction of the [Donzère-Mondragon] Canal.[14]

Close examination of this letter demonstrates the way the authors strategically challenged the CNR. First, consider how the authors set up their argument. Although they ultimately conclude the CNR is responsible for their troubles, the farmers describe their initial assumption that "last summer's terrible drought" is at fault. By doing so, they eliminate the gulf between locals' and CNR officials' explanations of these events. Several other authors adopted a similar tactic. Emile Lachaux, a schoolteacher, echoed the three farmers when he wrote to the CNR, "At first, I attributed all this to the extreme drought."[15]

However, Lachaux, like Roussin, Gilles, and Larmand, then marshaled evidence to challenge this conclusion. The three farmers stated that previous droughts had never caused problems with their wells. Other locals reiterated this point. For example, the mayor of La Garde-Adhémar sent the CNR a list of affected properties in October 1949. He concluded his letter politely but firmly: "I kindly ask you to note that there is no doubt that the lack of water in these farms is due to the construction of the [CNR's] canal because water has never been lacking, even during droughts much greater than this."[16] According to these locals, history demonstrated that nature had never before been this uncooperative.

Roussin and his neighbors then attempted to further undermine the CNR's account by pointing to recent rains. Since drought was by definition a shortage of precipitation, if recent "abundant rains" had not remedied the problem, then drought must not be the real issue. Seven other farmers agreed with Roussin when they complained that the "horrific situation" had not been solved "despite the rains and the Rhône's flood."[17] Locals therefore concluded that nature was not to blame.

Having built up their counterevidence to the CNR's explanations, Roussin and other locals reasoned that if current groundwater issues could not be caused by drought, then the CNR must be responsible. Unlike most letters, theirs cited a specific cause: the pumping of groundwater to facilitate construction of Donzère-Mondragon's diversion canal. The rhetorical tactics adopted by Roussin and his neighbors led these authors to end where most locals began: the ultimate responsibility of the CNR and its project.

The vast majority of locals writing the CNR may have blamed the agency, but they also sought its expertise to remedy their situation. Some letters simply asked for "an expert" without specifying a preferred institutional affiliation.[18] Yet a surprising number requested experts from the CNR itself. Charles Varennes, whose land had become so fissured that "a man could fall into certain crevasses," asked CNR officials to "please let me know what day the expert will be arriving so that I won't be absent then."[19] Only rarely did locals demand an outside expert.[20] Locals who readily blamed the CNR nevertheless granted the agency the authority to assess their claims.

When outlining their claims, most authors simply described the problems they faced without making concrete demands. At most, letters asked that the CNR "do whatever is necessary to remedy the situation."[21] Rarely did individuals request, let alone demand, specific fixes such as deeper wells or a certain amount of financial compensation.[22] These solutions can instead be found in the CNR's internal correspondence, in which agency officials discussed locals' claims and debated possible remedies. Locals therefore left the proper solutions to their problems in the hands of the CNR, even if they believed the same agency was responsible for their problems in the first place.

Most locals also refrained from accusations or condemnations, let alone protests or violence. Rather, they made straightforward claims of causality, simply linking Donzère-Mondragon to their drying wells and failing crops. G. Dumas exemplified this unemotional tone when he told the CNR "Our property, called 'La Mouette,' situated in the town of Pont St. Esprit (located in the department of the Gard), has been deprived of irrigation water because of the construction of the [Donzère-Mondragon] Canal." Locals also wrote polite letters to the politicians on whom they called to plead their case. Only a handful of letters made plaintive appeals. Because his wells had dried up, seventy-five-year-old Jules Albin had to supply his animals with water that he had carried "on my back." He characterized this necessity as a "terrible situation considering my age and the distance I must carry the water."[23] Although many residents waited years for financial compensation or for the CNR's technical solutions to become effective, their letters did not become progressively more accusatory.

The diplomatic, even hyperrational, tone of these letters is remarkable, given the widespread social unrest during the late 1940s and early 1950s, both throughout France and locally, where workers building Donzère-Mondragon repeatedly went on strike.[24] Despite this strife, rarely did the CNR receive letters expressing sharper sentiments or threatening legal action. In one of these unusual letters, L. Morand invoked Europe's recent fascist history when he compared Donzère-Mondragon's diversion canal to one "Mussolini had constructed to drain the Pontine marshes." His condemnation was blunt but, perhaps surprisingly, atypical. Moreover, only a few individuals threatened to contact judicial officials if the CNR did not take action.[25]

The seemingly patient tone of locals' letters masked frustration that emerged, however, in other ways.[26] Some extended families or groups of neighbors like Roussin, Gilles, and Larmand wrote joint letters on behalf of multiple properties, presumably in an attempt to bolster the validity of individual claims through collective action. These letters usually had between two and ten signatures, but occasionally locals submitted petitions for an entire town or even several neighboring communities. One included over six hundred signatures.[27] Locals also mobilized through new organizations that they established to help fight their cause. The "Federation of Those Expropriated by the Compagnie Nationale du Rhône in Bollène" was one of the most active. An even larger group, "Those Affected by the Canal of Donzère-Mondragon," publicly demanded information about the status of the CNR's repairs in *Le Dauphiné libéré*, a regional newspaper serving the Rhône-Alps region, in October 1953.[28] These organizations seem to have coordinated a letter-writing campaign, since several dozen letters sent to the CNR in 1952 appear to be form letters. Meetings between locals and CNR officials also demonstrated the degree to which these communities had mobilized around their shared predicament. At one meeting in February 1950, CNR engineer J. Rostagni noted locals' and elected officials' frequent protestations, recording their recommendations as to what steps should be taken.[29]

A poster that was plastered over the towns of Bollène and Donzère in April 1957 confirms that at least some locals had become extremely angry about what was happening and that their frustration may have peaked during the late 1950s. If locals correctly assessed that Donzère-Mondragon was the source of their groundwater problems, then by that point they had already been dealing with these issues and unsuccessfully lobbying the CNR for nearly a decade. Directed at "visitors" to the region, the poster's text attempted to rally support for the plight of local communities (see Figure 4.1).[30]

VISITORS

Before you

The Plain of the Canton of Bollène
Whose land has been dried by work from the canal of Donzère-Mondragon

Situation:

Before the project	After the project
Rich and fresh land	Dried land
Abundant and pure water	Non-potable water, sometimes contaminated
Raised and constant water table	Water table dropped to a level not beneficial for agriculture
Solid buildings	Fissured buildings, sometimes dangerous to inhabit

Results:

Dramatically diminished agricultural production
Leached land, sterile after a short period of time
Farmers in ruin

Conclusion:

Regarding agriculture: do not give any credit to the deceitful assertions of the Compagnie Nationale du Rhône. Go to the farms, taste the water, look at the buildings, and judge for yourselves.

Those Affected in the Canton of Bollène

Figure 4.1. *Challenging Donzère-Mondragon*. English translation of text of 1957 poster aimed to rally opposition to the CNR and Donzère-Mondragon in the towns of Bollène and Donzère.

This poster demonstrates that some locals rejected the celebratory discourse invoked by politicians, engineers, and writers who framed Donzère-Mondragon as the icon of Promethean France. No archival evidence suggests that these communities joined the conservative, antiindustrial, and antistate Poujadist movement founded in 1953, which quickly gained electoral representation in 1955 and 1956; none of the letters sent to the CNR referenced Poujadisme.[31] Although members of these communities did not appear to join political parties to express their frustration, other evidence, including this poster, conveys the difficulties and rage some locals experienced. After all, the poster depicted the CNR and its project as a destroyer of land and livelihood. Furthermore, it asserted that the landscape itself was proof of the CNR's responsibility. The poster's creators maintained that simple observation ("look at the buildings") and experience ("taste the water") yielded knowledge of the situation and its cause. According to the poster, locals and outsiders alike could read the landscape to determine the source of what its authors saw as a crisis: "farmers in ruin." No scientific expert or specialized expertise was needed. A transparent, legible nature spoke the truth. But CNR officials did not necessarily concur.

The CNR and the Politics of Knowledge

The CNR's administration and technical elites lacked the locals' clarity and unanimity about what was causing changes to the area's hydrology.[32] However, a close reading of CNR documents suggests that the intended audience shaped what (and which) story CNR officials told. Two documents by Delattre encapsulate tensions between the agency's public and private statements and are thus worth quoting at length.

On March 4, 1949, Delattre responded to the prefect of the Vaucluse, who had received numerous complaints from residents of Bollène, a small town located just south of the Blondel hydroelectric plant. Delattre began his reply by trying to reassure the prefect: "This question captures all of our [the CNR's] attention." Then he described the situation, its cause, and the CNR's course of action. Citing evidence from throughout France, Delattre blamed widespread drought for Bollène's water problems: "You know that many wells are very low or have been drying up throughout France because of the drought that has persisted for several years now:

this is certainly true for some wells in the Rhône valley, even in places where our work cannot be incriminated." Delattre proceeded to cast the agency as a good Samaritan, emphasizing how the CNR was helping those living near Donzère-Mondragon by supplying them with water even though, he claimed, the CNR was not actually responsible for their water issues: "Certain superficial wells in the region of Bollène will probably dry up in any event. Nevertheless, we have up until now supplied water at a relatively low cost to all of the houses that have lacked water in the region where we are working." Delattre even questioned whether groundwater subsidence was a consequence of the CNR's activities: "We do not think that continuing our work toward the south is tending to lower dangerously the groundwater level [there]; as you know, we are working by dredging on this side, thus without pumping out [groundwater]." However, as Delattre explained, "We are conducting experimental studies of artificial recharge of the water table at this time."

This account contrasts with a memo Delattre wrote to one of his senior engineers and the CNR's Research Division the next day. In this memo, Delattre explicitly references his letter to the prefect when he tells his subordinates the CNR had received numerous claims, especially from residents of Bollène and throughout the department of the Drôme. Delattre then asks the CNR's technical staff to conduct additional studies in these areas: "We need to see, particularly in the region of Bollène, if the lowering of the water table due to pumping for the [construction of the Blondel] plant is still continuing to drop or if it is reaching a stabilization point."[33] One day after informing the prefect of the Vaucluse that drought was to blame and "we do not think that continuing our work toward the south is tending to lower dangerously the groundwater level," Delattre privately informed his colleagues that the cause was the CNR and its pumping.

Had Delattre lied? Was this politically expedient double-talk? According to the 1957 poster, the CNR had made "false assertions." Indeed, at first glance, significant discrepancies between the two documents seem to indicate as much. However, a close reading of the CNR's public and internal documents tells a far more complicated story about the difficulties of modeling a dynamic environment, the uncertainties of knowledge, and the way power mediated the production and distribution of that knowledge.

The CNR's Local Public

The CNR's elites did not share locals' clarity about the cause of groundwater change, either publicly or internally. Moreover, the agency's public discourse was actually partitioned for two different audiences. The first included the farmers, mayors, and other residents living near Donzère-Mondragon. When writing to these individuals, CNR officials most frequently attributed problems to drought conditions. In November 1949, Rostagni explained to the mayor of La Garde-Adhémar that the CNR had "carefully" considered the claims of his town's residents and determined the town was located in an area "not at all influenced" by the agency's project, adding that the "worst drought in a century" was instead responsible.[34] Although CNR engineers would reverse their conclusions for this area six years later, agency officials told locals that drought played a substantial role in accounting for groundwater change in the central Rhône valley even into the 1980s.[35]

The CNR's officials did not absolve themselves from responsibility in their correspondence with locals, but comparing the agency's public statements with its internal documents exposes the extent to which its public and private discourses diverged. These officials assumed accountability in a few letters to individual claimants between the late 1940s and late 1950s, but in most cases held drought responsible.[36] Yet Rostagni estimated in a 1953 internal memo that two-thirds of the claims along the right bank of the tailrace were indeed legitimate. Internally, then, the CNR's leaders tended to characterize more claims as justifiable than they did publicly.[37] Furthermore, although CNR leaders declared it was "urgent to fix legitimate claims," they did not explain *what* or *whose* criteria might be used to determine legitimacy.[38]

Internal documents show that the CNR's top administrators and engineers so feared litigation that they were willing to repair damages for which they did not believe themselves actually responsible just to avoid further investigation. They worried that additional inquiries might reveal more extensive damages or, even worse, an undeniable link between the CNR's project and hydrologic change. For example, one week before the inauguration of Donzère-Mondragon, Rostagni warned his superiors that their agency needed to avoid "collective discontent," which might provoke "bothersome interventions" from local and regional governments.[39]

The CNR's leaders may have considered paying for a few "extra" repairs worthwhile, but they also hoped to marshal evidence to defend their agency against locals' claims. In February 1952, Tournier informed the CNR's Research Division that one resident, M. Bresson, had filed a complaint with his prefect. Tournier proclaimed to his colleagues that the "Affaire Bresson" was "extremely important" because "if we [the CNR] are eventually found guilty, [then] all of the property owners along the tailrace will present us with similar claims. Under these circumstances, we ask you [the CNR's Research Division] to proceed with a very detailed technical study of this question and let us know in a precise manner what, in your opinion, are the arguments we can make against the demands of M. Bresson."[40] As this internal memo suggests, the CNR's administration feared that losing one case might result in a spate of litigation because it would validate other locals' claims. Consequently, Tournier hoped his engineers could generate "arguments" to prevent such a dangerous and costly precedent.

During the late 1950s, however, an important shift in the CNR's discourse with locals occurred. Beginning in 1957, the CNR signed *conventions*, or legal agreements, with hundreds of residents. It seems likely the fear of litigation, mounting pressure from locals, and (as we will see) ineffective technical "solutions" together pushed the agency to make substantial concessions. The details of these *conventions* varied, but the agreements explicitly stated that Donzère-Mondragon had caused hydrologic change, thereby implicating the CNR. As one 1959 *convention* put it, the drop in the water table "following the initiation of Donzère-Mondragon's operational procedures" caused damages the CNR had to repair. The agency's *convention* with the extended Gilles family of Bourg-Saint-Andéol was typical of those signed in 1960: "The Donzère-Mondragon project provoked a lowering of the water table in the property of Erard, Anne, Baldus, Maurice, Jean-Marie, and Clément Gilles and Emmanuel, Maurice, Gabriel, and Jean-Marie Gilles, including the following parcels in the commune of Bourg-Saint-Andéol."[41] These *conventions* thus forged a clear link between Donzère-Mondragon and groundwater change, attributed unambiguous responsibility to the agency, and therefore validated locals' assertions.

The CNR's Public of Political and Administrative Elites

The CNR's explanation about the cause of groundwater change shifted when it addressed a second public composed of administrative and technical elites. This public included the prefects and Conseil Général of the four departments Donzère-Mondragon traversed, officials from state agencies such as the Génie Rural and the Conseil Générale des Ponts et Chaussées, and national politicians who became involved after they received letters from their constituents. For this audience, CNR leaders also attributed an important role to drought, but they assumed more responsibility than they did with locals.

Still, even to this elite audience, the degree to which the CNR's leaders admitted blame varied. The agency's officials seemed to accept the most responsibility in their correspondence with representatives from certain state agencies, especially those holding supervisory roles. For instance, in early 1950, Delattre reported to M. Pfahl, head of the regional division of the powerful Ponts et Chaussées, that the CNR and the Société Anonyme de Coordination des Travaux d'Aménagement du Rhône à Donzère (SACTARD), the agency charged with the actual construction of Donzère-Mondragon, had deepened wells "where the lack of water is clearly attributable to our work."[42] In letters to the Génie Rural and 6ème Circonscription Electrique, CNR officials brought up the role of drought but also stated that the agency played an important role.[43] They made similar arguments to prefects. As Delattre acknowledged to the prefect of the Drôme, "It is certain that our work has contributed to the drying [of the water table]," but then added a crucial caveat: "in certain cases and to a certain extent." Moreover, Delattre proceeded to chide the farmers, who, he said, "obviously overestimate the effect of the [CNR's] work and underestimate that of the drought."[44] That the CNR was at least partly accountable to some of these state agencies may help explain its position.

The CNR's correspondence with Édouard Daladier, prime minister three times during the 1930s and member of the National Assembly from 1946 to 1958, exemplify the agency's carefully crafted exchanges with national politicians. Daladier notified the CNR's administration of letters he had received from locals asking him to intervene on their behalf and hoping that he—born in Carpentras—had not entirely forgotten his Vaucluse roots. Responding to Daladier, Delattre acknowledged "damages to which certain property owners or those working the land have been subjected because of the draining effect of Donzère-Mondragon's tailrace." However, he also noted that the "delicate question" of evaluating these damages to determine "an equitable solution and to save the general interest" had not yet been calculated. Here Delattre appealed to the common good in the hope of defending the CNR against locals' claims, an argument he also made to Édouard Rastoin, a Marseille oil baron and member of the National Assembly.[45]

The CNR's officials seem to have accepted less responsibility in their correspondence with agricultural and hydraulic experts from other agencies. Locals' complaints to their political representatives, a handful of court cases, and the CNR's legal obligation to supply preexisting quantities of water to locals resulted in the limited inclusion of outside experts in groundwater discussions.[46] Minutes from meetings among these technical elites indicate that the CNR's representatives recognized "the damages caused by the [CNR]."[47] Yet the agency's leaders were quick to highlight inadequate data and inconclusive evidence. In a meeting in 1954, Roger Rougier, an agricultural expert who was defending local claimants, asserted that technical solutions and financial compensation should be determined by "a coefficient of responsibility accorded to the CNR based on the relative drop of the water table in different zones." Responding to Rougier's recommendation, CNR engineer A. Dagand remarked, "The CNR never contested that, at a minimum, the water table was going to be lowered in certain zones by the creation of the Canal of Donzère-Mondragon." However, Dagand warned, "It remains to be determined if the claimants have been subjected to damages from this drop and, if so, to calculate the indemnity." A month later, Dagand again underscored the CNR's ambiguous share of responsibility: "At this time, the principle of the CNR's responsibility is not actually known and it can not be admitted on the basis of simple hypotheses."[48] These statements suggest that CNR officials capitalized on partial knowledge and inconclusive causality to raise questions about their agency's responsibility.[49]

The CNR's multiple publics thus received different stories about the cause(s) of groundwater change in the central Rhône valley. The agency's political and economic relationship with each audience appears to have shaped its narratives. The CNR attempted to appease locals, the least powerful group, by pointing almost exclusively to drought. Until the late 1950s, letters, petitions, and protests did not sufficiently pressure the agency into making the same statements to locals that it did to many administrative and political elites. However, in their correspondence with national politicians and technical elites, CNR officials repeatedly added caveats about the difficulties of determining causality with any real certainty. The CNR's precarious financial situation and its need for state approval for each of its projects suggest that it may have sought to cast Donzère-Mondragon in the best light to preserve the future development of the Rhône and the agency's central role in that process.

The CNR's Internal Correspondence

On some level, multiple narratives constructed around imperfect knowledge and ambiguous causality also characterized the CNR's internal correspondence. The CNR's own experts had predicted significant hydrologic change even before construction of Donzère-Mondragon began.[50] They also asserted that drought played a significant role between late 1949 and the spring of 1952. For instance, Rostagni analyzed data on the level of the water table between 1947 and 1949 in an internal memo. He determined that "outside of other influences, the drought has caused an average drop in the water table of more than 1.5 meters in relation to its normal level." Although Rostagni did not make explicit the assumptions behind these calculations, other internal reports reiterated his findings.[51]

Because the CNR's technical elites blamed both drought and their agency's project, they attempted to determine the relative responsibility of each, just as Rougier had recommended. Calculating the CNR's precise share of responsibility could either strengthen or ultimately undermine the agency's position. It all depended on the calculation's outcome. An ambiguous result might also work to the agency's advantage. For example, in 1950 Marc Henry told his superiors that one of his engineers had conducted studies in an attempt to differentiate between the effects of drought and those of the CNR's actions, especially pumping out groundwater. After reviewing this report, Henry concluded, "It is *likely [vraisemblable]* that a large part of the drop in groundwater levels is due to pumping." However, when assessing locals' complaints about cracks in the ground and their houses eighteen months later, Henry stated that "it is possible" that groundwater subsidence had "provoked" these fissures, but notably, he did not identify the cause of that subsidence.[52] If subsidence was the cause and not the effect of another event or process, then the CNR's role was conveniently diminished. As Henry's comments suggested, the CNR and its supporters could use the language of plausibility and possibility to weaken attributions of causality and therefore lessen the agency's responsibility. It was in their best interest, then, to be in charge of the interpretation. Furthermore, it is worth noting that the CNR's discussion apportioning responsibility occurred primarily *within* the agency.

Imperfect knowledge exacerbated the uncertainty of whether and to what extent the CNR's project was responsible. The agency's officials stated that groundwater studies could only produce approximations.[53] As CNR engineer J. Mathian confessed to his superiors, "we have to admit that, in general, these hypotheses [i.e., those about the behavior of groundwater] are not verifiable by other, more direct methods."[54] For these technical elites, hydrologic change in the Rhône valley demonstrated the way ecological complexities could complicate scientific research, defying either simple or certain conclusions. In addition, as time passed, CNR experts found that hydrologic processes did not necessarily conform to the models they had developed. Moreover, problems worsened despite the CNR's efforts, only further exposing the discord between the Rhône and the knowledge the CNR "experts" possessed of it.

Yet this tension, along with the language of uncertainty expressed by Henry, Mathian, and other CNR officials, largely remained *within* the agency. Public discussion of it carried two risks: first, opening the CNR's findings to interpretation, and second, granting locals credibility. Democratizing the uncertainty debate might end up increasing the CNR's responsibility and thus spur costly mitigation, if not litigation.[55] Put more simply, publicizing and discussing uncertainty might not only undermine the CNR's authority as technical expert but also have serious financial and legal implications.

Uncertainty, generated partially by multiple causes and marshaled by the CNR to its advantage, therefore helped the agency downplay its responsibility. No archival evidence suggests CNR officials blatantly lied or used drought as a coverup.[56] Yet as Delattre's 1949 letter to the prefect of the Vaucluse and his subsequent internal memo together attest, CNR officials did appropriate complex causality and uncertainty to cast their agency in a better light. In another example, during early debates over the legitimacy of groundwater claims from the town of Donzère, Tournier remarked to a CNR division that most of the claimants lived outside the area affected by pumping. He continued, "under these circumstances, we are wondering if we are really responsible or if it would not be better to attribute the said damages only to the drought." Nine days later, Rostagni responded to Tournier's query. After revisiting the data, Rostagni concluded, "most of the claimants are located at the theoretical limits" of the zone of influence, and "a part of the damages about which the property owners are complaining is attributable to the drought."[57] That drought was at least partly involved allowed CNR officials to deflect blame to environmental factors. Invoking drought thus helped the CNR's leaders downplay or perhaps even minimize their agency's role, a role they admitted internally but only to a limited degree publicly.

The reliance of CNR administrators and engineers on descriptive, rather than causal, language offered another way to lessen the agency's role. The agency's officials titled several early documents "Drop in the water table due to pumping for the [Blondel] plant" but within eighteen months shifted their focus to the manifestation of the problem rather than its cause.[58] For example, Rostagni titled a memo to the CNR's Operations Division "Drop in the water table along the feeder canal." By doing so, Rostagni highlighted the problem—the drop in groundwater levels—rather than the potentially controversial cause or causes of that drop.[59] Also recall Henry's statement that subsidence had caused fissures in locals' land and homes. In some ways, CNR officials' descriptive language paralleled that of locals. While locals described their problems without asking for specific solutions, CNR officials described hydrologic change without identifying its cause(s). The CNR's descriptive language therefore performed powerful work for the agency through its passivity and apparent neutrality. It helped obscure the contentious and potentially expensive question of causality. In short, the drop in the water table, not the cause or causes of that drop, became the focus of debate. This rhetorical move helped recast what was cause and effect and ultimately deflected blame from the CNR to nature.

In November 1952, CNR engineers added another environmental variable to their causality equation: geology. That month, Mathian explained in an internal report to the CNR's administration that cracks in the land and buildings were "due to a local geological phenomenon and probably due specifically to the special formation of lime in these areas." Several months later, Mathian noted it was difficult to ascertain whether changes in the water table had indirectly "caused" these cracks by revealing these geologic formations. But after examining several claims, Mathian concluded that none was due to groundwater variations.[60] He thus absolved the CNR by neatly distinguishing natural and cultural processes, favoring environmental explanations and blaming geology alone. Or, to use Bruno Latour's term, Mathian transformed nature-culture into nature and culture.[61]

Mathian developed his geologic hypothesis several months after the initiation of normal operational procedures at Donzère-Mondragon. Probably during the summer and fall of 1952, CNR engineers began to suspect that the project's operation, not just its construction, was altering groundwater, a point reinforced by the numerous letters the agency received during those months. Consequently, faced with a problem that appeared unlikely to go away, the CNR's experts again attempted to identify its precise cause.

In this context of mounting local protest, Mathian continued to study the possible role of local geology. By June 1953, Mathian had refined his theory of microscale geologic variations to account for local differences in hydrologic patterns. He hypothesized that the relative permeability of certain geologic formations explained why some areas experienced greater groundwater subsidence than others. In particular, Mathian introduced the concept of "geologic discontinuities" to the CNR's model of the region and accordingly divided the central Rhône valley into fourteen zones. As with his earlier study, Mathian used hydrologic change to study and represent the valley's complex geologic formations. Because his research focused, however, on these geologic variations and not the CNR's project, which had motivated their investigation, Mathian effectively separated environmental and cultural processes, neatly framing changes in the region's hydrology as "natural" events and minimizing the CNR's role. According to Mathian's revised theory, the CNR was merely the secondary cause of groundwater change; geology was really at fault.[62]

Future studies refined the zones Mathian identified in his 1953 report, but after that year, his geologic discontinuity thesis fundamentally underlay the CNR's model of the central Rhône. Even more significantly, it was also used to assess locals' claims.[63] One of the most important assessment tools based on Mathian's research was a 1955 map titled "Influence of the CNR's work at Donzère-Mondragon on the behavior of groundwater." Mathian's zones, which explained hydrologic change by geologic discontinuities, were inscribed on the map. Yet the CNR also superimposed on Mathian's zones nine different kinds of hydrologic shifts based on the cause and extent of change. The map identified four causes: elevated groundwater levels "by the feeder canal"; lowered levels "by the tailrace"; lowered levels "by the Rhône River"; and finally, the ambiguous "modifications for diverse reasons." For each of these, the CNR designated regions as either being affected with "important" or "minor" modifications. Any area that remained unchanged was white. The map's color coding therefore indicated which areas were affected, the cause of those effects, and the extent of impact. It essentially communicated that the CNR accepted some responsibility for the Rhône's new hydrologic order; after all, the map's title was "influence of the CNR's work at Donzère-Mondragon on the behavior of groundwater." Recall, however, that Mathian's zones were also inscribed on the map. It literally illustrated, then, the CNR's two competing explanations: the colors attributed change to the CNR; the zones blamed geology. The 1955 map thus embodied CNR officials' own uncertainty about the cause(s) of hydrologic change along the central Rhône.[64]

Despite this ambiguity, and perhaps even the inconsistencies, of the 1955 map, it appears to have served as the guideline for the CNR's assessment of locals' claims. The particular location of a claim in the zone and color matrix determined whether the CNR deemed a claim legitimate or spurious. Agency officials did not explain, however, how they reconciled potential contradictions between the map's zones and colors, a tension both troubling and significant, foremost for locals themselves.[65] Even more incongruous was the fact that claims against the CNR were assessed by standards determined by the agency itself.

Ultimately, this map became more than a model of the region, and its assumptions became more than abstract ideas. In effect, the CNR's understanding of regional hydrology, however tentative and perhaps even erroneous, superseded the processes it was supposed to simulate. The principles underlying this model then took concrete form in the technological fixes CNR officials agreed or did not agree to make for affected properties. While the model may have been intended to simply depict the central Rhône, it in fact transformed that landscape by mediating the assessment and mitigation of locals' claims. In short, the CNR's engineers reproduced the model and its postulates in its technological "solutions" and therefore on the landscape itself. The boundaries between the natural world and models of that world blurred, as locals and CNR experts contended with a situation neither group seemed to understand completely.[66]

Although the CNR's experts eventually concluded the agency did play a role in transforming the area's hydrology, a point encapsulated by the 1955 map and outlined in the *conventions* of the late 1950s, they frequently depicted the new groundwater conditions as an improvement. They defended this position most vehemently between 1950 and 1954, the period in which Donzère-Mondragon was completed, claims proliferated, and the agency began to implement technical solutions to these claims.[67] For instance, in 1952, Henry explained that the CNR would not be obliged to "reestablish too strictly the previous status of groundwater" because "it was bad." In fact, whatever direction the water table went, CNR engineers seemed to assert that locals were better off. The agency's officials often made these convenient declarations to politicians and representatives of state agencies. One CNR official told the head engineer of the 6ème Circonscription Electrique, "The plain where the property owners reside was clearly too humid in many areas. . . . A moderate drop in the water table corresponds, therefore, often with an improvement for agriculture." Although he acknowledged that "the drop in the water table is desirable only within certain limits," he asserted that the CNR project "avoids an exaggerated drop."[68] If they had been asked, locals might have disagreed with his conclusions.

The CNR's ideology of improvement downplayed, if not dismissed, the consequences of groundwater change along the central Rhône. CNR officials' claims that they had improved, not destroyed, regional hydrology were expedient. Yet if they were successful in representing the valley's new hydrologic order as an asset, then perhaps they might not have to worry about such troubling issues as causality, responsibility, and mitigation. This argument was undoubtedly self-serving, both politically and financially, but it also fit within the broader ideology of development that guided the CNR's reconstruction of the Rhône. Agency publications portrayed it as improving nature and thereby the nation through its projects, a view shared by state engineers and an extension of the idea that the French engineering tradition served the public good.

Although the CNR's elites downplayed their agency's role, even in internal documents, by pointing to drought, emphasizing uncertainty and inadequate data, using descriptive language, and invoking geology, the CNR was eventually held culpable in the legal agreements signed with locals.[69] Little evidence explains why the CNR reached these agreements between 1957 and 1960. It is also unclear why the *conventions* included direct statements of responsibility after such a long period of contestation. It seems likely, though, that mounting local pressure coupled with a decade of largely ineffective solutions finally pushed the CNR into a corner. At that point, locals received concessions, but largely on the CNR's terms, on its timeline, and up to thirteen years after groundwater change had first become apparent—at least to locals.

The CNR's Technical "Solutions"

As locals, politicians, CNR officials, and state experts debated the cause(s) of hydrologic change along the central Rhône, what steps did these groups take on the ground? Despite the CNR's mixed admissions of responsibility, as early as October 1947 CNR officials had discussed the construction of an extensive system to fix problems they frequently denied or cast as improvements. If causality was so unclear, at least in the eyes of many CNR officials, why did their agency propose solutions to an ostensibly nonexistent problem? Contemporary laws placed some constraints on the CNR. The *cahier des charges spéciales*—a section of the by-laws governing Donzère-Mondragon passed in December 1953, six years after construction had begun and four years after groundwater change was first apparent to locals—outlined what the CNR would be obliged to do if groundwater was affected. Article 16 specified: "In case of groundwater subsidence caused by the reduction of the flow of the Rhône between the reservoir dam and the mouth downstream of the diversion canal, the concessionaire [the CNR] should undertake all useful measures in order to assure that people using groundwater have a supply at least equivalent to that to which they had access [before construction of the project]."[70] This provision required the CNR to provide communities with the same quantity of water to which they had access historically without specifying the source. Consequently, article 16 could be interpreted two ways. The CNR had to either ensure locals the same amount of groundwater or provide the same amount of water from another source. Technically, then, the CNR did not actually have to restore groundwater to historic levels, as long as users were provided with just as much water. While residents might have been guaranteed the same quantity of water, however, other consequences of shifting hydrology might make these mitigation efforts inadequate, if not unacceptable.

In principle, existing water law might have placed stronger limitations on the CNR, but there were no legal provisions regulating groundwater in the early 1950s. As Henry explained to Delattre in May 1950, "No legislation exists decreeing the conditions to which the modifications of groundwater must be submitted. The law of 1919 instituted various rights ([such as] submersion, aqueduct, [and] occupation), but it is silent with regard to subterranean water. In this regard, we find ourselves with relationship to the property owners and tenants in the same situation as the contractors of waterfalls who, before the 1919 law [which determined the legal status for hydroelectric development], wanted to obtain the constraints that this law subsequently instituted."[71] Regulating groundwater was thus uncharted legal terrain, just as the rapid development of hydropower three decades earlier had necessitated new legislation. Without a strong legal precedent, the CNR was subject to fewer constraints, and locals had no real legal precedent on which to ground their claims. Coupled with scientific uncertainty and the powerful push for modernization after World War II, the CNR found itself in a propitious position.

Despite the vague status of groundwater in French law, locals and state experts expressed concerns about the area's hydrologic order even during the planning stage of Donzère-Mondragon. These worries intensified in 1949, when work on the Blondel plant and tailrace began, and they continued throughout the 1950s. While the CNR feared costly precedents, it also hoped to avoid "collective discontent," interests that motivated the CNR's technical experts to address the agency's apparently problematic envirotechnical system. As a result, the agency discussed four possible solutions between 1947 and 1950, the first four years of Donzère-Mondragon's construction. Several options were thus proposed even before locals began to flood the agency with complaints. One proposed the construction of siphons under the diversion canal, an idea already abandoned by October 1947; the three other options remained under consideration for much longer. The CNR's technical elites assessed the effectiveness of a watertight wall or barrier that might slow, or ideally stop, the flow of groundwater toward the tailrace. They also considered expanding irrigation networks throughout the region to compensate for groundwater subsidence.[72] Finally, they studied a comprehensive water transfer system.

The CNR's engineers ultimately concluded that the last option, the water transfer network, was the best solution. Initially experts focused on drainage canals, a particular kind of "counter-canal" *(contre-canal)* that attempted to desiccate soggy land paralleling the feeder canal where groundwater levels were now elevated.[73] This was an old solution, since construction of the Canal du Midi had prompted the building of counter-canals three centuries earlier. In fact, that network had been a response to the complaints of neighboring property owners.[74] But when CNR officials concluded that Donzère-Mondragon would also affect hydrology downstream of the Blondel plant, the agency's engineers revised their proposal and envisioned a comprehensive water transfer system involving counter-canals along the entire length of the diversion canal.[75] The drainage canals along the feeder canal would send excess water from the area upstream of the plant to "recharge canals" *(canaux de réalimentation)* along the tailrace, where groundwater was now scarce. This water transfer network served as a micro-envirotechnical system that reconfigured canals and water flows in the hope of producing a new hydrologic equilibrium closer to what existed before the construction of Donzère-Mondragon (see Map 4.1).

The CNR's engineers concluded this solution was "best," meaning most economical. As CNR engineer J. Bonnier explained, the choice was "the least onerous," but even better, "it can function without great expense at least for a while."[76] Counter-canals were also solutions with which France's engineers had considerable experience. By the late 1940s, the CNR's administration charged its workers to begin building counter-canals along the feeder canal and soon turned to recharge canals along the tailrace.

But officials from the Génie Rural soon challenged the CNR's choice. This response manifested long-standing tensions between the two agencies and their envirotechnical regimes. Their conflict over groundwater followed on the heels of shifting engineering hierarchies under the Vichy regime and was only the first of numerous postwar debates. Each agency preferred different solutions: the CNR, groundwater recharge; the Génie Rural, expanded irrigation. Not surprisingly, each solution was to be built and managed by the agency favoring it.[77] While CNR officials condemned the Génie Rural for what they saw, probably rightly, as politically expeditious strategies, they did not admit their own self-interested position.

M. Bourgin, chief engineer of the 6ème Circonscription Electrique, attempted to mediate this conflict. He agreed that the CNR's solution to "try to battle directly against the drop in the water table . . . seems reasonable," but he also saw irrigation as a viable option. Consequently, Bourgin recommended that the Confédération Générale de l'Agriculture, a group founded during the war that tried to coordinate various agricultural organizations, designate several experts to study groundwater and its agricultural implications in the departments of the Vaucluse and Drôme.[78] Because Bourgin had authority over the CNR, the agency

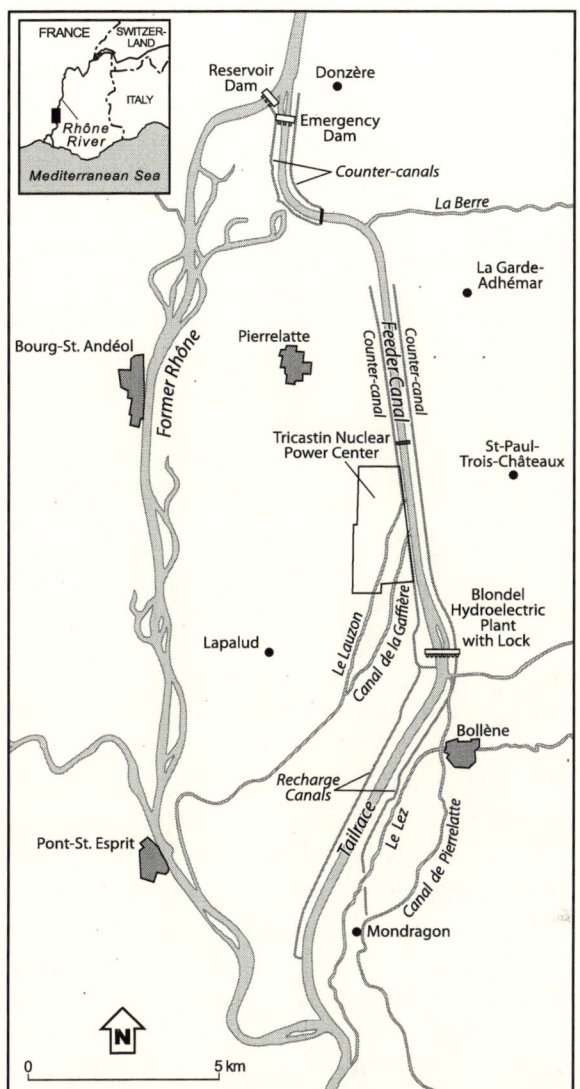

Map 4.1. *Donzère-Mondragon's counter-canal network*. This map shows the comprehensive water transfer network the CNR completed in the region surrounding Donzère-Mondragon. Counter-canals paralleled the feeder canal. They moved excess water from upstream of the Blondel plant to recharge canals downstream, where groundwater had become scarce. Not shown are dozens of groundwater pumps and test wells. (Map by Joseph W. Stoll, Syracuse University Cartographic Laboratory.)

could not keep outside experts from entering the debate, a fact that clearly annoyed the CNR's leaders.

After several months of deliberation, representatives from the Génie Rural changed their recommendation. Instead of supporting irrigation alone, they decided to advocate a "mixed system" that included both recharge and irrigation. Génie Rural engineer A. David acknowledged that the CNR's proposal might "permit the entire or partial neutralization of the reduction in the water table due to the tailrace," but he still recommended a two-pronged approach. Only this option, he claimed, would offer a "complete solution."[79]

The CNR, however, refused to compromise with a "mixed system" and invoked its expertise to defend its preferred method of mitigation: recharge alone. When Delattre told David to "leave the entire responsibility of the operation to the CNR, whatever the solution chosen," he implied that his agency knew best. Initially, representatives from the Génie Rural were unwilling to let CNR officials make that decision, but after several months of animated discussion, they finally backed down. The CNR had won a political and technical victory, but the celebration was short-lived.[80]

The CNR did implement its recharge program, but in the end, the agency's "solution" proved largely unsuccessful. As the unrelenting stream of letters from locals attests, their problems persisted. In fact, the CNR's own studies demonstrated that its recharge system was not working. A few months after test wells had been drilled, engineers found that borings had already begun to plug up. Surface wells sent only a third of expected flows into the water table. Meanwhile, a "central hole" along the tailrace seemed entirely unaffected by the CNR's recharge program, no matter how much water the agency pumped into the water table there.[81] In private documents, CNR officials revealed a great deal of pessimism about whether their fixes actually worked and how well. In June 1952, Bonnier even admitted in an internal memo, "If we were assured that recharge will be effective, I would suggest that we plan to reduce the deepening of [locals'] wells. But in the current state of affairs, it seems to me that the question has a speculative character."[82]

The CNR's response to its problematic test wells foretold the way the agency would remain committed to its preferred solution, despite growing evidence of its ineffectiveness. When CNR engineers studied these test wells, they ignored the fact that the borings were already plugging up. Instead, they made the generous assumption that the wells would remain unblocked. These engineers thus analyzed an idealized version of the area's hydrologic order, not one they—and locals—actually faced. Furthermore, by making this assumption, the CNR's technical elites calculated maximum recharge rates, which resulted in optimistic predictions of the agency's recharge program. These engineers also noted that their agency could always drill more borings. By ignoring sedimentation and focusing on the possibility of developing additional recharge sites, CNR engineers cast their recharge program as an effective, unproblematic solution.[83] Just as Mathian's 1955 map became reified in the landscape, the CNR's idealized assessment of the recharge program took on a life of its own.

Problems with recharge, verified by the CNR's own research, indicated that agency engineers did not fully understand the hydrology they managed. Yet the CNR administration defended its decision to the public in a proclamation a month after Donzère-Mondragon began normal operational procedures and two months before the project's official inauguration.[84] The timing of this proclamation suggests that the agency was attempting to shore up public support for its recharge program amid mounting evidence of a problem its "experts" seemed ill-equipped to solve.

Public pronouncements aside, growing uncertainties and proven complications with the recharge program forced the agency to reconsider alternatives. Irrigation was again at the top of the list, and Génie Rural officials led the call for it.[85] The CNR also considered buying up land in the problematic central hole, but this proposal divided the agency's administration.[86] Split internally, the CNR's elites returned to one of the agency's initial proposals: an underground barrier to block, or at least slow, the flow of groundwater toward the tailrace. Because exorbitant costs made construction of a continuous wall along the entire length of the tailrace unrealistic, the engineering firm, Solétanche, proposed a series of shorter walls, 0.5–1.1 kilometers long, to be located in areas with "the most influence." Solétanche engineers built a model and conducted studies to evaluate the proposal. They concluded that judiciously placed walls could have a "considerable effect" by reducing the amount of land without a captive water table. Depending on the length of the walls, it might be reduced to as little as 6 percent of current levels. Apparently, CNR officials found the Solétanche study persuasive, since they requested more detailed "technical and economic analysis." However, the costs of building a single wall one kilometer long, fifteen meters high, and five meters wide were extraordinary; needless to say, the costs of constructing several walls would have been astronomical. In an attempt to revive the proposal, Solétanche advised more studies, but the CNR's administration tabled the plan. The agency's finances were too precarious.[87] By the end of 1953, the CNR had closed the books on the wall solution for what would be the penultimate time.

Although CNR leaders renewed their commitment to the recharge program, they faced mounting evidence throughout the rest of the decade that the agency's preferred technical "solution" was failing. In the early 1950s, CNR technicians began to gather data at several hundred sites each year to track the Rhône's flow, the agency's injection rates, and the level of the water table. The CNR published an annual report based on these data, summarizing the state of groundwater in the central Rhône valley. By 1958, several years into the recharge program, CNR officials seemed optimistic that the water table was rising, but they also stated that maintenance of recharge infrastructure was "becoming more and more onerous." The following year, the agency reversed its optimistic conclusions. The agency's officials characterized 1959 as "one of the worst [years] since Donzère-Mondragon was put into operation." They had recorded significant drops in the water table, noting that this subsidence was due largely to the unusual absence of even minor floods that year. While CNR engineers acknowledged the role of precipitation, floods, and other "natural" factors in explaining the current state of the groundwater, they also pointed to human causes: "pumping done by those living along the river with the help of the borings that we constructed in 1957."[88] In light of a worsening situation, CNR officials seemed quick to blame local people.

If 1959 seemed bad, 1961 was even worse. That year, the CNR injected less water and for fewer days than ever before. The agency's change in policy resulted in a significant decline in the water table, but CNR engineers believed they could keep groundwater at its 1959 level by maintaining injection infrastructure and adding new borings. That they aimed to keep the water table at its 1959 level suggests that they lacked a higher standard than simply year-to-year maintenance, a goal that might result in ever-diminishing groundwater levels.[89] Recent agreements with locals may have weakened the agency's determination to undertake more ambitious, and perhaps ultimately more effective, solutions.

Only in the late 1960s did the CNR's leaders consider alternatives to the recharge program. They revived irrigation, especially when the CNR and state agencies in the Rhône valley sought to modernize irrigation networks. At a 1968 meeting in Bollène, state elites discussed the "eventual abandonment of groundwater recharge for the benefit of surface irrigation of property." The departmental director of agriculture (DDA) in the Vaucluse pointed out that shifting from recharge to irrigation had certain advantages, foremost for the CNR: abandoning recharge would lower the CNR's costs while increasing energy production. Although CNR officials had questioned irrigation in the past, this time they did not jettison the proposal, though they did not rush to implement the DDA's despite its benefits. Focusing on the legal implications of such a policy change, they stated that if the proposal was undertaken, the DDA would have to accept responsibility for any damages caused by ceasing the recharge program. Despite these reservations, the agency had moved to adopt the proposal by the summer of 1969. The CNR and DDA, however, did not have local support for it. The CNR's officials reported that locals had demanded "the immediate reestablishment of the water table [through recharge]." "According to the opinion of farmers," one official explained, "eventual sprinkler irrigation infrastructure does not eliminate . . . the absolute necessity to maintain the water table artificially."[90] By the late 1960s, then, it was locals who defended, and ultimately saved, the CNR's groundwater recharge program. To this day, the CNR continues to carry out this program along the central Rhône. The agency has not given up, however, the hope of finding a permanent solution. In 1999, the CNR initiated a new study of the construction of a wall that might replace a costly ongoing solution with a one-time fix.

The history of hydrologic change in the central Rhône valley since World War II is awash in irony. When technical elites debated the design of the CNR's projects during the 1930s and 1940s, they and politicians had argued that the diversion approach would save farming by avoiding the need for large reservoirs, which would have inundated sizeable parts of the valley. In general, the CNR's projects did circumvent this prospect, but the construction and operation of long diversion canals created its own set of problems. The CNR's archives suggest, however, that these problems were far from unexpected; and they were a lesson that might have been learned from the Canal du Midi three centuries earlier.

Hydrologic change in the central Rhône valley was a catastrophe for many local people, a point made poignantly clear by their correspondence.

No one bothered to tally its environmental costs. Yet out of the drying wells, dying crops, and cracking ground emerged a large body of knowledge about the Rhône and its hydrology. Studies proliferated. Scientists, engineers, and technicians working for the CNR and state agencies were sent to the Rhône, surrounding valley, and laboratories. They took data regularly and still do today. They developed models in an attempt to understand how groundwater flowed through the region. The CNR's comprehensive water transfer network, which aimed to modify hydrologic patterns that had already been altered by drought and Donzère-Mondragon, embodied this knowledge, however tentative or erroneous. These people, not to mention locals themselves, were intimately familiar with the river and produced the nature of the Rhône through their work, although locals might have also pointed out that it was a disaster that inspired this work and the knowledge it yielded.[91]

This understanding of the Rhône and its hydrology was certainly not the first. Since at least the seventeenth century, farmers had developed practices and artifacts with which to make the land more productive, regulating the amount of water in specific areas by building canals, diverting flow, and controlling sluices. Of course, these strategies did not always prove effective; both floods and droughts challenged the efficacy of farmers' envirotechnical systems, just as they would challenge those of the CNR. Nonetheless, residents developed knowledge that enabled them to make the land productive and profitable—and generally without causing a socioecological disaster. Donzère-Mondragon appeared to have undermined existing envirotechnical systems that locals had developed over centuries to manage the water around and under them.

As a result, control over that hydrology was transferred from farmers, their organizations, state agencies, and envirotechnical systems to the CNR, its engineers, and their projects. Ironically, the more problems Donzère-Mondragon engendered, the more the CNR extended its jurisdiction in a desperate attempt to fix the situation. The financial costs of the CNR's groundwater recharge program reflected the agency's expanding influence. Agricultural investment composed only 1.6 percent of Donzère-Mondragon's total costs, but without groundwater-related expenses, it came to just 0.5 percent.[92] This shift in control, therefore, took place not in spite of, but *because* of, the socioenvironmental consequences of the CNR's project.

Yet the CNR's control proved more illusory than real. The agency's inability to develop effective solutions made this clear. Once groundwater change was framed as a problem, due largely to locals' persistent complaints, CNR engineers developed complex models of regional hydrologic processes. In theory, the construction of new envirotechnical systems based on these assumptions increased the CNR's hold on both the social and natural environments. Counter-canals, recharge pumps, surveys, and the deepening of residents' wells literally extended the CNR's reach into and under the countryside. The CNR's authority was thus expanded as its models became reified in technological artifacts and the landscape itself. The apparent limits of this system indicated, however, that the CNR was less successful at regulating the natural than the social environment, but that too was a problem for the agency. After all, locals wrote dozens of letters, put up posters, created organizations, and lobbied their political representatives. Within the confines of laboratories, then, the CNR's models seemed to work, but in the (literal) field, they did not. Despite these problems, the agency remained wedded to its models. Only in the late 1960s did CNR officials briefly consider replacing recharge with irrigation. In many ways, whether these models were accurate or not is beside the point. By the late 1950s, the Rhône valley materialized the very models state engineers had developed in hopes of rectifying hydrologic change.[93]

The stories locals and experts told about these events came from different positions in the social and epistemological order. Locals tended to focus on drying wells and cracking homes: they were primarily interested in the effects of hydrologic change—albeit effects that had transformed their lives, communities, and land. In their view, perhaps it was less a crisis about groundwater than about what to do when their wells ran dry. In contrast, CNR engineers and politicians framed groundwater change in terms of recharge rates and "geologic discontinuities." These groups thus had quite different understandings of complex ecological, technological, and political events. While locals were concerned with the personal, immediate, and specific implications of dry wells, CNR and state officials tended to focus on the abstract, quantifiable environmental consequences of modernizing the Rhône in the pursuit of broader political and economic aims.

In the end, whose story do we believe? Delattre was probably correct in his assessment that locals overestimated the culpability of the CNR in their water problems. And locals were probably right in assuming that the CNR underestimated its own responsibility. Although their versions of causality differed, both locals and CNR officials tended to construct tidy boundaries between the natural and the human to bolster their arguments about what caused hydrologic change and what should be done about it. Yet a reductive classification of either environmental or human causality failed to capture the complexity of the situation, given the (in this case, literally) porous boundaries between ecological and technological systems. Furthermore, more important than the question of relative responsibility is that, in the end, the CNR's views were ultimately the most important. When wells ran dry, the CNR became judge in a trial where it was also the defendant. The rules of assessing claimants' cases and the basis of final decisions were derived from knowledge the agency's experts had produced in the context of the groundwater situation.[94]

Conflicts over the Rhône valley's hydrologic order offer a unique—and, notably, local—perspective on the processes of France's postwar reconstruction and industrialization. State building and debates over national identity did not take place solely within the halls of government or in official reports circulating among elites. They also played out in the farms and riverbeds of the central Rhône. Indeed, both locals and state officials experienced postwar modernization through their interaction with the river and the surrounding valley, giving new meaning to the phrase "history from the ground up."

As locals responded to the CNR—and indirectly to the broader industrializing impulse of the state of which the agency was a part—and forced the agency to contend with the consequences of its projects by modifying its envirotechnical system, they reshaped not only the landscapes of their own communities but also national agendas and wider envirotechnical regimes. The CNR tried to enroll the Rhône in the processes of modernization and nation building, but the agency's decisions, the material constraints of the Rhône valley, and locals' responses all undermined the coupling of nature and nation that was to serve as the foundation of France after 1945. And as these "outside" pressures began to fragment the new visions for the Rhône, it was revealed that the seeming unity of a state ideology wedded to large-scale technological and economic development masked widening fissures from within.

5

RETHINKING THE NATION

In 1968, Antoine Pinay, former prime minister and now head of Le Grand Delta, a new organization based in the Rhône valley, issued a stern warning. "Recent events have shown," Pinay cautioned, "how concentration, which taints the economic, social, and political life of our country, is a danger for all of France." As a result, "it is fundamental and urgent to rebalance [rééquilibrer] France." Achieving this goal, however, was not necessarily straightforward because, as Pinay put it, "we must avoid the same vice for which we reproach the national capital: creating polarized regions around a regional capital, which end up playing the role of a horrible devitalizing force outside its zone of influence." Yet in the Rhône valley Pinay saw a promising solution, not only for France but also for Europe and beyond: "The geographic situation of Provence and the development of the great Rhône corridor require intimate cooperation among all regions that are interested in the creation of a major north-south axis for the Common Market as well as its [potential] extension into Africa."[1]

Offered amid tremendous social strife that year, Pinay's appraisal echoed that of the postwar geographer Jean-François Gravier's famous book *Paris et le désert français* (1947), while referencing two more recent challenges: decolonization and European integration. In his own study, Gravier had painted similarly stark contrasts between the country's center—an urban, wealthy, and "modern" Paris—and its rural, impoverished, and "backward" periphery, which he dubbed "the French desert." Like Pinay, Gravier believed that the provinces drained France of its prosperity, and he denounced the vast disparities between the capital and the rest of France for threatening the health of the entire nation. Gravier's condemnation of the country's uneven development was trendy among the general public and widely read by bureaucrats at the time, and to this day the image of France's periphery as "the desert" remains a shorthand for the discrepancies between Paris and the provinces.[2]

Gravier's training as a geographer undoubtedly shaped his representations of France's postwar political economy in both illuminating and problematic ways.[3] His spatial lens brought the nation's inequalities into sharp relief, but it is worth noting that his concerns were not entirely new. Instead, they built on established critiques of centralization and urbanization while foregrounding some of the special difficulties facing France after World War II. Immediate postwar reconstruction that had repaired basic infrastructure and restored key industries had failed to rectify the disparities between Paris and its hinterlands. Moreover, as Pinay's assessment twenty years later indicated, these issues remained unresolved. In fact, they were ever more urgent, given the Cold War, decolonization, and European integration. On some level, then, the anxieties Gravier and Pinay expressed about "the French desert" reflected the nation's tenuous position, both domestically and internationally.

At the same time, their neocolonial portrayals of provincial France were highly charged, given the country's situation at home and abroad. After all, several politically freighted dichotomies underlay these representations of France: developed and undeveloped, urban and rural, modern and traditional. Such binaries simplified and ultimately polarized the country's far more complex socioenvironmental landscape. In many ways, Gravier and Pinay's arguments exemplified post-1945 development discourse and modernization theory, which usually targeted the colonial and subsequently postcolonial world.[4] Yet they were describing France's provinces, not its colonies or former colonies, suggesting that French intellectuals and politicians were struggling once again to rearticulate France and to reassert control over the nation's domestic periphery at the very moment its empire was falling apart. Economic, political, and cultural distinctions between Paris and its hinterlands may have been legitimate concerns for France's postwar leaders, and the ways Gravier and Pinay framed the "problem" of the provinces embodied these anxieties about France's place in the world, but these writers essentially reproduced the unequal power dynamics of colonialism at home, what Michael Hechter has called "internal colonialism."[5] This was not the only connection between the French metropole and imperial periphery, however. As Gravier's desert metaphor already suggests, elements of the "colonial declentionist environmental narrative" that geographer Diana Davis has explored in her study of French North Africa seem to have been applied to provincial France as well.⁶

Between the early 1950s and 1964, the French state responded to these issues by implementing both economic and political reforms that attempted to correct the country's "disequilibrium." More expansive *aménagement du territoire* (regional planning) came to supplant the narrower demands of postwar reconstruction, especially following passage of the Second Plan (1954–1957). As Eugène Claudius-Petit, minister of reconstruction and town planning, explained in 1950, "By placing France within a geographical framework, regional planning explores a better division of human population based on natural resources and economic activity."⁷ Because *l'aménagement du territoire* aimed to meet regional needs within France, intellectual and political elites viewed the fates of regions and the entire country as fundamentally entwined. Furthermore, the economic integration of Europe, epitomized by the creation of the EEC, only intensified apprehension about the nation and its future. On the political front, the centralized model of the French state became formally contested in 1964, after years of tempered critique, when officials approved the creation of twenty-two official "regions," or interdepartmental polities that were each to mediate between the national government and a regional collection of communes and departments. Together these reforms suggest that the spatial framework of the region had begun to challenge the postwar nation-state.⁸

The reemergence of this regional frame, for which Herriot and Perrier had argued so vehemently during the interwar years, reshaped how contemporaries saw not only the Rhône but also France itself. As this chapter shows, these were in fact closely linked. Likewise, the meanings and goals of development began to shift, a trend that would accelerate in subsequent decades, as the final two chapters of this book also demonstrate. By the mid-1950s, a regionalist vision of the Rhône began to call into question the national, even nationalist, perspective that had materialized after World War II. In fact, France's political geography became more complex, as regional and European concerns both influenced "national" imperatives—and vice versa. In particular, *l'aménagement du territoire* recast the Rhône's role within the national economy. Remaking the Rhône thus aimed to strengthen southeastern France within the context of the nation and the continent alike. Decolonization and European integration only amplified the importance of this process. Moreover, advocates of development saw the Rhône as the primary means of "regional" regeneration, a process that actually had regional, national, and transnational implications simultaneously. They explicitly represented the river and its flow as technologies of development that would transform Gravier's "desert" into what became known as "Le Grand Delta." This chapter traces how the spatial dimensions of France's political economy shifted during the 1950s and 1960s, a shift that ultimately led to a rethinking of the Rhône as the nation's river and set the stage for the rhetorical and material remaking of a regional Rhône.

Situating Gravier's and Pinay's Critiques

As their commentaries suggest, both Gravier and Pinay perceived centralization and urbanization as two significant challenges plaguing southeastern France and the nation as a whole. Postwar imperatives help account for these issues, but they were also emblematic of concerns dating to at least the late eighteenth century.

Consider the opening epigraph to Gravier's *Paris et le désert français*, which quoted Alexis de Tocqueville's declaration that, by the early days of the French Revolution, Paris "was already France itself."[9] In reality, it was only the result of complex historical processes. Other cities such as Lyon, Marseille, and Montpellier once rivaled the northern metropole that eventually became the center of political, economic, intellectual, and cultural life in France. But it took absolute monarchy to consolidate power in the Parisian basin. Even the revolution of 1789, which had overthrown the king and ended the monarchy (at least temporarily), shared with its royal predecessor a tendency toward centralization. Not all revolutionaries sanctioned this perhaps surprising continuity. Girondins, whose support came from the countryside, demanded greater local autonomy, but it was the Jacobins, advocates of statism and centralization, whose views eventually reigned.

The revolution of 1789 was also significant in that it helped frame the problem of power within the state: the question of centralization versus decentralization. The revolutionary ideals of unity and equality came to be pitted against decentralized power and liberty. This formulation of the debate favored proponents of centralization, who asserted that alternative political geographies such as regions imperiled the revolution by signaling a return to the provincial federalism of the ancien régime that had just been overthrown. These arguments, however problematic, were far from short-lived. Before the Third Republic, there was little opposition to Jacobin centralization.[10]

Although other political geographies were not sanctioned during the late eighteenth and nineteenth centuries, they did persist into the twentieth. For instance, regionalism experienced a modest revival during the interwar years. While Herriot and Perrier were defending a regional Rhône, former prime minister Georges Clemenceau proposed the establishment of regions. Only after World War II, however, did a sustained and ultimately more persuasive challenge to centralization take hold. The postwar economic boom known as *les trente glorieuses* (the thirty glorious years) appeared to aggravate France's existing disequilibrium and thus bolstered the cause of decentralization.[11]

The popularity of *Paris et le désert français* evinced concern about these growing disparities on the part of planners, politicians, and geographers like Gravier. A geographical perspective on France's political economy proved especially damning. Conceptual tools such as regional economics and the discipline of geography yielded the dire conclusions that the nation's domestic periphery was politically weak and economically sluggish, bringing down the country as a whole. A spatial lens thus revealed deep, persistent inequalities that belied France's revolutionary ideals, eventually forcing its leaders to confront the nation's uneven distribution of power.

Building on these earlier arguments, postwar social scientists began to question how political and economic power were distributed and organized, but they also challenged contemporary alternatives to the existing system.[12] Gravier, for example, believed that a laissez-faire economic approach only worsened regional inequalities: the French state and its experts could—and should—regenerate regions experiencing economic stagnation by articulating the philosophical ideals of development and proposing projects that would achieve these goals. Such state intervention became even more important as postwar urbanization and industrialization accelerated, threatening to worsen existing disparities. In short, Gravier proposed state-sponsored regional development to reduce the power of Paris, renew provincial France, and foster a more balanced economy. His vision reflected broader postwar ideas about the proper relationship between the state and the economy, including state planning, that aimed to avoid either a command economy or unrestrained economic liberalism.[13]

These critiques and their associated correctives were intended as breaks from the systems they condemned, but inherent tensions weakened claims of rupture. First, professionals in social scientific fields such as geography helped formulate the conceptual framework underpinning regional planning. In so doing, they both legitimated the field and bolstered their own roles in these solutions. As Gravier's own career illustrates, geographers worked for several central government agencies responsible for quantifying, analyzing, and addressing inequalities within the national body. They held key positions in developing postwar political-economic policy, and many state programs institutionalized their spatial approach.[14] Second, as their methods might already suggest, these analysts were unquestionably technocratic: they sought "better" decision-making, not wider participation or more democratic representation.[15] Regional planners and sympathetic bureaucrats may have lamented France's inequalities, but ironically, they perpetuated unequal power by asserting their exclusive expertise. Postwar regional planning was not, then, as distinct from its precursors as Gravier and others maintained.

The impact of these reforms was also tempered by the fact that some politicians within the central government worried about undermining France as a nation through major political restructuring. Even in 1963, just one year before the state did approve the establishment of regions, de Gaulle's former prime minister Michel Debré admitted a "fear of seeing competition from twenty regional parliaments, which might be capable of preparing the disintegration of the nation with the complicity of European organizations."[16] European integration clearly compounded these anxieties. As Debré himself had noted, regions were key vectors of European integration, thus heightening apprehension about French national autonomy precisely because "France," the "nation," and "autonomy" all appeared at risk.

Notwithstanding these persistent reservations, the geography of political power did change in the mid-1960s. The central government's creation of the twenty-two "regions" in 1964 was a move to redistribute power so that the sociopolitical order was less concentrated and hierarchical. But long-standing resistance to decentralization continued after the reform had passed: if the national government did not transfer considerable decision-making powers to the regional level, then the reform's actual impact would be limited. Indeed, the regionalization of France remained largely symbolic for almost twenty years. Only after the Socialist Party won electoral victory in 1981 did the regional turn become realized in more substantial ways. One scholar has described the eventual result of these reforms as a shift from "*le* territoire" to "*les* territoires."[17] Just as critics had feared, regionalization signified a waning of the nation-state. However (as I discuss below), the region, the nation, and a transnational Europe were not entirely incompatible.

The timing of the regional turn in France suggests recognition of the limits of the national idea(l)—limits embedded within the revolutionary precepts of unity and equality. Proof of rising inequalities, especially after 1945, particularly exposed these contradictions. Gravier's analysis not only exemplified but also contributed to this growing evidence. Moreover, the fact that regionalization came right on the heels of decolonization suggests another limit to the French nation-state. The central government formally constituted the region at the end of France's imperial experience. Its preoccupation with (French) regions crystallized just as the nation's empire had disintegrated. Southeastern France, which was centered on the Rhône, appears to have been an early site of postwar regionalism, including the formulation of an official policy of regional development in France.

If centralization was one of France's fundamental challenges, then urbanization was the other. Embedded in Gravier and Pinay's analyses was a second critique: of cities and their harmful influence on French citizens. This perspective maintained that urban life and spaces threatened not only the individual body but also the social body as a whole; yet this view was also rife with contradictions. The material conditions of modern cities signified progress, but critics lamented the physical pathology and moral degradation they associated with urbanization. They believed that physical, mental, and social diseases afflicted the inhabitants of modern cities at disproportionate rates, and that urban environments played significant roles in causing such problems. State officials intervened through public health campaigns and sweeping urban renewal efforts such as Haussmannization. While large-scale processes of the nineteenth century, including industrialization and urbanization, did alter the physical and social landscape of France, they exacerbated rising anxieties about class, gender, and race.[18]

The idealization of rural life, as epitomized by the wartime Vichy government, only bolstered this discourse of urban degeneration. During the early 1940s, the regime celebrated bucolic France, glorifying farming and the peasantry. Vichy leaders favored agriculture, which became a guiding principle of government policies. In 1943, they created the Délégation Générale de l'Equipement National, essentially France's first state agency for *l'aménagement du territoire,* which attempted to counter the dominance of Paris and the industrial centers of the north and northeast by reviving rural life in the hinterlands, including the Crau and Camargue.[19] It is worth noting that Gravier published *Paris et le désert français* only three years after the end of this turbulent and controversial period in French history, suggesting an important, if not also contentious, continuity between the Vichy government and the early Fourth Republic.

After the war, Gravier echoed these older concerns about the unhealthy relationship between city and hinterland while seeking to solve them. Regional centers had the potential to offset Paris, but they might end up dominating their own periphery. State experts therefore attempted to plan and then to construct (literally) a hierarchical but ultimately harmonious urban structure in the country's provinces. Claudius-Petit's Plan National d'Aménagement du Territoire, the 1950 successor to Vichy's Délégation Générale de l'Equipement National and a precursor to the 1963 Délégation à l'Aménagement du Territoire et d'Action Régionale (DATAR), largely oversaw these ambitious efforts.[20] Its first report defined "large provincial agglomerations," explaining that they "exercise a true directing role in the economic and social life of a regional or multiregional zone of influence, thus avoiding all generalized recourse to the capital [Paris]." It then identified nine *métropoles régionales,* notably also called *métropoles d'équilibre,* which comprised a second tier after Paris in the urban hierarchy. Lyon and Marseille, both located in the Rhône valley, were the two most important cities. Officials then filled in the rest of the urban matrix by classifying ten "regional centers" and three more subsequent levels of towns. Neighboring regional metropoles and centers could also be united into urban networks. For example, Marseille, Aix-en-Provence, and the Rhône delta together formed the Mediterranean southeast, while Lyon and Saint-Étienne were the heart of the inland southeast.

The state's First Plan began altering the uneven relationship between Paris and the rest of France by channeling funding into these *métropoles d'équilibre.* Subsequent Plans reproduced this urban hierarchy and proposed projects for specific cities. Gravier described this conscious attempt to reengineer the relationship between urban and rural spaces through regional planning in one of his later works, *L'aménagement du territoire et l'avenir des régions françaises* (1964), devoting chapters to the themes of "cities and countrysides" and "metropoles and regional organization."[21] As both *Paris et le désert français* and his subsequent book suggested, regional economic development organized France spatially, paying particular attention to the distribution of urban and rural spaces, rather than dividing the country by industry or economic sector. Regional economic planning thus offered a geographical representation of France's political economy. In this new model, the southeast, centered on the Rhône, was key.[22]

Regional Development and the Rhône

As its long name suggests, the BRL (Compagnie Nationale d'Aménagement de la Région du Bas-Rhône et du Languedoc) and its precursor not only reflected but also forged the Rhône's ties to the state's emerging policy of regional economic development. According to advocates, regional development—flowing from the ample waters of the Rhône—promised to turn the "desert" into an oasis. Furthermore, this region played a critical role in the formulation of nationwide policies of regional development. In other words, the laboratory of the Rhône valley during the late 1940s and early 1950s ended up having far-reaching implications for the rest of France.

Philippe Lamour, the BRL's first president, deserves some of the credit. Lamour was born in rural Landrecies in 1903. During World War I, his family moved to Paris, and in 1922, he started advanced studies in law. As a young Parisian lawyer, he specialized in real estate law, defending poor clients. By the time war broke out again, he had married and was supporting a family. In 1940, he rented land in the department of the Gard as part of Vichy's program of agricultural repatriation, but he complained about the difficulties of farming there and began to advocate agricultural reforms. By the end of World War II, Lamour had started organizing farmers in the area and created a federation of winegrowers in Languedoc-Roussillon. Having become a leader of agricultural interests, Lamour became general secretary of the recently formed Confédération Générale de l'Agriculture (CGA) and promoted investment in agriculture as part of postwar reconstruction. As Chapter 3 showed, industrial priorities ultimately outweighed those related to farming, especially during the immediate postwar era, but Lamour eventually managed to pursue his goal of agricultural reform in southern France in other ways.

Lamour's ambitious vision for the lower Rhône valley and Languedoc likely emerged in part while accompanying Jean Monnet on his tour of the United States during early Franco-American discussions of reconstruction and what eventually became the Marshall Plan. Not only did Lamour read David Lilienthal's celebratory account of the TVA, but French bureaucrats also visited the project during their stay in America. With Monnet's support, Lamour proposed several new projects as CGA general secretary, proposals that soon became state policy. For example, a July 5, 1946, decree by the Ministry of Agriculture established "pilot regions" for postwar agricultural reform, the very first of which targeted the lower Rhône valley.

Lamour hoped to create an organization modeled on the TVA to oversee construction of a large irrigation canal on the Rhône's western bank and thereby help revive this part of rural France. Historically, farmers and state agencies had diverted some of the Rhône's waters into lower Provence, but communities in Languedoc barely tapped the river. Initially the 1946 pilot program aimed to divert 25 m^3/s of water from the Rhône for agricultural purposes. It also assessed existing groundwater supplies and the possibility of expanding irrigation. Rural electrification was also one of the program's first goals. Languedoc had some of the lowest electrification rates in all of France. Once constituted, the program was administered by thirteen farmers, ten bureaucrats from the Ministry of Agriculture, and five scientists, with Lamour serving as president.[23]

When recounting the program's early history, Lamour explained that "after the war, we were looking for the most efficient means to achieve reconstruction and most of all, modernization. Rather than having general and abstract ideas, I told [Minister of Agriculture François] Tanguy-Prigent that it was necessary to have long-term projects based on natural regions. We came up with a pilot program. This was the lower Rhône." Fading memories and subsequent events may have colored Lamour's account, but his story is noteworthy for several reasons. First, from the outset, the program's supporters hoped that their efforts would go beyond mere reconstitution, as we already saw in Chapter 3. By 1947, its administrators had formulated a much broader mission: the modernization of agriculture, which would serve as the basis for the revitalization of southeastern France.[24] Second, Lamour remembered asserting that projects should be "based on natural regions," suggesting that political boundaries should flow from natural ones. He believed, then, that a regional development company's concession should reflect a certain ecological unity.[25] Lamour therefore tied the effectiveness of the projects to their supposed grounding in nature, thereby naturalizing these new organizations. Such arguments implied that postwar modernization was simply fulfilling the design of nature, which conveniently downplayed its transformative effects.

In 1951, Plan officials began to turn the pilot program from a small laboratory for rural renewal into a regional mission with potential implications for the entire nation. That year, they established a commission, the Société d'Etudes des Canaux de la Rive Droite du Rhône, to study the comprehensive development and modernization of the lower Rhône and Languedoc. In ensuing months, the Société completed preliminary studies of a proposal to irrigate the plains along the western bank of the lower Rhône. Moreover, these officials appeared primed to replicate such schemes throughout France; that year, they began to lay the groundwork for the creation of regional development companies across the country.[26]

The Société immediately clashed, however, with locals who lived in the area under growing scrutiny. Vintners and farmers opposed the state's intervention and questioned the Société's activities. Local and regional agricultural unions staged protests, and one anonymous pamphlet lambasted the organization: "Why are we going to waste twenty billion [francs] on the irrigation of the Bas-Rhône-Languedoc?"[27] Meanwhile, graffiti such as *"pas de canal en Costières"* ("no canal in the Costières") stamped critics' views on the landscape itself. Elections also demonstrated the extent of local opposition. During the mid- and late 1950s, many Languedoc residents backed Poujadiste candidates. In fact, National Front politician Jean-Marie Le Pen was an outspoken early critic of nascent development efforts in the southeast.[28]

Despite local opposition, the government moved to transform the Société into a fully fledged regional development company. In 1953, Plan officials sketched out the organization's objectives.[29] The Conseil d'Etat began to draw up legislation, and by February 1955, Decree 55-253 established the legal foundation for the creation of regional development companies, outlining general rules governing the administration of concessions that would promote "increasing the value of certain regions." It also charged the Plan and interested government ministries with the authority to determine exactly how to carry out this mission. Another decree (55-254) formally created the BRL, France's first regional development company.

What was the BRL's mission? As 55-254 decreed, "the improvement of the Bas-Rhône and Languedoc region will be realized in order to promote the development and the agricultural reconversion of this region as well as supply potable and industrial water for certain rural and urban collectivities."[30] This mission applied to an area totaling 250,000 hectares with agricultural "reconversion" as the agency's primary but not sole objective. State agricultural and economic experts maintained that monoculture dominated too many of the region's farms, with low-quality wine production the biggest culprit. BRL experts and state elites saw modernized irrigation as the best solution to this problem and eventually constructed an extensive network that tapped the Rhône's waters and sent them into the countryside through a complex system of canals, pumps, and pipes. The decree's second article conferred the concession and responsibility for these projects to the mixed economic organization that was the BRL. Just over half of the BRL's *actionnaires* came from public entities such as communes and departments, with the rest from private parties.[31]

Just as Plan officials had hoped, the BRL became only the first of several regional development companies created, perhaps ironically, by the central government during the 1950s and 1960s. Subsequent Plans established additional organizations to improve other regions perceived as stagnant and backward. For instance, a government study in the 1950s investigated the possibility of using water to modernize rural Provence. Approved in the early 1960s, this project, the Canal de Provence, eventually supplied nearly sixty thousand hectares of farms with irrigation water while helping to meet Marseille's growing demands for water.[32] Other regional development companies included the Compagnie d'Aménagement des Côteaux de Gascogne, Société pour la Mise en Valeur Agricole de la Corse, Société d'Aménagement des Friches et Taillis de l'Est, Société de Mise en Valeur des Régions Auvergne-Limousin, Société d'Aménagement et de Développement de la Région Authion-Loire, and two more for the Landes de Gascogne and the Marais de l'Ouest.[33] The structures and objectives of these organizations paralleled those of the recently founded BRL, but the CNR should also be seen as a predecessor.[34]

By the same token, *l'aménagement du territoire* also transformed the missions of existing agencies like the CNR. In 1962, the CNR began construction of Pierre-Bénite, a project in southern Lyon. Completed four years later, Pierre-Bénite expanded the Feyzin industrial site, part of the industrial heartland of suburban Lyon sometimes called *la Vallée de la chimie* (Chemistry Valley) thanks to its preponderance of petrochemical manufacturers. The project was intended to be the first stage in realizing the modernization of the Rhine-Rhône liaison. Its goals and "technical" features demonstrated how the CNR modified its multipurpose mandate to reflect the new politics of development.

Indeed, the activities of organizations like the BRL helped remake the socioenvironmental landscape of southern France (see Map 5.1). On

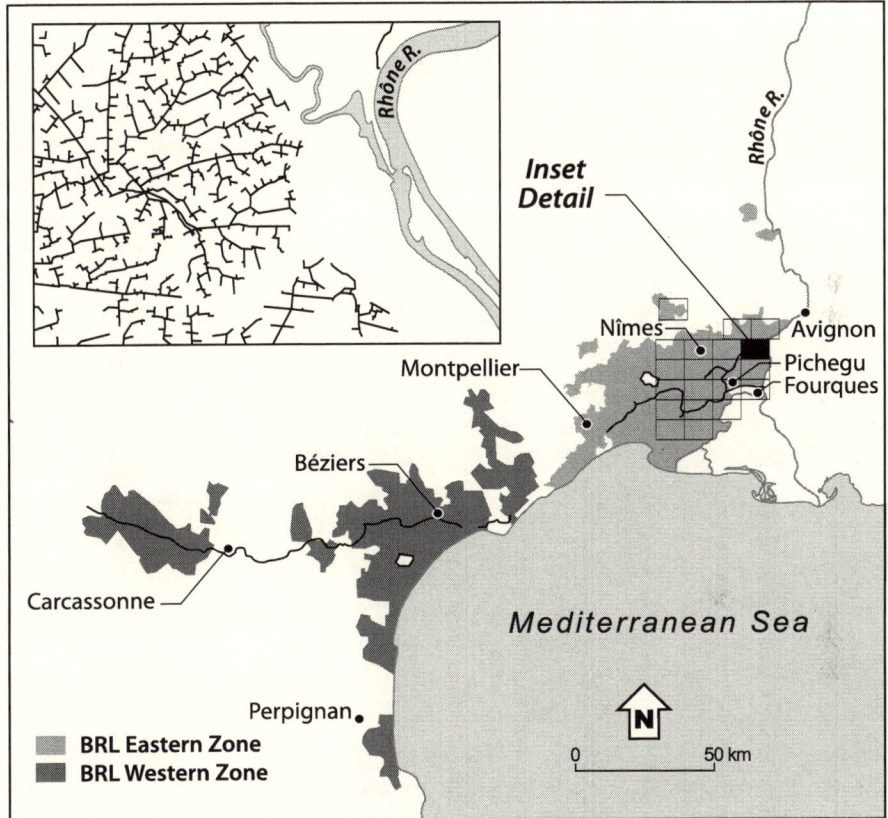

Map 5.1. *The BRL's irrigation system in southern France.* The Rhône feeds the BRL's eastern zone, while several now-dammed rivers supply water to its western zone. The dark lines show the primary canals for both zones. The inset helps convey just how extensive this irrigation network is: it is composed of primary canals, secondary and tertiary canals, pumps, and other infrastructure that together send part of the Rhône's flow into the countryside. Even so, only one small part of the BRL's entire system is shown. Extrapolation from this inset to the entirety of the BRL's domain suggests how the landscape of the lower Rhône valley and Languedoc became literally gridded by such extensive irrigation infrastructure from the mid-1950s. (Map by Joseph W. Stoll, Syracuse University Cartographic Laboratory.)

September 14, 1956, the Conseil des Ministres officially granted the BRL the right to develop and irrigate 230 communes in the departments of the Gard, Hérault, and Aude. The BRL's officials divided the agency's concession into *casiers* (literally "compartments" or "pigeonholes"). Essentially, the eastern and western zones of the BRL's concession were broken down into small, clearly defined areas of development in an attempt to facilitate each one's needs. Interestingly, this approach seems to have been tested in French Indochina for several decades before being implemented extensively after World War II, apparently both in Vietnam and southeastern France.[35] The BRL's initial efforts focused on a thirty-thousand-hectare part of the Gard known as the Costières, but at least some state officials hoped the program might be extended even further.

In the early 1960s, the state expanded its authority by extending the western boundary of the BRL's concession. Two amendments authorized the BRL to build the Avène dam on the Orb River and to develop the Lauragais Audois in the eastern foothills of the Pyrenees. Tapping the Hérault, Orb, and Aude rivers, all of which drain into the Mediterranean, these projects fostered irrigation and agricultural reconversion while promoting flood control and drainage. During the 1960s and 1970s, the BRL also became involved in the development of the Mediterranean littoral between the Rhône delta and Béziers. By the 1970s, then, the BRL oversaw water management in an area that stretched from the Rhône to the medieval city of Carcassonne and from the coastline of the Mediterranean to the Pyrenees and Massif Central.[36]

The eastern zone of the agency's extensive irrigation network shared a common origin: a single intake of 75 m^3/s of water from the Rhône near Fourques, a village located along the right bank of the river north of Arles. When the state enlarged the BRL's concession, the agency was authorized to tap water from other rivers farther to the west, but the intake at Fourques fed all of the BRL's projects in the department of the Gard and some of those in the Hérault. A primary canal channeled the water from Fourques approximately four miles to the main pumping station at Pichegu, where the BRL's network divided the Rhône's waters. Pumps moved a smaller share of the water uphill over sixty meters with an irrigation system distributing it throughout the Costières. Most of the water continued in the BRL's primary canal from Pichegu to the southwest near Montpellier. Meanwhile, the BRL's network in its newer, western zone channeled waters from the Hérault, Orb, and Aude rivers for distribution to farms along the parched western edge of the French Mediterranean coast.

The BRL's engineers and technicians thus built a complex and large-scale yet also diffuse envirotechnical system to distribute this share of the Rhône's flow to orchards, vineyards, and fields located sometimes dozens of miles from the river. Secondary and tertiary canals along with pumps and regulators acted as the highways and engines of the BRL's water delivery system. Thousands of pipes then sent the water from these canals into individual fields. A remarkable illustration of the BRL's distribution network revealed agricultural engineers' idealized vision of these pipes, depicted perfectly perpendicular to one another and neatly interspersed among orchards and fields.[37] Officials from state hydraulic and agricultural agencies had conducted numerous studies in an attempt to determine the most effective design for maximizing irrigation efficiency. After passing through the BRL's distribution system, farmers irrigated their crops through a pressurized sprinkler system. One author, undoubtedly echoing the views of state experts, called this technology a "revolution" in the history of irrigation.[38] By the 1960s, the BRL's envirotechnical system had substantially transformed the agricultural landscape of the Bas-Rhône and Languedoc.

The BRL's irrigation network was a central feature of its agricultural reconversion and regional development efforts, but the agency oversaw other projects as well. Some were directly linked to agriculture, from promoting polyculture and building an experimental agricultural research station to completing, with the American firm Libby's, a new tomato processing plant in the commune of Vauvert. Others—constructing rural housing projects, potable water systems, and tourist facilities and setting aside green space—fell under the broader rubric of rural development.

As the eventual expansion of the BRL's concession attests, the national government also aimed to improve coastal tourism in the Bas-Rhône and Languedoc. This goal hinted at the rising demand for vacation destinations by both French and other tourists. Visitors packed the famous beaches of the Côte d'Azur, but the Mediterranean coastline west of the Rhône was far less visited. Admittedly, it had fewer roads, limited amenities, and more rustic facilities. Moreover, the relatively undeveloped littoral environment was hardly conducive to tourism. Swamps punctuated the coastline, mosquitoes harassed human visitors, and limited supplies of fresh water dampened prospects for a vibrant tourism industry. State officials sought to overcome these limitations by building new tourist complexes along Languedoc's coastline beginning in 1963. These efforts eventually resulted in many now famous destinations, such as the popular beaches near Sète and Montpellier. In the process of transforming the Mediterranean littoral into an environment more suitable for tourists' consumption, the BRL helped bring water to booming coastal towns just as repatriation surged in the early 1960s.[39]

Responses to the BRL, its envirotechnical system, and envirotechnical regime varied. As we saw earlier, the activities of the BRL's precursor had generated intense local opposition. Protests continued once the Société became the BRL. Just as the siting of the CNR's diversion canals in the central Rhône valley had inflamed local communities, the BRL's projects sparked conflicts with neighbors. The location of the BRL's intake along the Rhône and the path of its primary canal proved especially controversial. Locals feared expropriation and worried about the impact of the canal on their farms. Given the CNR's recent work at Donzère-Mondragon and the EDF's at Tignes, it was understandable why many locals were wary of large-scale projects.[40]

These disagreements may also help explain why farmers failed to adopt the BRL's irrigation program as quickly as state officials had predicted. These experts had conducted countless studies in an attempt to understand the region's hydrology and historic water usage. Technical elites hoped to use this information to maximize irrigation efficiency and raise agricultural yields, as well as pay off the costs of the new irrigation infrastructure. When few farmers signed contracts with the BRL, bureaucrats and engineers formulated alternative rate structures to encourage adoption of the new technology. Technical experts and farmers therefore perceived the new irrigation technologies quite differently, and the discourse of rural modernization clearly failed to persuade many locals.[41]

In addition to these disagreements between state elites and local communities, conflicts among state agencies' envirotechnical systems and regimes emerged as well, just as they had in the central Rhône. This time, BRL and CNR officials sparred over the river's lower reach. Although the concerns of CNR officials were generally distinct from locals', agency representatives also worried about the location of the BRL's intake. They hoped that the BRL would locate the intake downstream of the CNR's planned project, Vallabrègue, lest its hydroelectric production be diminished by the amount of water diverted to the BRL's irrigation network. Just as the CEA's and EDF's demands for water had worried CNR leaders, so too did the BRL's requirements for demanding modifications to the CNR's regime.

Disputes also surfaced over how much water should be allotted to the BRL. The CNR's representatives, state elites, BRL officials, and locals eventually agreed on 75 m³/s of water, but only after considerable debate over who had rights to how much of the Rhône's flow. The CNR's officials hoped to limit the BRL's allotment because it would decrease energy generation at all CNR projects located downstream of the BRL's intake. Moreover, the CNR officials emphasized that their agency's concession predated that of the BRL. Meanwhile, some communities along the lower Rhône complained that the BRL's project allowed inland residents to receive "their" water, a grievance that echoed Rhodanians' objections in the late nineteenth century, when Parisians and northern industrialists had sought access to Rhône-produced electricity.

Despite these concerns, the BRL and its envirotechnical system became an integral part of the state's program for *grands aménagements régionaux* (large-scale regional planning) throughout France. This program oversaw regional development companies like the BRL and was housed within the Ministry of Agriculture. The close relationship among regional planning, economic development, and the Ministry of Agriculture suggests the importance of rural France to regional development more broadly. For state elites, the rural and the regional were virtually synonymous.[42]

The founding of the BRL and its subsequent efforts demonstrated the state's shifting priorities during the postwar era and the emergent politics of *l'aménagement du territoire,* a goal of the Second Plan that became institutionalized even more with the founding of DATAR in 1963. Indeed, it embodied new interpretations of "development" that came to be reflected in the Rhône itself. Moreover, the BRL and CNR played formative roles in shaping regional development nationally. These two organizations, both centered on the Rhône, offered models that were eventually implemented nationwide. Key figures such as former BRL president Lamour also held influential positions in these national ventures. Lamour headed DATAR's predecessor and played a decisive role in the organization's early years. He also advocated *l'aménagement du territoire* to the general public through a television program, *60 millions de Français,* and a popular book of the same title.[43] Lamour's experience with the Sociéte and then the BRL likely shaped the formulation of regional development programs throughout France.[44] Perhaps ironically, the regionalization of the Rhône during the early 1950s actually became a model for the nation.

France's colonial experience also indirectly, though significantly, influenced how lessons learned along the Rhône were extended across and beyond France—and vice versa. Experts who had worked in French North Africa made up a notable share of the BRL's technical division. The BRL's second president, Henri Pommeret, also a former CNR official, explained the demographic composition of the BRL's workforce by claiming that "France did not have a great tradition of hydraulic development."[45] It is unclear why Pommeret did not connect the BRL's activities to a long history of canal building, not to mention more recent efforts by the EDF and CNR. However, Pommeret's somewhat cryptic reference to the sizeable presence of scientific and technical elites who had worked in North Africa had important implications. These places appeared to offer valuable insights to the metropole's hydraulic engineers. Colonial administrators in Algeria, Morocco, and Tunisia had undertaken large water management projects there. The combination of this recent experience with the fact that southern France and French North Africa shared some environmental similarities made it logical to draw on the knowledge of colonial experts, especially as the political situation in the colonies worsened during the 1950s and early 1960s, eventually leading to a spike in repatriation. In the end, it appears to have been not entirely by chance that regional economic development in provincial France, including the Rhône valley, flourished at the very moment the nation's empire came to a violent end. This example provides only further evidence of the way "colonies" fundamentally shaped the "metropole."[46]

From *"le désert"* to *"Le Grand Delta"*

By the 1960s, proponents of *l'aménagement du territoire* viewed the Rhône as not only the heart of regional renewal but also the primary means of regenerating the southeast. As Jacques Soustelle, anthropologist, parliamentary deputy of the department of the Rhône, and former governor-general of Algeria, put it, "the Rhône is the central artery" of the entire Rhône-Alpes region and even of "[Le] Grand Delta," or all three regions in southeastern France that had been created by the 1964 law: the Rhône-Alpes, Languedoc-Roussillon, and Provence-Côte d'Azur.[47] Other contemporaries called the Rhône the "spine" of "Le Grand Delta."[48] These evocative metaphors forged vital connections between the Rhône and the economic health of the national body.[49]

This term, "the Great Delta," held multiple meanings. It was at once a strategic representation of a (constructed) place, an organization, and an ideology. When contemporaries invoked "Le Grand Delta," they were referring not only to the river's delta itself but also to the entire Rhône valley. Some advocates of development stretched this already loose interpretation to argue that "the Great Delta" spanned the entire French Mediterranean and reached inland as far as Lyon. Le Grand Delta was also the name of an association that promoted regional economic development within the territory of "the Great Delta." Finally, state elites and business leaders, including those involved in this new organization, adopted the idea as a conceptual framework for the economic regeneration and expansion of the southeast. It was therefore also an ideology that the association's members promoted to the French state, European businessmen, and the region's residents. As the phrase itself suggested, all three "Great Deltas" seamlessly but strategically conflated hydrologic, political, and economic boundaries.

Le Grand Delta, a convenient shorthand for the new organization's lengthy official name, Association pour le Développement Économique et Social du Sud-Est Français, was founded in December 1966 by politicians, business leaders, and public figures. The association's *conseil d'administration* included officials from departmental governments, chambers of agriculture, ports, regional economic expansion committees, regional development companies such as the BRL, and the CNR. As its formal title indicated, Le Grand Delta aimed to promote economic development in the three neighboring regions of southeastern France. Joined together, this "Great Delta" did indeed form a large triangle. It included one-fifth of the country's surface area and, at the time of the organization's creation, was home to over ten million residents, or one-fifth of France's population.[50] Given these figures, this delta did seem to have great potential.

Although the constituencies within "Le Grand Delta" were actually quite diverse, officials from this organization mobilized a unified conception of this multiregion territoriality under its singular mantra of "the Great Delta." With its capital of Lyon, the second largest city in France, the Rhône-Alpes was the wealthiest and most populated of the three regions. Languedoc-Roussillon, a coastal region that stretched along the Mediterranean from the Rhône to the Pyrenees, was dominated by agriculture, especially the wine industry. Spanning the Mediterranean coastline in the other direction, from Marseille to Nice, beaches, tourism, and products such as olives and lavender made Provence-Côte d'Azur world-renowned. Politicians and planners argued that to find solutions to the problems these regions faced, they needed to be considered collectively. If adopted, this interregional "solidarity" would mediate between the newly created region and the French nation-state.[51] Officials from Le Grand Delta and their supporters thus integrated these three areas into an interregional political entity, "the Great Delta," that the hydrologic metaphor conveniently naturalized. By doing so, this concept reproduced historic slippages between political and natural boundaries while legitimating both the relatively new political unit of the region and the organization that shared the same name.

Representatives from Le Grand Delta claimed the right to speak for the interests of the territory for which their association was named. This new association promoted the southeast to companies and politicians not only within France but also across Europe. Le Grand Delta was in effect a well-funded marketing campaign that promoted this region, which became inseparable from the concept of "the Great Delta" and its tremendous potential. Even the title of its journal, *Delta, Revue d'action régionale*, captured the organization's objectives. This publication advanced the multilayered idea(l) of "the Great Delta" in articles as well as in copious advertisements from towns and departments seeking economic opportunities.[52] And in June 1972, at an exposition for twelve hundred European business executives organized by the association, representatives from villages, cities, departments, and regions organized tours, sponsored wine tastings, and coordinated many other events for attendees.[53]

At first glance, this "Great Delta"—as both territorial description and ideal—seems to mark a significant shift from Gravier's representation of provincial France as "the desert." Its "great" promise, at least according to Le Grand Delta's supporters, appears to contrast sharply with Gravier's image of a barren and backward desert. But they are, in fact, complementary, with the idea of a delta flowing naturally, so to speak, from that of the desert. Neocolonial representations of provincial France as economically stagnant and in desperate need of reform underlay both the ideology and the organization of "the Great Delta." This ideal underlined the desert's hidden but true possibilities, as the association endorsed these changes and positioned itself to realize these reforms.

Furthermore, the historical emergence and juxtaposition of these two metaphors for southeastern France—the desert and the delta—are revealing. Water, specifically water from the Rhône, was what promised to transform this difficult arid region into a viable, if not flourishing, political and economic oasis. Gravier's memorable desert metaphor, emblematic of other neocolonial representations of the southeast, therefore helped lay the conceptual foundation for substantial socioenvironmental interventions in that area, precisely the kinds of changes that members of Le Grand Delta advocated. Just as the representations of the Rhône as a powerful but unruly river helped pave the way for large-scale hydroelectric development after World War II, the metaphor of the provincial desert legitimated regional development in rural France, beginning in the 1950s with agencies like the BRL and continuing into the 1960s with associations like Le Grand Delta.

Proponents of regional development ultimately highlighted three of the southeast's strengths (its "nature," its envirotechnical potential, and its people), assets Gravier had perhaps overlooked in his pessimistic portrayal two decades earlier. First, members of Le Grand Delta represented the region's environment as its greatest asset, with the waters of the Rhône's flow as the southeast's best resource. After all, it was the river that promised to turn the desert into a productive delta. "Nature has given us a magnificent instrument," Soustelle avowed. "It is the Rhône, and it has to be said that until the present, it has not played the role that it could have."[54] Soustelle's remarkable proclamation was an unabashed representation of the Rhône as a technology that historically had not reached its full potential. In addition, by claiming that nature "has given us" that "instrument," Soustelle naturalized regional development efforts and the associated socioenvironmental changes that attempted to fulfill the river's supposed promise.

By emphasizing the Rhône's, and by extension the region's, vast possibilities, proponents of regional development like Soustelle underscored the way materializing "the Great Delta" could reverse the southeast's historical marginality. Backers claimed that the river's potential for improving transportation and communication—economic sectors critical to the regeneration of the southeast—was enormous. State officials celebrated the Rhône's "almost limitless quantities" of water. Furthermore, regional development companies, including the BRL, had already begun to divert some of the river's flow into the valley's more distant hinterlands. Their efforts were on the cusp of remaking the entire French Mediterranean coastline. Advocates of regional development even compared southeastern France to California, but highlighted one crucial difference: unlike the parched American state, they believed that thanks to the Rhône, *"la Californie européenne"* had abundant water. Although many French politicians, intellectuals, and citizens feared Americanization, perhaps the Golden State's postwar economic boom offered a vision of the southeast's potential, for better or worse.[55]

As these bold visions of the power of the Rhône's waters suggest, while the southeast's "nature" may have been a major asset and proponents of development eager to highlight it, they quickly shifted their focus to the region's "second" nature—that is, its envirotechnical possibilities. This potential for an intensely managed Rhône was the region's second strength. The importance of the Rhône proved debatable, but as a river remade, advocates of development agreed, there was even greater promise in its future.[56] The Rhône valley may have had "its handicaps," but its intersection with a variety of transportation methods and convenient geopolitical location boosted its potential as a "fluvial axis." This reference, a theme to which I will return, recalled postwar invocations of the Rhône as a historic crossroads and corridor of civilization.[57] Now, two decades later, state officials hoped that Fos, a new deepwater port and oil refinery on the outskirts of Marseille, might become *"l'Europort du Sud,"* a reference to Rotterdam's commercial dominance of northern Europe.[58] Many pinned their hopes on the Rhône for improving the prospects of southern France and even the entire nation. This required, however, substantial investment in new envirotechnical systems.

Finally, in addition to the southeast's "natural" and envirotechnical qualities, contemporaries viewed the area's human population as not only augmenting but also fundamentally enabling the realization of its potential. Unlike many areas of France that suffered from population declines, the southeast experienced demographic growth during the 1960s. All three regions that made up "Le Grand Delta" had growth rates between 1962 and 1968 that surpassed those of Paris. Provincial metropoles such as Lyon and Marseille saw even higher growth rates. Consequently, by 1968, the population of "the Great Delta" had surpassed that of Paris, reversing the circumstances from only six years earlier. Development supporters concluded that the southeast had a "critical population mass," which made it competitive even Europe-wide. This reference to the region's transnational context was hardly accidental.[59]

Decolonization played a major role in accounting for these demographic shifts, both numerically and spatially. The repatriation of French citizens from former colonies, especially Algeria, and higher immigration rates contributed to the country's population increase after 1962, a trend magnified in the southeast. In fact, Soustelle advocated even more growth to make "the Great Delta competitive with other historically powerful regions of Europe." Yet such demographic increases had to be managed carefully, he warned (as had Gravier and Pinay), since they might aggravate existing inequalities.[60]

In outlining what could be achieved by taking advantage of the southeast's untapped promise, members of Le Grand Delta and their supporters expressed three goals. First, they explicitly aimed to revive Marseille. For centuries, this city had served as France's primary Mediterranean port, linking the country to European, Eurasian, and even global trade. Both Lyon and Marseille once rivaled Paris, but when power eventually shifted to the north, Marseille became a regional metropole. While the city was on the literal margins of France and its North African empire, Marseille's doubly peripheral status ensured that it was actually a central node in the country's colonial enterprise.

As a result, Marseille and its residents particularly felt the loss of France's colonies in North Africa. Soustelle described the city's precarious state in stark terms, a depiction that also hinted at his break with de Gaulle over Algerian independence and anger at the loss of French Algeria: "Given that Marseille no longer has an empire in front of it, that it no longer has North Africa, that it no longer has French territories on the other side of the Mediterranean Sea, it is therefore necessary that Marseille now have a hinterland." Nonetheless, this bleak assessment served the purposes of those advocating regional development. By the mid-1960s, and thus only a few years after decolonization, proponents framed southeastern France as Marseille's internal empire. As Soustelle stated, the city "must have an interior territory, and it is here that the Rhône valley takes on its importance." Others reproduced Gravier's metaphor when they called for the southeast to be transformed from a "semidesert economy" into Marseille's promised hinterland.[61] This idea of an "interior territory" evoked internal colonialism, a forging of quasi-colonial relations within the boundaries of the nation-state.[62] In short, rural, provincial France resumed its status as the country's periphery with the fading of more distant margins of empire beyond national borders.

Second, in addition to regenerating Marseille, contemporaries hoped that regional economic development might restabilize France as a whole.

Like Gravier two decades earlier, Pinay proclaimed that "it is fundamental and urgent to rebalance France." Large economic centers that could "constitute an indispensable counterweight to the Parisian mass" were imperative. Pinay perceived the southeast as the country's most promising ballast because of its "economic weight, the dynamism of its actors, and its relative distance from the capital." In fact, it might be the only area of France with such potential. Pinay asserted that such a restructuring would return "vigor and dynamism to what can be called, admittedly with a certain level of exaggeration, but with a powerful force of expression, 'the French desert.'"[63] In making such claims, Pinay was essentially applying modernization theory to France itself, without fully recognizing the contradictions in doing so.

Lastly, supporters of the southeast hoped that it might secure a vital role for France within Europe. The economic and political integration of Europe posed serious challenges, but it also created possibilities for "the Great Delta," including the creation of "an entirely new regional entity, deliberately conceived in a European context and at the scale of an enlarged European Community."[64] The relevance of the southeast was, then, as much about Europe as about France. Promotional literature associated with Le Grand Delta's 1972 exposition declared simply, "Le Grand Delta: Un rendez-vous avec l'Europe."[65] Similarly, government officials envisioned not only the "equilibrium between Paris and the provinces" but also "the double equilibrium of national and European development." They recommended integrating the Rhône corridor "within the framework of the major European axes" to "make sites in the French southeast as attractive, and—why not?—even more attractive than the sites of the traditional heavy triangle of the Rhine."[66] Uneven development may have plagued France, but Europe faced similar troubles. "The Great Delta" seemed to solve both problems at once.

International pressures intensified these growing concerns over regional disparities. Without a doubt, the creation of the EEC heightened the stakes of staying competitive. Commentators lamented the fact that, like much of southern Europe, "which is falling behind," France continued to lag behind Germany. Numerous references and images of the industrial powerhouses of the "northern delta" and the "heavy triangle," centered on the Rhine, explicitly compared the relative importance of these two river-centered regions. But Germany was not the only worry; northern Italy was also becoming a formidable rival.[67]

Given the political geography of industrialization across Europe after

World War II, France's future appeared increasingly precarious. The country risked becoming to Europe what the southeast was to France. Advocates of regional development, members of Le Grand Delta, and bureaucrats repeatedly invoked and thus reproduced these entwined relations among region, nation, and a transnational Europe. Staying competitive in Europe meant improving the economic state of France, which in turn depended on modernizing the nation's "backward" areas in particular. By creating the "union of three regions" in southeastern France, advocates hoped that the southeast might "counter the smashing power of the north."[68] The goal was, therefore, to balance not only the excessive power of Paris but also the threat of the German-Dutch economic engine.[69] Otherwise, one politician warned, "it is clear that for us, the French of the south, there is a danger that we will remain on the margins, and thus be marginal to the economic development of Europe."[70] In short, France, and especially the southeast, had to catch up. If successful, the country might join the industrial and economic leader, Germany, but if it failed, former BRL president and now DATAR official Lamour cautioned, France would wind up a weak player like Spain.[71]

As these declarations suggest, many contemporaries framed the Rhône's "Great Delta" in terms of its regional, national, and European implications simultaneously, as concentric circles centered on the river. Although this multifaceted spatial construction of the Rhône's transformation, with its multiple geographical foci, may seem at odds with the emphasis on the nation-state immediately following World War II, it is clear that contemporaries did not see them as either wholly discrete or necessarily in conflict with one another. Advocates of regional development appeared to move seamlessly from one political geography to another, each one justifying and reinforcing the others. Government officials and development proponents argued that national and European economies ultimately depended on local ones. Moreover, the Cold War, decolonization, and European integration all heightened such concerns and bolstered the southeast's relevance. In short, the region mattered. Given these arguments, we might think of these spatial frameworks as nested and interlocking political geographies through which politicians and planners sought to understand the continued transformation of the Rhône. At the same time, as local and regional economies became increasingly tied to national and transnational contexts, they became subject to growing political pressures commensurate with the high stakes of France's development—or lack thereof.

But these nested political geographies were also defined in relation to what was no longer there: empire. This gaping absence spoke as loudly as the explicit new realities of European integration. As development proponents, state officials, and others argued, the periphery of France, including the southeast, had to be bolstered because the colonies could no longer serve this role. Given European integration, regional weaknesses potentially threatened the future of the entire nation, and without an imperial periphery, France's provincial hinterlands now appeared all the more important.

Several challenges did jeopardize these attempts to advance regional development for the purposes of the region, the nation, and a transnational Europe. First (as seen in Chapter 3), government officials aimed to strike just the right balance between industrial and agricultural improvements. Second, politicians and development backers hoped to avoid reproducing the domineering effect that regional metropoles had within their territories, just as Paris had on the national scale. Cities could serve as anchors or as menaces to regional economies, and the relationship between the urban and the rural within regional development was clearly marked by anxiety.

On the one hand, government officials and intellectuals framed urban development as critical to this new geography of power within the southeast, across France, and beyond. They argued that regional metropoles had to be "equipped" *(équiper)* in order to assure their roles as *métropoles d'équilibre* against the centralizing vortex of Paris. In fact, failure to do so might undermine the entire project. Supporters believed that the "success of developing the Rhône corridor and the Mediterranean coastline in the future depends in large part on solving the excessive retardation of the infrastructure in the two metropoles of Lyon and Marseille."[72] Not only was it important to "equip" lagging regions, then, but the "urban equipment of balancing metropoles" was particularly important.[73]

On the other hand, commentators like Lamour and Pinay warned that provincial centralization should not reproduce national centralization, essentially becoming the "same vice" on the regional scale.[74] Because regional centers, often called *villes-pôles*, played structuring roles for their respective regions, they might end up dominating them as much as Paris dominated France. Regions could therefore not escape the pull of cities, even if those cities were not Paris.[75] Once again invoking Gravier, development advocates also recognized that they faced the constant risk of creating "new economic deserts, and new population deserts," because large regional industries and cities had the potential to disrupt local economies.[76]

Politicians and planners attempted to strike the right balance, then, between urban and rural, industry and agriculture, so that the economy might achieve a state of "symbiosis," both regionally and beyond.[77] If the cultivation of these "balanced regions" was successful, they would serve as harmonizing rather than destructive forces.[78] Through a diversity of activities, it might be possible to foster "the harmonious development of this Great Delta, which," Soustelle argued, "seems to be an absolutely indispensable element to the economic riches of France in the Common Market."[79] As part of regional economic development, the Rhône's "Great Delta" could potentially redeem both the region and the nation, thereby securing for France a prominent place in a newly integrated Europe.

Furthermore, contemporaries maintained that such a balanced approach to regional development had important sociopolitical consequences. French citizens might experience greater mobility as the nation's rural hinterlands urbanized and modernized, but Pinay asserted that prudent development would permit mobility *"sans dépaysement ou déracinement"* ("without disorientation or rootlessness").[80] This implicit condemnation of the ways urbanization removed people from the land may have echoed Vichy critiques of industrial cities, but it also alluded to France's recent experiences with decolonization, repatriation, and immigration, all of which had, in fact, uprooted thousands of people. But the phrase was even more telling because it again linked political, social, and natural orders.

Government officials and others involved in overseeing regional development explained that systems thinking and *"la prospective"* (future-oriented planning or forecasting) offered powerful tools to guide the realization of this important objective.[81] Systems thinking proved central to respatializing France's political economy and altering the geographical parameters of economic development. Politicians emphasized *"l'ensemble"* (the group or whole)—from *"l'ensemble de la France"* and *"ces ensembles régionaux"* to Europe's northern and southern *ensembles* that reached from the Rhine to the Rhône.[82] At the same time, *la prospective* also shaped decision-making, orienting it very much to the future, as Olivier Guichard, the first head of DATAR, put it in his 1965 book, *Develop France: Inventory of the Future*. Similarly, Pinay emphasized that regional development "must function with respect to the future and not the past. This is why it seems essential to form regions on the European scale."[83] Meanwhile, the Ministries of Equipment, Housing, and Transportation clarified that their officials analyzed any potential project based on a multiyear time horizon. In 1973, state officials were already looking ahead to "Horizon 1985."[84] Future-oriented systems analysis thus attempted to offer a holistic response, both temporally and spatially, to contemporary anxieties over the relationship between the urban and the rural, industry and agriculture, and the three interlocking geographies of contemporary political economy (region, nation, and a transnational Europe). In particular, systems thinking and *la prospective* provided ways to make these daunting challenges manageable—hence solvable—for government leaders.

As these hopeful visions suggest, contemporaries perceived the Rhône as the literal center of this comprehensive system of "Le Grand Delta" and thus the core of the region's future. "After having served its role as a divider," one government report declared, the Rhône "should now become the great uniter."[85] Proponents recommended that the river could—and should—forge links across Europe, thereby rebalancing the economy of the entire continent. For instance, Pinay situated "the Great Delta" beyond the boundaries of France, claiming that it should be connected to Spain and Italy, perhaps extending from Catalonia all the way to Lombardy. In fact, simply following national boundaries and ignoring historic, transnational links denied the possibility of "Mediterranean unity."[86] Soustelle viewed such an opportunity to forge connections across the Mediterranean as "a very interesting and promising proposition."[87]

Moreover, the European potential of "Le Grand Delta" was "great" because the Rhône valley was a true crossroads. This assertion echoed postwar references to the Rhône as the corridor of Gallic civilization, but decolonization and European integration inflected this allusion in new ways. Soustelle asserted that southeastern France was located at the intersection of two major axes of development: a north-south axis connecting the Mediterranean to the North Sea, largely via the Rhône, and an east-west axis that spanned the greater Mediterranean as far east as the Danube. In addition, supporters of an improved Rhine-Rhône liaison argued that expanding and modernizing the link between the two major rivers and creating a single, uninterruped "axis" from Marseille to Rotterdam would unite the "two European deltas." Given this potential, proponents argued that it was "an urgent necessity" "to construct it as it should be, this great valley, this communication route between the north and the south of Europe, to relink the north and south of our continent by the intermediary of the Rhine and the Rhône, these two great rivers, and thereby to resolve some of the most important problems that our country faces today." Officially, then, "the Great Delta" conceived during the 1960s might fall within the boundaries of France, but from the outset politicians, bureaucrats, and developers believed that it was critical far beyond the nation's borders. Moreover, as Pinay suggested in the passage that opened this chapter, "the development of a major north-south axis," ostensibly for the EEC, could also make inroads "into Africa."[88] France's imperial history was over, but the country's ties to former colonies and other nascent nation-states of the global South still held vast economic potential, and a regional Rhône might actually make this possible.

By the mid-1960s, a regional Rhône offered an alternative spatial framework to the nation-state that had dominated the early postwar era. This regional turn engendered new political divisions within the nation-state, as well as political geographies that might, it seemed, actually transcend national borders. Given thinking such as Pinay's, decolonization and integration probably accelerated the revitalization of regionalization efforts begun a decade earlier. State-approved regional development companies like the BRL and their programs had first endorsed this regional frame during the mid-1950s. Moreover, the regionalization of the lower Rhône valley appears to have served as a model for the rest of France. These institutions and programs therefore predated both the creation of the region as a political intermediary and organizations like Le Grand Delta. The Rhône valley was, then, not simply an *object* of regionalization during the 1950s and 1960s but a key *laboratory* for the formulation of this increasingly influential policy nation-wide.

Even in the mid-1950s, then, some proponents of development had begun to advocate the regionalization, rather than the nationalization, of the Rhône: that is, they reinterpreted "development" as specifically *regional,* framing economic growth and modernization in particular political-geographic terms.[89] By doing so, they viewed and ultimately legitimated the Rhône's continued transformation through this alternative political geography, at the same time constructing and justifying the geography of the region. However, as we have seen, this regional frame was not entirely antithetical to the nation-state. In fact, by the late 1960s, three political geographies—the region, the nation, and a transnational Europe—had become nested and were perhaps surprisingly mutually supporting. Regions within France such as the southeast, centered on the Rhône, promised not only to modernize domestic hinterlands, but also to revive the nation and assure the country's role within Europe. At the same time, the regions did have the potential to transcend national borders, thereby challenging the primacy of the nation-state within Europe, just as the contemporaneous processes of political and economic integration did. The Rhône was not simply "between nation and region," as one recent French scholar put it, but at the confluence of regional, national, and transnational political economies simultaneously.[90] As such, the river was crucial. After all, advocates of regional development during the 1950s and 1960s had framed the Rhône (specifically, the remade river) as a technology; or to recall Soustelle's words, "a magnificent instrument." In other words, these were envirotechnical means to profoundly political ends.

Decolonization and European integration help explain the two aforementioned patterns: the emergence of a regional framework and the construction of the Rhône as a technology. These complex and often contentious political processes renewed interest in the river and its remaking, although they simultaneously fostered new aims and new constituencies. As Chapter 6 will show, regions continued to matter within France and across Europe into the late twentieth century. Thus, in rethinking the Rhône, these regional, national, and transnational frameworks persisted into the 1970s and beyond. However, this period also witnessed stronger challenges to and new definitions of "development." Not only did the spatial framework of development shift, then, but activists, locals, and even certain state officials began to question the goals and perhaps even the idea of development itself.

6

RETHINKING THE RHÔNE

In an inconspicuous chapter of *The Identity of France,* published in 1986, celebrated Annales School historian Fernand Braudel discussed the place of the Rhône in the history of France. Braudel closed the chapter with a description of current plans for developing the river. Historian Braudel even became politician Braudel by advocating the modernization of the Rhine-Rhône liaison, an ambitious scheme to dramatically enlarge the canal linking the two continental rivers. By including these contemporary proposals in his historical study of "the identity of France," Braudel intertwined into a single narrative the Rhône, questions of identity, France's past, and the country's potential future. By endorsing the enhanced liaison, he outlined his vision of the proper relationship among these elements.[1] The CNR's officials and their supporters were undoubtedly thrilled that the esteemed Braudel championed the project.

At the time of the book's publication in France, Braudel was optimistic that the liaison's modernization would soon be realized, but other events that year portended its eventual fate. In 1986, CNR leaders cancelled the last project they had originally planned for the Rhône, Loyettes. Then in 1997, the CNR's administration, experts, and employees were forced to face the agency's second cancellation in just over a decade. That June, Dominique Voynet, leader of the French Green Party (Les Verts) and newly appointed minister of the recently merged Ministère de l'Aménagement du Territoire et de l'Environnement, called off the Rhine-Rhône liaison proposal within days of taking office. The fates of these two projects marked a new chapter in the history of the transformation of the Rhône and of France.[2]

Until the late twentieth century, the ultimate demise of Loyettes and the liaison would have been politically and culturally unimaginable. As we have seen, during the interwar and postwar eras, a broad coalition of politicians, experts, and even left-leaning political parties and labor unions supported the Rhône's remaking. By the late 1960s this alliance and its vision had begun to fracture, as critics envisaged the Rhône in new ways. In particular, environmentalism joined in the debates over the future of the river and of the nation, again altering the landscape of development. The state itself was not immune to these changes, as it became responsible for simultaneously promoting industrial development and ensuring environmental protection. By the mid-1980s, only two projects in the CNR's blueprint for the reconstruction of the Rhône, originally conceived in 1935, had not yet been completed. One of them, Sault-Brénaz, was realized by 1986, but the other, which was the project at Loyettes—where the Ain River meets the Rhône—was not; eleven years later, the modernization of the Rhine-Rhône liaison also met its demise. Within four decades, the drive to transform the Rhône, begun with a firm commitment to the state-sponsored industrialization of nature for national reconstruction and modernization, reversed itself in cancellation of this effort because of probable environmental costs. In the end, Braudel's prediction was proved wrong.

Echoes of the Postwar Rhône

After the 1973 oil crisis, the French state paved the way for the CNR's remaking of the upper Rhône, the river's mountainous reach between Geneva and Lyon. The agency, not to mention state officials, hoped that hydroelectric projects there would augment the country's domestic supply of energy amid escalating oil prices. As Chapter 7 will show, the CNR's technical division proposed five projects along this reach of the river, receiving approval from the state in 1975 to conduct preliminary studies. The CNR ended up building four of these projects between the mid-1970s and 1986.

Many of the metaphors used by the CNR's administration and its supporters to describe the upper Rhône and these projects echoed those deployed by the previous generation when building Donzère-Mondragon. Backers of development continued, for instance, to represent the Rhône as a "wild" river that needed to be "disciplined." As one CNR administrator explained in 1974, the "complete development of the river" would "replace the wild and devastating river with a useful river."[3] Throughout the 1970s and 1980s, journalists also peppered their coverage of the CNR's latest efforts with comments about the unruly Rhône. A journalist from *Le progrès* essentially quoted Michelet when he described the river as a "furious bull flowing toward the sea." Other reporters emphasized the "violence" and "tumultuous" nature of this "torrent." Until the CNR completed its projects, the Rhône would remain a "proud and torrential river."[4] Since each CNR project still required a lengthy process of state approval, a process now even more bureaucratic, thanks to a web of environmental regulations, advocates of river development justified the agency's projects with such language well into the 1980s.[5]

Not surprisingly, then, CNR officials, as well as some politicians and writers, still sanctioned the agency's "taming" of this insubordinate river. One scholar recounted that for twenty-five years, the CNR "had carried out ... a patient and obstinate undertaking: to domesticate the river." Development proponents celebrated what the CNR and its projects had achieved, characterizing these accomplishments as a "victory." After all, the Rhône had been "subdued" and "purified."[6] Although the CNR's supporters applauded its efforts to "harness the Rhône for all time," other writers noted that the agency's work was far from finished.[7] They expressed confidence, however, that the river "would soon be harnessed."[8] According to the CNR's advocates, man and technology would ultimately conquer the Rhône once and for all.

Employing a utilitarian vocabulary that represented the Rhône as a resource for human betterment, development proponents welcomed the river's role as a servant of human society, and specifically of "man." Such gendered language harkened back to the masculinist portrayals of the agency's attempts to master the Rhône during the postwar era. In 1973 Tournier, now retired but still informal publicist of the CNR's endeavors, praised "industrious man" for his efforts in achieving the "profound changes" in the Rhône's "natural regime." The writer André Castelnau concurred that this "most majestic" and "most grandiose" of rivers should be put in the "service of man."[9] Profound links among technology, development, environmental change, and gender had persisted into the late twentieth century.

In mobilizing support for their projects on the upper Rhône, the CNR's political and technical elites reproduced several of the key rhetorical strategies of legitimation invoked from the 1940s through the 1960s. Again speaking for the river, agency officials and their allies reiterated that they were logical partners, as when one journalist avowed that "in the Rhône valley, industry and environment want to go together." Representatives of the CNR even asserted that "sometimes, nature decidedly has the need of man."[10] The gendered and perhaps sexualized character of such expedient declarations remained pronounced.

Development backers continued to collapse distinctions between natural and human-induced changes to the river, which echoed the postwar idea of a "former" Rhône. According to CNR representatives, humans and their technological systems could recreate ecological processes like flowing water just as effectively as the river itself.[11] This was a declaration of confidence in human abilities indeed, but CNR officials strategically conflated nature and culture in other ways. They diminished, if not masked, the agency's responsibility for environmental change, asserting that since the Rhône was historically dynamic, the CNR was merely replicating natural processes. As one journalist argued, "Under our very eyes, the Rhône transforms itself again. It has not finished being born and giving birth to energy [and] human communication."[12] In this rather confused reiteration of the earlier birth metaphor, the river itself was being (re)born; yet it was simultaneously the parent of projects that yielded electricity, commerce, and transportation. Of course, given the long-standing gendering of the Rhône as male, it was another very unusual birth. Overall, by blurring nature and technology, CNR officials and their advocates conflated naturally occurring ecological changes and those carried out by humans, and in fact framed river management technologies as emerging organically from the Rhône itself. This naturalization of the CNR's projects aimed to reduce opposition to them, an objective that increasingly met with mixed results. (See Chapter 7.)

Engineers, politicians, and journalists also continued to historicize the CNR's efforts along the upper Rhône, presumably in an attempt to diffuse or disguise this growing hostility. Numerous coffee-table books published during the 1970s and 1980s recounted the history of France through the lens of the Rhône and its own history. Like some postwar writers, these books' authors naturalized the river's transformation by deploying a narrative strategy in which textual and visual lineages seamlessly linked past and present. They followed the Rhône's course from its source to the Mediterranean while creating a parallel temporal narrative from the geologic past to the present. Many seminal figures of French history were woven into these accounts, which generally culminated with the CNR's most recent projects. *Le Rhône de Genève à la Méditerranée,* by J. M. Delettrez, France's general inspector of finances, typified this approach. It juxtaposed images representing France's Roman and medieval history with those depicting the country's contemporary landscape. One photograph, for example, captured dilapidated ruins on a hillside overlooking the Rhône, with Donzère-Mondragon serving as an imposing backdrop.[13] By situating the CNR's projects within a long history and visually weaving the different eras together, development proponents suggested that the CNR's projects in fact reflected, rather than challenged, the country's rich human and natural histories. As Prime Minister Raymond Barre proclaimed in 1980, the CNR's projects had been conceived and built out of a "respect for our natural and historic patrimony."[14] The portrayal of the CNR's program as a steward of, not a threat to, national *patrimoine* established the transformation of the upper Rhône as the inevitable next step in that history.

This rhetorical move helped proponents represent the CNR's program as "the destiny of the Rhône," as Daniel Faucher entitled the final chapter of his 1968 history *L'homme et le Rhône.* In it, Faucher described the CNR's projects, both past and proposed, and the promised growth they would bring. Among other things, they would help Lyon achieve "its international radiance."[15] Like Braudel and Delettrez, Faucher painted the CNR's efforts, including the vision of the modernized Rhine-Rhône liaison, as the climax of the river's history. By framing the Rhône's recent history as its intended destiny, these authors naturalized historical contingency and thereby depoliticized an inherently political project.[16] Historicization thus continued to provide a powerful rhetoric of legitimation, especially amid growing challenges to the CNR's final projects.[17]

Once again, CNR leaders and their backers touted the agency's latest plans as a critical step in ensuring the future prosperity of France. During the early 1970s, press coverage referred to the Rhône as the "corridor of the future," implying that the river's political and economic potential flowed from its hydrologic channel. Delettrez also echoed Virenque's postwar criticism of those who romanticized the river's past, calling on his fellow citizens to instead admire the impressive technical marvels that would help to realize this productivity.[18] Although Delettrez and other supporters justified development through historicization, they hoped to focus attention on the country's future, rather than its past.[19]

In particular, development proponents presented the upper Rhône as an instrument of regeneration in yet another era of economic difficulty.

The oil crises of the 1970s exacerbated the worldwide economic downturn brought about by slowing growth and rising inflation. France's fiscal woes help to explain why CNR administrators marketed their agency's projects during the 1970s and 1980s as an effective solution to the problems the country faced. Advocates of continued development hoped that the Rhône would again become a powerful tool of industrialization and modernization. Echoing Soustelle, one CNR administrator wanted the river to become "the instrument of new economic development."[20] Other CNR officials argued their agency's most recent projects would help stimulate the regional economic growth discussed in Chapter 5. As CNR General Director Gemaehling explained, the "water of the Rhône is capable of facilitating the industrialization of the southeast [of France]."[21] Facing another era of economic crisis, CNR officials enlisted the upper Rhône in the larger project of economic revitalization. For them, the river promised to refurbish an ailing economic order, especially at the regional level. Apparently, the agency made a persuasive case; in 1975, the state decreed that the CNR should move ahead with its proposed projects on the upper Rhône. Development proponents may have cast the upper Rhône as a technology of regional development and national regeneration, but in doing so they generally focused on its "natural" qualities while minimizing the significant infrastructure, not to mention political commitment, it would actually take to fulfill the river's ostensible nature.

Rethinking the Nation's River

Despite the centrality of the upper Rhône to the latest economic revitalization of the country, proponents of the CNR's program articulated a much weaker ideology of national prestige than their predecessors had during the postwar era. They made fewer connections between *grandeur* and the Rhône's remaking, especially when compared to the flurry of nationalist exhortations that had surrounded Donzère-Mondragon. Prime Minister Barre counted the Rhône among France's "national riches" but did not see the continued development of the river as conflicting with the country's "natural and historic patrimony."[22] The political unit of the nation still had currency, but such declarations were infrequent, to say nothing of the lack of reference to Notre Dame or other icons of French radiance.[23]

In one of the few allusions to *grandeur* after the late 1960s, Vincent

Simon actually criticized the connection between nature and nation forged by his predecessors when he wrote "the Pharaohs had their pyramids, Édouard Herriot his dams. Proof that all great men leave tangible traces of their passage, memories of their greatness, or, at a minimum, their political influence."[24] Simon's ironic tone condemned the link the earlier generation had made between *grandeur* and the river's domestication. According to Simon, the CNR's projects testified not to the impressive state of the nation, but to the excessive power of self-interested (male) politicians who built enormous monuments to themselves. His article not only hinted at a waning in nationalist claims about the fundamental bond between the state and nature but also served as a strong critique of that relationship.

Simon's comments illustrated a reworking of the postwar coupling of nature and nation. This shift both reflected and enacted wider political changes, particularly after the founding of the Fifth Republic in 1958 and the events of May 1968.[25] These changes emerged from decades of domestic political instability and the reconfiguration of international politics during the Cold War. The constitution of the Fourth Republic was approved in 1946, but over the next two decades, citizens and their political representatives would debate the proper balance of power between the president and Parliament, with the Fifth becoming a more centralized presidential form of government. The eventual stabilization of the domestic political order by the early 1970s, coupled with the generational distance from World War II, may have diminished the need for frequent invocations of national glory.

Similarly, nationalist *grandeur* no longer fit as comfortably within the emerging politics of European integration that had strengthened during the Cold War. In fact, advocates of a more international politics within France mobilized the CNR's program for the upper Rhône in support of their cause, thereby generating support for the CNR's proposals and, in turn, offering a way for the agency's leaders to justify them. This move required and produced new spatial understandings of the Rhône and its continued transformation (as Chapter 5 detailed). The Rhine-Rhône liaison proposal, which Braudel had advocated in *The Identity of France*, epitomized these shifts.

Proposals to link the two rivers had been put forth since Roman times. Several small-scale projects linking tributaries of the Rhine and Rhône had been completed by the early nineteenth century. Charles de Freycinet, minister of public works in the 1870s, oversaw the modernization of these canals so that they could accommodate larger barges as part of the Third Republic's efforts to industrialize and incorporate provincial France. Calls to improve this canal network again began in the early 1960s, undoubtedly as part of regional development efforts, but some had bigger ambitions: basically to turn the existing Rhine-Rhône liaison into an enormous canal capable of handling seagoing vessels between the North and Mediterranean seas. At that point in time, the CNR was in a good position to undertake the latest incarnation of the initiative because it had managed the Rhône since 1934.

Politicians, CNR officials, and the press mobilized around this bold plan, framing the modernized liaison within the context of the emerging economic and political integration of Europe. In 1975, Paul Ribèyre, president of the Conseil Régional of the Rhône-Alpes region, alluded to these changes when he suggested that "this grand idea" to modernize the connection between the two rivers "can represent for today's youth the symbol of the new ambitions of France." The recently elected president, Valéry Giscard d'Estaing, made these aims even more explicit by declaring that completing the enhanced Rhine-Rhône liaison would make "France the meeting point of Europe." Similarly, Tournier touted the Rhône as the "corridor of Europe," and press headlines described the new liaison as "a great European undertaking."[26] This Europeanization of the Rhône and its continued development marked a distinct shift in the representation of the river, its transformation, and meaning from the postwar era.[27] Indeed, enrolling the CNR's projects into a transnational political context bolstered the case for European integration, as the agency's envirotechnical system would make this new technopolitics literally concrete.

Yet the CNR projects that advocates of European integration marshaled to their cause were also defined and debated in terms of shifting state policies toward economic development within France. As noted, Gravier's *Paris et le désert français* had alluded to deep-seated anxieties over inequalities between regions of France. The state had begun to propose projects of political and economic decentralization, including the creation of regional development companies in the mid-1950s and DATAR in 1963. These trends only accelerated under the Socialist government of François Mitterrand.

Officials of CNR, and the press, began to negotiate, even embrace, the Rhône's role in this regional, rather than national, economic redevelopment. Journalists from *Le progrès* admitted that the Rhône was no longer a river of poets, alluding to Mistral's famous ode to the Rhône, but instead an instrument promising to economically rejuvenate southeastern France. In 1981, other journalists cast the CNR's first project on the upper Rhône, Chautagne, as critical to the economic development of northeastern France. Even Simon, who had so sharply criticized Herriot and the CNR, believed that the river, once remade, would serve as "an incomparable factor of progress for the economies of regions through which it runs."[28] These commentators suggested that with the help of the CNR's engineers, the Rhône had the capacity to realize the state's new politics of regionalist development and revitalize lagging regions of the country. By the 1970s, development proponents had recast the Rhône, cast for earlier purposes as the nation's river, as a regional and European river.

Environmentalism, Environmentalists, and the Upper Rhône

The river's new political geographies would ultimately contrast with the agenda of environmentalists who fought the transformation of the upper Rhône beginning in the late 1960s. Although the CNR received state approval to initiate the development of the upper Rhône in 1975, the agency's final series of projects was debated and came to fruition in a much different cultural and political milieu. Not only had a series of domestic and international shifts weakened the river's relationship to nationalist *grandeur*, but new voices had also begun to offer alternative visions of the river and of France.

A comprehensive history of environmental movements and environmental politics in twentieth-century France is beyond the scope of this study, but these issues strongly shaped events along the upper Rhône and thus the river itself. Like similar movements in western Europe and the United States, French environmentalism was partly rooted in the social critiques of the 1960s and 1970s, which tended to share antistate, antiinstitutional, and antiestablishment sentiments.[29] Although the student protests of May 1968 helped catalyze environmental organizations, especially among the younger generation, a 1972 survey showed that environmental issues had in fact become important to France's general population as well. Furthermore, membership in these movements was certainly not exclusionary. At the same time, while the concerns of environmental organizations often overlapped with those of the antinuclear movement, frequently they disagreed over energy policy. Nor were members always in agreement even among themselves. French environmentalists experienced an early split between "militants" and "organizers" over the proper means to enact change. Despite initial antistatist inclinations, certain factions of the environmental movement sought to effect change internally, aiming to institutionalize "green" representation at all levels of elective office.

These ideological disjunctures within the environmental community contributed to widening institutional fractures in the movement, especially as it matured over the final third of the twentieth century. Between the early 1970s and 1980s, environmentalists created formal political organizations such as the Green Party, in addition to dozens, if not hundreds, of local, departmental, and regional grassroots groups focused on specific environmental issues. By the late 1980s, the Greens had emerged as the strongest of France's "environmental" political parties, but the emergence of Génération Ecologie in 1990 introduced a viable alternative. Despite their new rival, the Greens made important electoral gains during the 1990s, and in June 1997, after lending critical support to the Socialist Party in the national elections that year, they joined the central government as part of its ruling coalition for the first time.[30]

The plan to develop the upper Rhône occurred, then, just as French environmentalism crystallized and began to attain greater political authority. Given the timing of the CNR's proposals and the fact that the state had already approved preliminary studies of these projects, activists immediately began to question the state-sponsored transformation of the upper Rhône, whether for regional, national, or transnational goals. They formed organizations, foremost among them the Fédération Rhône-Alpes de Protection de la Nature (FRAPNA), to challenge the ongoing development of the river.[31] Set against the backdrop of significant social and cultural change in France beginning in the late 1960s, these new constituencies now played a more active role in shaping the specific redesign of the upper Rhône (see Chapter 7).

Protests against the Rhône's continued development during the 1970s and 1980s were not limited, however, to *écolos,* the somewhat derogatory term for French environmentalists *(écologistes).* Press coverage and CNR publications indicate that scientists, intellectuals, and a diverse array of local organizations, including agricultural groups, town commissions, anglers, and outdoor clubs, lobbied against the CNR's efforts on the upper Rhône—albeit at different times, to varying extents, and for many reasons. Some questioned the findings of the CNR's preliminary studies or wrote editorials to the press; others participated in the projects' formal decision-making process.

Indeed, the boundaries among the identities of various members of this antidevelopment coalition were far from clear. Particularly complicated is the story of scientists who often played multiple roles as experts, environmentalists, and local residents. Biologists and ecologists especially, and those who worked at universities and research institutes in Lyon, Saint-Étienne, and other cities along the Rhône, are most relevant here. Some became involved officially in the debates over the CNR's proposed projects on the upper Rhône because of environmental impact statements (EISs) required by a July 10, 1976, law for large development projects.[32] On the basis of their expertise, these scientists had to predict what might happen to the river if the CNR's proposed projects were actually built. By extrapolating from existing data and projected designs, they produced knowledge on which policy decisions would be based, including the fundamental decision of whether to implement the projects.

Many of these scientists took stances on the various policy recommendations. A number of them thus served as both experts and activists.[33] Some who contributed research to the state's mandated EIS even passed along their findings to environmental organizations in order to bolster the activists' claims against the very projects the scientists were studying. Others adopted more direct tactics to challenge the projects.[34] Meanwhile, the CNR's administration hired its own scientists whose conclusions supported the agency's position.

The legal framework of the EIS ultimately allowed the CNR's administration to downplay some of the concerns that other, non-CNR scientists raised. Rather than requiring a third party to synthesize the evidence and the conclusions of the two competing camps of scientists, French law permitted the CNR itself to present research related to a project's potential environmental impact to the state commission in charge of assessing and determining the future of the proposed project. Not surprisingly, the agency's report to the state usually cast the project under consideration in a better light. Yet the dual role of at least some scientists as both experts and activists helped undermine the CNR's attempts to minimize the environmental impact of its projects. Armed with data from these sympathetic scientists, environmental groups marshaled evidence against the CNR. Moreover, by comparing the CNR's documents with scientists' original reports, these activists portrayed the CNR as an agency willing to sacrifice public welfare in an attempt to develop the upper Rhône at any cost. The inclusion of some activist scientists in the decision-making process, both officially and unofficially, thus altered existing power dynamics for any potential projects and ended up shaping how—even *if*—the upper Rhône would be transformed.[35]

Remembering the Rhône

Although this emerging coalition of scientists, environmentalists, outdoor enthusiasts, and locals opposed the CNR's plan, activists and agency officials actually shared a common vocabulary for describing the upper Rhône. Both groups characterized it as a "wild" and "capricious" river, although the environmental coalition viewed any vestigial wildness after three decades of development favorably.[36] Activists argued that certain sections of the upper Rhône were the last "natural" and "wild" parts of the river, and because "there still remains about fifty kilometers of the 'natural' Rhône" north of Lyon, the CNR's projects should not be built there.[37] These conclusions reflected many environmentalists' strategies for defending the river and critiquing the CNR and its envirotechnical system.

Other journalists and activists focused less on what part of the Rhône remained supposedly untouched than on the vast majority of the river that was now "dead." In a 1978 editorial in *Le Monde*, the writer Bernard Clavel proclaimed dramatically, "I saw the Rhône die." Other opponents declared similarly, "a great river is dead."[38] Some writers rejected these depictions of the Rhône because they masked humanity's role in its decline. These activists asserted that the river had not simply died; the government and the CNR had "killed," "assassinated," or "massacred" it.[39] As a result, they fought to prevent any future "murder" of the Rhône. Members of FRAPNA even asserted that if the CNR completed the modernization of the Rhine-Rhône liaison, the agency would "massacre" the Doubs, a tributary they wished to rescue from the Rhône's unfortunate fate.[40] These pronouncements all implied that technologies of development resulted in the death of nature.[41] Their division of the Rhône into the "wild" and "dead" river bolstered environmentalists' ambitions to save what little of the "natural" river still remained.

Environmentalists drew on Christian metaphors to describe the upper Rhône and its history. They retold biblical stories of humanity's fall from grace. In a lengthy article about the transformation of the river, photographs of flamingos, the famous bulls and feral horses of the Camargue, and spectacular sunsets painted the Rhône as a utopic Eden. Pierre Gascar's accompanying poem called the place "another Genesis, a landscape belonging to the most secret nostalgia of man."[42] However, the "wild" Camargue was actually a cultivated, managed landscape: ranchers raised the bulls pictured in the article, and other parts had been productive rice paddies for over a century.[43] Moreover, even less managed reaches of the river like the upper Rhône had been affected by what had occurred upstream. It was impossible to separate entirely the various parts of this complicated envirotechnical system, much as opponents of development may have tried.

The romanticism of environmentalists and sympathetic journalists, therefore, idealized the river, its history, and its ecology. Haroun Tazieff, mayor of a small town in the region and former anchor of a nationally distributed children's television program about the environment during the 1970s, characterized the river of the pre-CNR era as "magnificent" and "wonderful."[44] Such characterizations minimized, if not ignored, the risks the river had historically posed to its human neighbors, basically whitewashing (or perhaps "greenwashing") destruction such as flooding. Activists' romanticism thus relied on a selective memory of the river's natural and human histories, one often cast through a Christian lens.

Saving the Rhône, Saving the Nation

These romantic versions of the Rhône's histories provided the narrative framework in which many activists called for the "saving" of the "last sixty wild kilometers of the French Rhône."[45] In 1982, for instance, a coalition of environmentalists, farmers, fishers, and locals had circulated a petition to "save the upper Rhône and the Ain rivers" by fighting the construction of Sault-Brénaz and Loyettes, the CNR's last two projects.[46] Such new organizations brought together local communities while transcending potential social, political, and educational divides, as a variety of interest groups shared a common goal: to preserve what remained of the "natural" river.[47]

United in their objective to "save" the Rhône, some of these activists invoked themes of sacrifice, loss, and finality to justify the river's protection. During the early 1980s, after the CNR had built the first two of its five proposed projects along the upper Rhône, the writer Françoise Holtz-Bonneau characterized the agency's projects as "backward dams" because they would channelize the "last sixty wild kilometers of the French Rhône." She hoped the CNR would be forced to cancel its final projects while the last portion of the wild river still remained. Similarly, writing in *Le progrès,* R. Michel suggested that the costs of losing the last vestiges of the wild Rhône far outweighed the benefits of development when he asked "is it necessary to exchange several dozen kilometers of the natural Rhône for a hundred or so megawatts?"[48] The implication was clear: if the CNR succeeded in building its final projects, what little remained of the natural Rhône would be gone forever. These advocates hoped that this last stretch of the wild Rhône instead might serve as an antidote to the bad effects of the industrialization of the greater part of the river. In making this argument, they neatly separated humans from the Rhône, the river from technology, and even one stretch of the river from the rest.

Environmentalists articulated other rationales for "saving" what remained of the "natural" Rhône. Some maintained that its ecological value needed to be analyzed, quantified, and ultimately included in the state's official decision-making process. FRAPNA president Philippe Lebreton, a chemist, argued that the agency's initial project assessments inherently favored development because CNR officials had not, and ultimately could not, incorporate the value of the wild river into their calculations.[49] Although Lebreton and other activists often criticized the CNR's tendency to objectify the Rhône, they hoped this strategic move might serve alternative ends. For these environmentalists, the quantification of nature validated protection, rather than enabling development.[50]

Other environmentalists fought to save the Rhône for scientific and ecological reasons. Rejecting the CNR's claims that the river should serve human betterment, these activists mobilized the ecological sciences to their cause. They cited the inherent ecological value of the Rhône and the potential damages the agency's projects might cause to the environment. They particularly feared that the Rhine-Rhône liaison's modernization would have permanent, irreversible environmental impacts.[51] Indeed, FRAPNA members asserted that the CNR would destroy wilderness, landscape, flora, and fauna unique in Europe by "massacring the confluence of the Ain [and Rhône rivers]" if the agency completed the final two of its five planned projects.[52] Several leading scientists lent authority to environmentalists' claims about the probable ecological impacts of the agency's projects.[53] Not surprisingly, CNR officials disagreed with their conclusions.[54]

It was not only the CNR's projects along the upper Rhône that incited these concerns. Activists were also motivated by the potential environmental consequences of planned nuclear expansion along the upper Rhône, especially as France realized its "all nuclear" policy during the 1970s. Those in the Ain department, located northeast of Lyon, protested nuclear development in their communities. In particular, locals, activists, and scientists voiced deep concerns about thermal and radioactive pollution of the Rhône. Early problems with the new reactors only seemed to justify, if not amplify, these fears.[55]

As writers and members of FRAPNA had already suggested, these critics hoped to save the nation by saving the Rhône. They asserted that large-scale projects were not a mark of success but of loss and failure. Several authors characterized the CNR's efforts as a tragic mistake. Gilles Morel evoked striking imagery when he pronounced that bulldozers had silenced the river. When environmentalists compared the ecology of the upper Rhône to other sites in Europe, they also enrolled nationalist claims in their defense of the river. Cancelling the modernization of the Rhine-Rhône liaison would benefit, then, not just nature but also the nation, since such projects "risk causing us to lose a patrimony of high ecological and historical value." According to these views, the landscape of the "subjugated Rhône" marked all that was wrong in France. The country had industrialized too extensively and sacrificed agriculture too much.[56]

Among other reasons, critics condemned the CNR's program for the upper Rhône because of the agency's inordinate power and *gigantisme*, *grandeur*'s vile twin, or what Paul Josephson has memorably called "gigantomania."[57] In an article in *Voix de l'Ain*, Alain Gilbert characterized the CNR as a modern-day Gargantua, invoking Rabelais's famous all-consuming gourmand. Other writers feared that the *gigantisme* of powerful agencies like the CNR threatened the nation's future. In 1982, Holtz-Bonneau claimed the CNR was powerful enough to imitate the tactics of the influential EDF and to "impose" its will on the state. Leaders of FRAPNA even accused the CNR of attempting to get projects approved simply to employ their technicians, thus placing the agency's interests above those of the public.[58] According to these critics, large, powerful agencies, their gigantic projects, and above all their technocratic visions threatened the future not only of the Rhône but also of France itself. These accusations also carried an anti-American subtext, hinting at fears that the nation's culture, and perhaps its nature, might be Americanized.[59]

Furthermore, environmental activists asserted that such gigantomania marked the incompatibility of development with the environment. In 1975, several articles in the activist magazine *Stop Pollution: Vallée du Rhône,* emphasized that large projects like those of the CNR embodied a certain *gigantisme,* which did not fit with the river valley. These environmentalists condemned the way these projects recast the relationship between human society and the natural world, challenging the envirotechnical regime that had shaped the postwar reconstruction of the Rhône. They hoped that preserving what remained of the wild Rhône would help redirect France onto the right path. Mobilizing the nation to their cause, environmentalists attempted to ground the country in the late twentieth century on the absence, rather than presence, of development. Although these activists did not win all their battles, they did eventually place sufficient pressure on the CNR to alter the design of its envirotechnical system in the upper Rhône valley and even to cancel its last two projects, something that had been basically unimaginable a generation earlier.

It is not surprising that representations of nature and technology in France during the late twentieth century diverged from those of the 1940s through 1960s. The CNR projects on the upper Rhône that were eventually built during the 1970s and 1980s were debated within different political, social, cultural, and even ecological contexts from those of the postwar era. By then, postwar reconstruction had ended, and nearly three decades of economic prosperity had helped to create an urban middle class that demanded cheap consumer goods and idyllic rural vacation retreats alike. The 1960s and 1970s also marked a period rife with public debate, social struggles, and increasing dissent. Political activism on behalf of a variety of causes, including the environment, had intensified. Meanwhile, the Cold War, de Gaulle's departure from power, relative political stability after 1969, and the EEC together contributed to a new articulation of French identity and new political geographies of development. Finally, the actual landscape of the Rhône in 1968 little resembled that of 1921, when the state had founded the CNR. Perhaps it was easier to appreciate a "wild" river when so little of it remained. In fact, what might be more surprising are the resonances of the postwar representations of nature and technology during the late twentieth century despite all of these shifts.

By the 1970s, then, two competing but legitimate narratives about the Rhône and the nation had come into tension with one another, as the CNR's latest projects entered into public inquiry and encountered state bureaucracy. On the one hand, development advocates continued to lobby for the CNR's program, but the meaning of development came to be reinterpreted and ultimately respatialized. The crystallization of the presidential republic by the early 1970s offered one of the more stable governments since the beginning of the twentieth century—perhaps since the late eighteenth. Greater political stability helped diminish the need to constantly invoke national prestige in political and popular culture. Meanwhile, the Cold War further undercut nationalist *grandeur,* while European cooperation offered a new context in which to situate the Rhône and its remaking. As a result, the CNR and its supporters began to market a European Rhône to serve as a means of economic unity and a symbol of transnational community, rather than as an object of French pride and a means of national reconstruction. While some looked beyond the nation to the continent, other development proponents narrowed their focus to the region by framing the Rhône in terms of the economic revitalization of the country's poorer areas. Reformulating the connections between the political order and nature, both European integration and regional renewal offered different political geographies in which to frame the meaning of the Rhône and its continued transformation, while the CNR's projects provided key sites for considering these pressing questions.

On the other hand, the combination of evolving state priorities, internal reforms, and external protests resulted in significant shifts in power relations in the Rhône valley that helped to create new images of the river. By the 1970s and especially by the 1980s, a growing coalition of activists with a fresh political voice connecting the future of the nation to a wild Rhône proved relatively successful in challenging older narratives about the river and its remaking. Environmentalist discourse had therefore taken a legitimate place alongside development within the French state and society.

As Chapter 7 demonstrates, these activists reshaped the design, even the completion, of envirotechnical systems in the upper Rhône valley during the late twentieth century. They were, however, less radical than they may appear at first glance, and their achievement did come at a cost. Critics generally maintained a distinct divide between "nature" and "technology" in their representations of the Rhône and especially its upper reach.

This strategic depiction of the upper Rhône served to challenge how it had been managed for decades and strengthen arguments to protect it. In many ways, environmentalists' stance on the upper Rhône is surprising, given the general French propensity for hybrids, including nature-culture, and comfort with, even preference for, cultivated—that is, fundamentally envirotechnical—landscapes.[60]

Three factors may help explain the activists' position, as well as the seeming contradiction between their arguments about the upper Rhône and their tendency to embrace technological modernity, which Michael Bess has ably traced. First, transnational environmental discourses may have influenced "French" environmentalism, especially by the 1980s. Second, throughout the second half of the twentieth century, CNR leaders and their supporters constantly blurred the boundaries among nature, technology, and culture in an attempt to legitimize development. However self-serving their claims that the CNR's projects were simply extensions of nonhuman nature, the agency's elites acknowledged the inextricable interdependency of the cultural, technological, and natural worlds—connections that were, in fact, tightened by the agency's construction of its envirotechnical systems. In contrast, environmentalists and other activists in the Rhône valley rejected these links. Perhaps it was because the CNR and its backers continually conflated nature and technology in the name of development that environmentalists and other antidevelopment constituents held these concepts apart and focused their efforts on what supposedly remained of the "wild" Rhône. In short, if development proponents repeatedly merged "nature" and "technology" to legitimate development, then these activists responded by insisting on the separation of these concepts in an attempt to justify preservation.

However, one of the ironic consequences of this stance is that although environmentalists expressed many antimodernist arguments in their efforts to defeat the CNR, here they replicated some of the most fundamental analytic categories of the modernists they so strongly criticized. Moreover, framing the upper Rhône as a last wilderness in an attempt to protect this reach of the river offered little guidance as to how one should address places that had been transformed by humans. In reality, by the 1970s, the Rhône was an intensely managed, envirotechnical landscape. Yet the way these activists imagined the Rhône tended to mask its very hybridity—its status and fundamental identity as a blending of ecological and technological systems. By trying to save the "natural" and "wild" Rhône, they risked giving up on most of the river. The upper

Rhône's story suggests, then, that environmental critiques of the Rhône's late-twentieth-century development were selective and uneven.

Finally, a geographic or spatial lens may also help to explain the way defenders of the "wild" upper Rhône embraced both environmental protection and technological modernity, despite their seeming contradictions. Although activists eventually did seek to modify existing projects like Donzère-Mondragon by, say, increasing its reserved flow or planting more vegetation along the diversion canal's banks, they generally targeted the "undeveloped," "natural" reaches of the Rhône. In other words, during the 1970s and 1980s, they concentrated their efforts on what had not yet been built, rather than what had already been completed; they focused on the remaining spaces of undevelopment, rather than existing spaces of development. This differential spatial geography of environmentalism ended up curbing the full extent of its critique and therefore its impact. This would have serious consequences for humans and nonhumans alike.[61]

Bess has characterized post-1945 France as a "light-green society," arguing that while environmentalism eventually pervaded much of civil society and the state, ultimately it was a shallow critique. In many ways, the Rhône became a "light-green" river in the final three decades of the twentieth century. As I will show, environmental critics did protect a few areas from further development, and their protests led to modest modifications in the management practices and features of existing envirotechnical systems. Overall, however, they did not question what the previous generation had achieved by proposing, for instance, the possibility of dam removal. In the end, environmentalists largely reconciled their goals with existing envirotechnical systems. Through their spatial differentiation of a "wild" and a "developed" Rhône, they ended up accommodating both environmentalism and development.

7

A NEW MODERN

On March 10, 1982, French minister of the environment Michel Crépeau proclaimed, "The national interest, which consists of protecting a beautiful natural region near the metropolis of Lyon, is even greater than the energy benefits of [Sault-Brénaz and Loyettes]."[1] Crépeau's appraisal of the CNR's last two projects on the upper Rhône then under consideration embodied some of the key shifts in cultural attitudes toward the river discussed in the previous chapter: from constructing its upper reach as natural to assessing its value within a national framework. Crépeau's pronouncement is even more remarkable given France's energy emergency less than a decade earlier. The oil crises of the 1970s prompted the nuclearization of the upper Rhône as part of France's "all nuclear" energy policy, but it also revived interest in its hydroelectric potential, a fact often overlooked, given the close ties forged between nuclear power and French national identity. In fact, between the mid-1970s and early 1980s, the CNR built the first three of its five proposed projects on the upper Rhône. Yet despite this intense push to increase France's domestic energy supplies, opposition mounted with each project. By 1982, Crépeau declared that "a beautiful natural region" served the national interest "greater than the energy benefits" of the CNR's final two projects. Four years later, the CNR cancelled Loyettes. These were important changes in just thirteen years.

Even as the state authorized the CNR to undertake preliminary studies for its first projects on the upper Rhône in late 1975, internal reforms and external pressures already began to split the coalition that had backed the river's remaking since the end of World War II. Internally, environmental issues had become part of the central government's institutional landscape and agenda. The creation of the Ministry of the Environment in 1971 established a discrete government agency to address these concerns.[2] Then, on July 10, 1976, the government passed a law mandating EISs for large development projects to which some of the CNR's projects on the upper Rhône became subject.[3] By the mid-1970s, then, environmental commitments had joined development objectives as part of state bureaucracy and became a consideration in government policy-making, thereby intensifying existing intrastate conflicts. Although the French state was "greener" than in previous years, disagreements among state agencies illustrated its unevenness.[4]

As parts of the state took up these environmental initiatives, various groups of residents, environmentalists, scientists, intellectuals, and outdoor enthusiasts increasingly questioned the ongoing development of the Rhône. Locals and activists had leveled criticism at the CNR and government agencies before the mid-1970s, but these protests and their influence now reached unprecedented levels. Furthermore, the critiques intensified between the mid-1970s, when the state granted the CNR the authority to conduct preliminary studies, and the mid-1980s, when the agency sought approval for its last planned project. Combined with state-directed environmental agendas and expanded public input regarding the proposed projects, these activists played a greater role in the redesign of the upper Rhône than their predecessors had in the remaking of its central and lower reaches during previous decades.

The transformation of the Rhône over the second half of the twentieth century, therefore, demonstrates several important shifts in the goals and power structures of the envirotechnical regimes that sought to remake the river. These changes occurred, then, not only between World War II and the 1970s but even between the early 1970s and mid-1980s. They eventually became embodied in the CNR's projects on the upper Rhône—those that were actually built and those that never materialized. The designs of the completed projects and Loyettes's cancellation thus illustrate a shift from the high modernist, developmentalist ideology of the postwar era to a more environmentally informed approach in the late twentieth century. Yet it is important to note that state officials, and even environmentalists, did not call for more drastic measures such as dam removal. By 1986, the river had become a product of both historic and more recent approaches to river management. A new modern had begun to govern the Rhône.

France's Energy Crises and the Development of the Upper Rhône

France's energy crises during the 1970s triggered the contentious debates over the future of the upper Rhône. The outbreak of the Yom Kippur War in October 1973 brought about a rapid change in France's energy policy and a concomitant shift in the transformation of the Rhône. When OPEC quadrupled oil prices that winter, France, like many other Western countries that relied on Middle Eastern oil, faced an energy emergency. At the time, 75 percent of the country's energy came from oil, three-fourths of which was from the Middle East. In response, the central government issued an "all nuclear" *(tout nucléaire)* energy plan. This policy, outlined in the Messmer Plan of March 6, 1974, promoted substantial political, financial, and technical investment in nuclear technologies in an attempt to shift the country's primary energy source from petroleum to nuclear-generated power. It aimed to supply 70 percent of the country's energy needs with nuclear-generated power by 1985. Although the state did not quite meet its goal, such energy did provide approximately two-thirds of France's electricity in 1985, compared with 31 percent in West Germany and 19 percent in the United Kingdom. By the mid-1980s, then, France was one of the world's most nuclear nations when it came to the production and consumption of atomic energy, an achievement many of its leaders and citizens celebrated. Over the past two decades, France's reliance on nuclear power has only increased; about 75 percent of the country's electricity is produced through nuclear generation today.[5]

The new energy context sparked by the 1970s oil crises mediated the transformation of the upper Rhône in at least two ways. First, the government approved Creys-Malville, France's largest fast breeder nuclear reactor, thirty miles upstream of Lyon. Also known as Superphénix, Creys-Malville incited some of the country's most violent antinuclear protests. In 1976, fifteen thousand protesters attempted to put an end to the reactor's construction; by July 1977, the number of demonstrators had swelled to an estimated eighty thousand.[6] Despite these protests, France's energy needs and the state's atomic ambitions succeeded in reshaping the upper Rhône into an envirotechnical system more suitable for the production of nuclear energy. Second, although the state had declared an "all nuclear" energy policy, politicians and bureaucrats in fact also turned to the hydroelectric potential of the upper Rhône. As an article in the newspaper *Bref Rhône Alpes* portended in December 1973, "the current energy crisis, including the notable increase in oil prices, could make the development of the upper Rhône between Lyon and Génissiat, which was part of the original mission of the CNR, actually feasible."[7]

The "feasibility" of the hydroelectric development of the upper Rhône hinged on the *rentabilité,* or economic viability, of the CNR's proposed projects there. This assessment relied, in turn, on the EDF and its analysis of the estimated costs of hydroelectric generation. During the 1950s, EDF officials had shifted their primary goal from the quantity of electricity produced *(rendement)* to the costs associated with electricity production *(rentabilité)*: the agency's objective changed from producing as much electricity as possible to producing electricity as cheaply as possible. This repositioning reflected France's move away from the imperatives of postwar reconstruction, but as a result, EDF officials began to compare any proposed electricity-producing plant to the theoretical economic baseline elaborated in the famous "blue notes" their agency periodically updated.[8] They determined this baseline by calculating the cost of producing electricity at an "average" coal-fueled plant. Because the oil crises had dramatically increased the price of petroleum, the EDF's calculation of energy production costs after October 1973 made historically financially marginal hydroelectric sites viable. The new context of energy after 1973 thus recast the hydroelectric potential of the upper Rhône.[9]

The CNR's leaders, hoping to make the most of this new energy calculus, took steps to present their agency's projects not only as economically feasible but also as effective solutions to France's pressing energy needs. The agency's engineers had already begun to compile data on the potential hydroelectric development of the upper Rhône for the agency's technical committee by February 1974, just a few months after the Yom Kippur War broke out and, notably, one month before the government announced its "all nuclear" policy. In internal reports that revealed the extent of France's latest energy crisis, CNR officials stated frankly that this reach of the river "would be developed solely for energy," suggesting that the country's energy imperatives might override the agency's multipurpose mandate. That April, members of the CNR's technical committee attempted to advance the agency's plans when they toured possible sites along the upper Rhône. However, the Messmer Plan, which had been released the month before, was undoubtedly weighing on their minds, since they explicitly compared the advantages of hydroelectric and nuclear power along the river's upper reach. CNR leaders expressed fears that politicians might favor the EDF's nuclear reactors over the agency's hydroelectric plants. Their apprehension was not unfounded. The CNR's own calculations showed the EDF's facilities were far more *rentable* than the CNR's plants. Yet CNR administrators ultimately claimed it was an unfair comparison, since water was a national resource and uranium was not.[10]

These arguments obviously served the CNR's interests, but the agency's elites had a point. Growing dependency on uranium, a natural resource different from petroleum but also "foreign," might simply replace reliance on one imported commodity with another. Such reasoning exposed crucial contradictions inherent to France's "all nuclear" policy. Nevertheless, the central government remained unswayed, and during the final months of 1973 and early 1974 it moved ahead to lay the foundation for the rapid expansion of nuclear power. Despite this seeming paradox, the fact that hydroelectric plants could be framed as national projects, as CNR officials had done, did offer a compelling reason to foster the development of additional nonnuclear energy sources even amid a supposedly "all nuclear" policy. The CNR's leaders expressed cautious optimism, believing that state officials would eventually agree with their logic and support the agency's projects. By emphasizing the inevitable growing French dependency on "foreign" uranium that would result from the "all nuclear" policy, CNR leaders reproduced the nationalist rhetoric that had celebrated hydroelectricity after World War I and again after World War II. In effect, France's latest energy crisis revived the historic connections between hydropower and nationalism. As the rest of this chapter shows, the reforging of this connection would have important implications for the management of the upper Rhône during the final third of the twentieth century.

Indeed, in April 1974, only one month after the Messmer Plan was issued, members of the Senate and National Assembly discussed how to maximize France's own energy resources, particularly hydroelectricity, to reduce the country's dependency on foreign energy. That spring, Minister of Industry Michel d'Ornano established the Pintat Commission to study possible sites for future hydroelectric development and asked CNR administrators for their recommendations.[11] Because the agency had decades of experience and its engineers had already gathered pertinent information, the CNR's leaders were in a good position to advocate the hydroelectric development of the upper Rhône.

In November 1975, the Pintat Commission, echoing the CNR's likely self-interested counsel, recommended to the minister of industry that five additional hydroelectric plants be constructed along the upper Rhône at

Chautagne, Belley, Brégnier-Cordon, Sault-Brénaz, and Loyettes (see Map 7.1). The Commission also asked the CNR to submit preliminary dossiers for each of these projects. Just four months later, the CNR requested public hearings for the first three: Chautagne, Belley, and Brégnier-Cordon. With d'Ornano's approval, the agency began detailed studies of Chautagne and Belley shortly thereafter.[12] Fueled by both ambition and optimism, the CNR's president, Max Moulins, confidently informed d'Ornano's successor in April 1978 that his agency not only planned to start construction of Chautagne the following year but also intended to complete one project each year for the next five years.[13] Moulins may have had grand plans, but actually carrying them out proved far more difficult than CNR and state officials had initially envisioned; in reality, they masked a widening chasm in development politics in late-twentieth-century France.

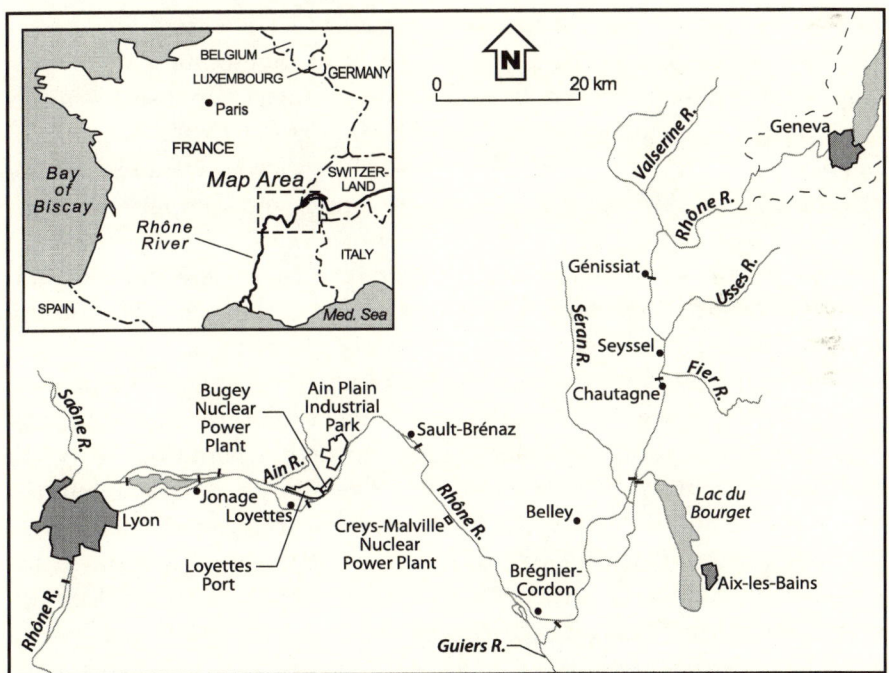

Map 7.1. *Developing the upper Rhône.* The CNR hoped to build five multipurpose projects on the upper Rhône between Génissiat-Seyssel and Lyon after the 1973 oil crisis. After mounting protest, the agency built only four: Chautagne, Belley, Brégnier-Cordon, and finally Sault-Brénaz; the state cancelled Loyettes. Note the location of Bugey, a nuclear power plant built in 1964. The earlier nuclearization of the upper Rhône ended up mediating the design of Sault-Brénaz. (Map by Joseph W. Stoll, Syracuse University Cartographic Laboratory.)

The Uneven "Greening" of the State

Understanding the bureaucratic hoops, both legal and administrative, through which projects like those of the CNR were required to jump, is critical to grasping why it became more difficult for the CNR to carry out its transformation of the upper Rhône. State institutions, policies, and laws are never static; this was especially true in the 1970s and the 1980s—precisely when the CNR aimed to complete its projects. Although CNR officials and their supporters found these changes frustrating and viewed them as impediments, critics believed they leveled a playing field that had historically favored large-scale development.

One key to this bureaucracy was the state's "declaration of public utility" *(déclaration d'utilité publique;* DUP). If the state issued a DUP for a proposed project, it had effectively authorized construction of the project according to the technical plans outlined in the proposal submitted by the project's sponsoring agency. Because the DUP was a critical turning point in the history of a proposed project, obtaining it involved a lengthy review process that started with the project's sponsors—in this case, the CNR—putting forth a preliminary dossier to the Ministry of Industry and requesting public hearings. The sponsors were required to advertise these hearings to affected communities and compile comments from locals and other interested groups, which they then presented to an "inquiry commission" *(commission d'enquête)* for review and consultation with the appropriate technical experts, when necessary. The inquiry commission then summarized its findings and issued its conclusions. The entire "inquiry dossier" was then transferred to the head of the Direction Interdépartementale de l'Industrie, who invited responses from the project's sponsors. After the sponsors had a chance to defend their proposal, the interdepartmental group released an overall opinion and submitted it to the minister of energy. The project then continued its tour through the government offices of Paris, where an interministerial committee discussed the proposal. If this committee reached a unanimous decision, only then might it decree that the project merited a DUP and subsequently guide the proposal through the Conseil d'Etat for signature. Not surprisingly, the protracted process usually required twelve to eighteen months to complete.[14]

The creation of the Ministry of the Environment in 1971 altered existing power dynamics within the state relating to projects like those of the CNR and decision-making processes like the DUP. Although this new ministry affected Chautagne and Belley, the CNR's first two projects on the upper Rhône, it was the agency's third project, Brégnier-Cordon, that made this shift most apparent. The CNR's officials had initially submitted one preliminary dossier for all three projects, but early in the DUP process, the minister of the environment ordered the CNR to separate Brégnier-Cordon from the other two, asserting that it merited its own study.[15] Because complying with this demand delayed the state's consideration of Brégnier-Cordon, the project became subject to the 1976 law requiring an EIS.[16] By mandating a separate DUP for Brégnier-Cordon, the Ministry of the Environment effectively put this and all subsequent CNR projects through the EIS process, over which it presided.

Although this ministry had less jurisdictional authority over Chautagne and Belley, its participation in interministerial meetings during the final stages of the DUP process alone destabilized the state coalition that had historically advocated the Rhône's transformation. Because the interministerial commission was required to reach a *unanimous* decision, the Ministry of the Environment had to "bless" the CNR's projects before they could receive a DUP.[17] Consequently, other ministries within the state, including the historically powerful Ministry of Industry, became at least partly beholden to it. By the mid-1970s, then, the new Ministry of the Environment had become a key player in the new dynamics of the French state. Moreover, these dynamics reflected the fact that the state now had a more complex and ultimately ambiguous relationship to industrial development and environmental protection.

On the one hand, environmental politics had explicitly become incorporated into state infrastructure through the creation of the Ministry of the Environment and the passage of laws like the EIS. Environmental concerns also played an increasingly influential role in the political system at the ideological and institutional level. For instance, during the elections of May 1974, presidential candidate Valéry Giscard d'Estaing circulated his "texts and statements" on the topic, *"Pour un environnement 'à la française.'"* The theme of the booklet alone suggested how interrelated Giscard d'Estaing considered environmental politics, political authority, and national identity to be; he even went on to assert that one of the fundamental objectives of "our" civilization was the right of all individuals to access nature and to outline an eighteen-point program to carry out this goal. Furthermore, he recommended that *"une politique d'ensemble de l'environnement"* (a systematic environmental policy) serve as the foundation of the state's Seventh Plan (1976–1980), signifying an important shift in the relationships among economic development, industrialization, and environmental management.

These "texts and statements" were put in print in 1977, three years after Giscard d'Estaing's successful presidential campaign, along with the measures his administration had already undertaken and those it planned to pursue in the near future. The publication particularly highlighted the 1976 EIS law, which he called "a real charter of nature" that would ensure environmental protection and establish a policy for creating natural reserves while guaranteeing the "effective participation of citizens."[18] His invocation of a "charter" *(charte)* and use of the language of historic rights and privileges were not accidental. Both alluded to the long-contested history of the relationship between the government and citizens of France under the monarchy and especially following the French Revolution. As France experienced a revolving door of political regimes between 1789 and the Third Republic, government officials had often referred back to *chartes* in an attempt to forge legitimacy for their often tenuous rule. Giscard d'Estaing thus articulated French environmental politics in a double sense. The state would address environmental issues, to be sure, but legislating the proper relationship between society and nature would in fact serve as a cornerstone of the nation's political order.[19]

Not everyone agreed with Giscard d'Estaing, of course. Caught in the middle of the state's new environmental bureaucracy, CNR administrators complained about the EIS, voicing frustration at its protracted procedures and lamenting frequent delays. They were also aggravated that their historically amicable relationships with supervisory state agencies appeared to have ended. In December 1976, five months after the state had passed the EIS law, CNR president Moulins bemoaned the "heavy toll of formal preparations associated with the public inquiries that sometimes seem out of proportion with the projects themselves."[20] CNR leaders like Moulins may have grumbled about these bureaucratic hurdles and attempted to circumvent them by pleading with sympathetic superiors, but the agency could not get around them, even when its administrators apparently tried to do so. In November 1976 and again in March 1977, the Ministry of the Environment informed the CNR that the agency's DUP dossier for Chautagne and Belley was unacceptable because it did not adequately outline the projects' "environmental effects." In fact, the Ministry had refused to approve the projects in 1976 because CNR leaders had ignored these concerns.[21] Consequently, the CNR had to go back to the drawing board not once but twice.

On the other hand, despite proclamations such as those by Giscard d'Estaing and laws such as the EIS, the French state was not uniformly "green." Conflict over Chautagne and Belley suggested that intrastate conflict worsened once the Ministry of the Environment had destabilized existing power relations within the state with respect to environmental management, whether under the rubric of "development" or "protection." In general, the debates over the transformation of the upper Rhône suggest that political allegiance from the early 1970s to the mid-1980s generally broke down along the following lines: the Ministry of the Environment on one side with several other, historically powerful state bureaucracies, including the Ministries of Industry and Finance, on the other.[22] The "greening" of the French state thus tended to be concentrated within a single ministry, thereby exacerbating tensions among these government ministries and their envirotechnical regimes.[23]

At times, however, officials from other state agencies and government ministries did raise environmental concerns, suggesting that Giscard d'Estaing's call for "a systematic environmental policy" did extend beyond the Ministry of the Environment. For example, in 1976, the Ministère de l'Aménagement du Territoire recommended higher reserved flows at Chautagne and Belley than those initially proposed by CNR leaders. Also that year, even the CNR's technical committee privately acknowledged that the agency would have to consider the environmental repercussions of its projects.[24] This admission may have been pragmatic, rather than an expression of any significant ideological shift; nonetheless, it signaled a change. By the mid-1970s, deep institutional and ideological conflicts over the proper balance between economic development and environmental protection were playing out within the state—and on the Rhône—and the envirotechnical systems completed during the 1970s and 1980s reflected the compromises that were made.

The "Greening" of Society

As Chapter 6 showed, the proposal to develop the upper Rhône also emerged during the blossoming of France's environmental movement. Yet it was precisely projects like those of the CNR that catalyzed the mobilization of environmental organizations and spurred the articulation of environmentalist discourse. An uneven greening of French society thus simultaneously accompanied the uneven greening of the state.[25] Undoubtedly, some locals welcomed the CNR's projects just as some government agencies did, but (as the previous chapter demonstrated) activist organizations, especially those concerned with environmental issues, proliferated during the 1970s and 1980s.[26] Public mobilization against the development of the upper Rhône commenced almost as soon as the idea was first proposed. In May 1974, one month after the Pintat Commission was formed, a top CNR administrator reported that numerous groups had already organized against any project located near Belley.[27] Within months, the agency had become the target of protests, petitions, letter-writing campaigns, and countless public meetings.

Environmentalists were certainly some of the most vocal critics. Among environmental organizations, FRAPNA was one of the most active.[28] Others included Comité Ecologie, Association pour l'Environnement de la Vallée du Rhône, Fédération Française des Sociétés de Protection de la Nature, Groupe Ain Nature, and Coordination Pour la Défense du Fleuve Rhône et de la Rivière d'Ain (CODERA).[29] As their names suggest, at least two of the groups had been established specifically to fight the CNR's development of the upper Rhône. By 1980, when the state's inquiry commission synthesized its observations in Brégnier-Cordon's DUP, the report's authors listed nine different environmental organizations and "assorted individuals" who opposed the project.[30] In addition to environmentalists, other constituencies established their own groups. Fishers' associations wrote letters to the CNR and attended public meetings, and farmers expressed concerns over the compatibility of agriculture and the CNR's projects.[31] Many local, departmental, and regional leaders and bureaucrats became embroiled in the debates as well. These efforts convey a sense of the diversity of the opposition and the degree of their mobilization.

Environmentalists often invoked these multiple constituencies as a way to bolster opposition to the CNR and its projects. One FRAPNA petition claimed that farmers, hunters, politicians, locals, naturalists, fishers, and hikers all supported their organization.[32] Indeed, as animosity toward the CNR grew, oversight committees were formed in an attempt to bring these diverse perspectives to the negotiating table so they might reach an amicable agreement. In early 1977, the "Working Group on the Upper Rhône" met at the state's request to negotiate approval for Chautagne and Belley. A prodevelopment stance united the constituents of this group, but the membership of other committees reflected the highly contested nature of river management by the late twentieth century. For instance, a 1979 meeting for Brégnier-Cordon included dozens of representatives from state agencies, the CNR, professional associations, and local organizations. By the late 1980s, the region even held an annual meeting "on the environment." This new forum brought together the prefect of the Ain and representatives from various organizations, communities, and all levels of government administration to discuss environmental issues.[33] These organizations did not necessarily have significant regulatory authority, but they did at least offer a form of political power to locals who had historically lacked a real voice in the project approval process.[34] In effect, community mobilization based on shared critiques of the CNR was a by-product of the proposed development of the upper Rhône.

Many of these activists were united in their opposition to the CNR's projects, but specific critiques diverged. Some highlighted the costs locals would bear, rather than the benefits they might see. For example, in March 1976, the prefect of the Ain argued that communities near Chautagne, Belley, and Brégnier-Cordon had already paid a heavy toll, thanks to the construction of the Bugey nuclear plant, the expansion of the national highway system and the Train à Grande Vitesse, and the creation of a large industrial park. Now they faced the hydroelectric development of the upper Rhône in their backyards.[35] Other activists criticized the bureaucratic process for proposed projects, arguing that it lacked sufficient accessibility and genuine public participation. Echoing earlier complaints of *gigantisme,* environmental groups such as Amis de la Nature, Groupe de Nature d'Isère, and Nature et Vie Sociale implied that antidemocratic tendencies within the state's approval process disadvantaged their constituencies and that the system inherently worked against environmental concerns. Some critics' complaints were thus social and political, and their laments became more frequent beginning with the DUP hearings for Brégnier-Cordon.[36] Meanwhile, other groups raised a number of environmental concerns, citing specific ecological changes associated with the CNR's proposed development program to challenge it. Framing their apprehension in terms of "environmental" or "ecological" "effects," "consequences," or "impacts," they not only pointed to concrete problems such as the volume of water allotted to the former Rhône but also issued broader proclamations about the need for an entirely new politics founded on *"la protection de la nature."*[37]

The projects' reserved flows, the water that passed through the Rhône's original riverbed, was one of the major points of contention from the outset. Local groups, such as the Comité de Défense du Rhône Savoyard-Bugiste, recommended higher reserved flows than what CNR officials proposed. Even some organizations that supported the CNR's projects hoped to increase the reserved flows.[38] Over the next few years, they made even more exacting demands for the CNR's later projects.[39] The Rhône's fish populations in particular fueled many of the concerns over the projects' reserved flows, whether from individual fishers and locals or anglers' organizations such as the Groupement des Pêcheurs Sportifs. Some fishers opposed Brégnier-Cordon because it would destroy several *lônes,* a specifically Rhodanian term for the small, temporary islands created by the river's meanders that created especially rich fish habitat. They maintained that these *lônes* helped make the Rhône's ecology valuable and ultimately unique.[40] Numerous letters to the CNR merged discussions of reserved flow, fish, and the projects' overall environmental "effects," reflecting the various connections among these issues.[41]

Finally, many members of these organizations, especially those affiliated with FRAPNA, articulated romantic arguments to challenge the CNR's plans for the upper Rhône. Some highlighted the beauty of the river, arguing it would be lost forever if the state permitted the CNR to proceed. Others invoked the notion of a last, wild river, even comparing the Rhône to the Amazon.[42] In 1979, Robert Hainard, an esteemed artist from the region who had published a series of 150 engravings of the river entitled "When the Rhône Ran Free," sent an impassioned letter to Brégnier-Cordon's DUP inquiry commission. On behalf of FRAPNA's chapter in the department of the Isère, Hainard expressed his "desolation at the possibility that the magnificent landscape of the Rhône would be destroyed by a dam."[43] Other activists proclaimed dramatically that "the upper Rhône at Sault-Brénaz is in danger of dying," rallying a romantic cry to justify their "defense of the environment" and to ensure the "protection of nature."[44]

As environmentalism gained both strength and legitimacy locally and nationally, CNR officials began to cast the agency's latest round of projects as environmentally friendly—although cynics at the time might have simply viewed it as greenwashing. Around 1980, the CNR's elites began to adopt selectively environmentalist discourse to reframe their agency's projects. Criticizing *Le progrès*'s coverage of Brégnier-Cordon, CNR General Director Claude Gemaehling maintained that his agency had in fact protected the environment, or at least the part of it that "merited" protection. Similarly, in 1985, CNR officials insisted to the prefects of the Isère and Ain, their longtime opponents, that the agency pursued "an active policy of nature protection in diverse sectors."[45] Mounting pressure from environmentalists and other activists may have pushed CNR officials to adopt such rhetorical strategies to improve the odds of their projects being built.

The greening of the French state and society may have posed challenges to the CNR's program on the upper Rhône, but significant changes in France's energy production patterns during the late 1970s and early 1980s also worked against the hydroelectric development of the river's remaining reach. The agency's leaders and their supporters had mobilized the oil crises to legitimate new hydroelectric projects, just as these crises had served those committed to expanding nuclear power. In time, however, "white coal" composed an ever diminishing share of the country's energy supplies, particularly as more and more nuclear reactors came online following the implementation of the Messmer Plan. By the 1980s, these reactors produced about three-fourths of France's electricity. Furthermore, the country has become a net exporter of electricity to the newly integrated European network.[46] During the 1970s and 1980s, cultural values may have shifted and the French state may have become greener, but there were also fewer incentives to build costly new hydroelectric facilities. With adequate electricity and much of the river already harnessed for these purposes, the country's energy context in the mid-1980s called into question the continued hydroelectric development of the upper Rhône.[47]

To Build or Not to Build

Given the new institutional and political context of development during the 1970s and 1980s, progress slowed on the CNR's first two projects, Chautagne and Belley. Recall that the CNR had begun preliminary studies in 1976, but the Ministry of the Environment twice rejected the agency's dossiers due to inadequate assessment of the projects' environmental impacts. Facing an increasingly divided state and mounting protests, CNR president Moulins wrote the minister of industry in April 1978 pleading to end "this real and incomprehensible impasse" blocking the projects' approval.[48] The delays over Chautagne and Belley would be only the first of many worsening "impasses."

Moulin's letter was also the first of numerous appeals by CNR leaders to the Ministry of Industry for assistance. The agency's officials and Ministry representatives exchanged frequent correspondence in an attempt to get the agency's projects out of interministerial deadlock. In June 1978, one state representative, M. Legrand, lamented the demands the

Ministry of the Environment had placed on the CNR, including the Ministry's assertion that the agency had failed to supply adequate information about the environmental impacts of Chautagne, Belley, and Brégnier-Cordon. Although Legrand sympathized with the CNR, he told the agency's administrators they could not simply discount the Ministry's concerns; the CNR would have to gain its support if the projects were going to move forward. Legrand even warned the CNR's leaders that their refusal to comply with the Ministry's demands might jeopardize the projects' DUPs and therefore the projects themselves. If the state did not approve Chautagne and Belley before the December 1978 deadline, the prolonged inquiry process would have to start all over again. It is unclear whether the CNR's leaders heeded Legrand's advice, but the Ministry of the Environment did informally approve the projects in July 1978. However, it did so with a series of reservations, including the volume of their reserved flows. Finally, at the eleventh hour, Chautagne and Belley received official DUP status in late November 1978, just days before the deadline.[49] The contentious DUP process for Chautagne and Belley did not bode well for the CNR's three remaining projects.

Indeed, debates had already delayed Brégnier-Cordon, the CNR's third project, especially once the EIS law boosted the Ministry of the Environment's authority. In December 1978, it rejected the CNR's studies for Brégnier-Cordon and ordered that the agency complete a new EIS. Once again, CNR officials wrote to their allies in the Ministry of Industry, hoping they could expedite approval of the CNR's last three projects, in particular helping to resolve a number of "delicate" questions with the Ministry of the Environment. These persistent disagreements between CNR and Ministry of the Environment officials suggest that the Ministry had more power because it oversaw the mandated EIS. Facing yet another environmental review, Brégnier-Cordon was already far behind Moulin's idealistic schedule of one project per year, although it finally received a DUP in March 1980.[50]

Although this hard-won decree had been issued, the CNR's leaders began to express concerns that a DUP might no longer guarantee a project's actual completion. That July, CNR official J. Lecornu warned the agency's general director of "the political turn" ongoing debates over Brégnier-Cordon had taken, attaching a lengthy report on the political implications of canceling the project for environmental reasons.[51] That CNR officials had essentially developed a policy paper on the possible cancellation of Brégnier-Cordon despite having received the project's

DUP suggests that they foresaw waning, if not rapidly diminishing, support for their program. Just as CNR leaders had perceived the "Affaire Bresson" as a potentially threatening precedent during their postwar groundwater debates, a cancellation at Brégnier-Cordon appeared to place the agency's future at risk. Their anxieties for Brégnier-Cordon eventually proved groundless, and the project was completed. However, the growing apprehension about the future of the agency and its remaining projects was well founded. Only six years later, Loyettes failed to move beyond ambition and blueprint.

Sault-Brénaz and the "Failure" of Loyettes

Recounting the histories of the CNR's final two projects is difficult, especially for Loyettes. The agency's documents pertaining to the project have not been centralized, compiled, or organized. Perhaps the relative silence of the CNR's archives on the history of Loyettes reflects the agency's attempt to forget a project that most of the agency's administrators and employees still regard as a failure. In contrast, environmentalists and other critics viewed its very nonexistence as a resounding success.

Debates over Chautagne, Belley, and Brégnier-Cordon revealed mounting pressure against the continued development of the Rhône, but the CNR's final two projects opened the floodgates of intense, explicit opposition. In December 1980, the vocal mayor of Sault-Brénaz described his community's concerns in a letter to the French president's chief of staff. Other politicians, residents, and environmentalists also began turning to Paris to lobby their case, a strategy that replicated the approach of some locals in the groundwater controversy of the late 1940s and 1950s and thereby suggests that allies within state bureaucracy had begun to emerge.[52] By January 1982, the CNR reported that environmental organizations had launched a "vigorous" campaign against its remaining projects. As one perceptive CNR official observed that month, "the major problem now facing the two projects located at the confluence of the Ain [Sault-Brénaz and Loyettes] is the convergence of two negative opinions: farmers and environmentalists."[53] According to this author, political confluences seemed to be working against the CNR and its proposed projects at the intersection of the Ain and Rhône rivers.

But views of the two proposed projects had also started to diverge. A vote at the Commission Départementale des Sites, Perspectives, et Paysages in January 1982 encapsulated this trend. That month, the thirteen voting members of the commission cast their ballots for or against the CNR's two final projects, which were considered separately. This turned out to be quite important. In the end, Sault-Brénaz received ten favorable votes and three opposed. Loyettes met with a more splintered judgment: with one member abstaining, five favored the project, and seven voted against it. On the basis of these results, the commission decreed an official opinion backing Sault-Brénaz and abandoning Loyettes.[54] This split vote portended the eventual fate of the two projects.

Tensions mounted in February and March, as locals and environmentalists stepped up their protests and newspaper coverage of the growing controversy expanded. Members of CODERA began to circulate a petition, hoping to gather twenty-five thousand signatures and submit this demonstration of widespread opposition to France's prime minister and minister of industry. Unfortunately, no evidence remains as to whether or not the group met its goal. Nonetheless, this petition and other documents hint at the intensity of mobilization, especially during the first quarter of 1982. In March, CNR official R. Bichet characterized this activism as "sometimes rather lively hostility" that "certain communes, especially those situated along the confluence of the Ain [and the Rhône]," displayed toward the CNR.[55] It seemed the agency was swimming against the tide of popular opinion.

This growing opposition to the CNR's final two projects was not limited to locals and activists. Indeed, Minister of the Environment Crépeau handed the CNR an enormous setback when, in a long letter to the agency, he sided more with critics than with the CNR: "In my opinion, this part of the upper Rhône is one of the last wild sections of the great rivers of western Europe and this region has already paid a heavy tribute to energy because of nuclear power plants and a number of hydroelectric projects. Furthermore, a partial amputation of the confluence of the Ain [River] with the Rhône cannot leave me indifferent." After this striking description, Crépeau noted "the difficulty in knowing the appreciable consequences of the Loyettes project on the ecology of the confluence of the Ain with the upper Rhône, a point that is obviously an essential element in the ultimate decision." Perhaps most remarkable was the assertion by Crépeau that opened this chapter: the "national" value of protecting the Rhône exceeded the "energy benefits" that might be generated by the CNR's remaining projects. His declaration, made in March 1982, marked a dramatic shift in the state's priorities since the postwar era and even since the 1970s oil crises. Crépeau proclaimed that the nation should *sacrifice* energy in order to save the Rhône, not the reverse. Moreover, his avowal echoed the split vote of the departmental commission two months earlier in his focus on the greater ecological, hence national, value of an undeveloped Loyettes.[56] While early critics may have complained about the volumes of the CNR's proposed reserved flows or quibbled with other elements of the projects' design, here Crépeau dealt an apparently far-reaching challenge to the continued development of the upper Rhône.

Although Crépeau had not finalized his decision—he planned to consult the Haut Comité de l'Environnement and ask the opinion of the Conseil National à la Protection de la Nature before definitively announcing his judgment on the affair—the general concerns and preliminary view outlined in his letter did not bode well for the CNR. By that spring, state opinion over the remaining development of the upper Rhône had fractured into two positions: Sault-Brénaz was still viewed favorably, as long as the CNR carried out "compensatory measures" for the project's environmental impacts, but state support for Loyettes was rapidly fading. These increasingly diverging views determined both projects' ultimate fates.[57]

Officials from other state agencies also began to oppose construction of Loyettes, providing further evidence of the fragmentation, if not deterioration, of state support for the continued development of the upper Rhône. Responding to the probable environmental impacts of both projects, officials from the Ministry of Urbanism and Housing concluded that Sault-Brénaz "seems acceptable for the environment" as long as measures were taken, they said, to "diminish landscape and ecological impacts. In contrast, the chute at Loyettes is much more unfavorable to the environment, because it involves the destruction of the remarkable natural site that is the Ain-Rhône confluence."[58] It came as no real surprise, then, when Crépeau announced, in November 1982, "the site of the confluence of the Rhône and the Ain will be preserved"; or as one headline declared, "M. Crépeau and the High Committee on the Environment Pronounce Themselves against the CNR's Project Loyettes."[59]

Loyettes's end appeared settled, but CNR officials waged a fierce, if desperate, campaign to combat the idea that their final project was doomed. Throughout 1983 and continuing into 1984, they defended both Sault-Brénaz and Loyettes amid mounting opposition. Although Sault-Brénaz had already received a DUP, FRAPNA members still continued to protest the project, even demanding an annulment of the

DUP.⁶⁰ Meanwhile, in an attempt to revive Loyettes, CNR engineers began to assess alternative designs so they could locate the project upstream of the confluence of the Ain and Rhône, the site of greatest opposition to the project.⁶¹

The agency's political and technical elites also mobilized green rhetoric, presumably in an attempt to bring the projects to fruition. Writing to Crépeau's successor in January 1984, Gemaehling asserted: "For all the projects that the CNR has built over the past few years, the agency has always sought to respect the environment and to research the proper ways to compensate, in the best fashion possible, the costs that have been borne by the natural landscape. The agency continues to work in a direction, especially for the development of the upper Rhône, that responds to the legitimate needs of the river's neighbors. Furthermore, the most recent projects in this region, notably that of Belley, constitute an example of this policy to preserve the environment."⁶² Belley, like Chautagne, had been built according to what the CNR called "integrated" development *(l'aménagement intégré)*. The project's basic features were still modeled on the "Rhône formula" of the earlier projects, but trees and shrubs literally shrouded the new hydroelectric plant's concrete structure in greenery. The plant itself was also smaller, largely due to new technologies that allowed hydroelectric generation through a horizontal rather than vertical turbine; this orientation produced more energy from a shorter drop and therefore required less excavation. Gemaehling asserted that by literally blending nature and technology, the CNR was a green agency, but his claims seemed more concerned with saving the project than with saving the river.⁶³

Meanwhile, the attempt by activists and state officials to preserve the environment of the upper Rhône, or at least its most valued reaches, provoked more conflicts between farmers and environmentalists. Farmers relied on flood control measures. Yet building, operating, and maintaining these structures often transformed the same places that environmental activists hoped to preserve. Struggles over the upper Rhône's floodplains consequently intensified, with farmers accusing the CNR's projects of worsening flooding and locals bemoaning the delays caused by environmentalists' frequent protests and the Ministry of the Environment's voluminous paperwork. As CNR official Lecornu put it, "the environmental impact study [for proposed flood protection measures near the town of Sault-Brénaz] will probably show a more or less intense conflict between 'nature' and 'agriculture,' depending on the technical solutions."[64] Although many locals and environmentalists had together fought against the CNR's new envirotechnical system on the upper Rhône, their own envirotechnical regimes increasingly came into conflict with one another. The two groups did end up "saving" one reach of the river by rejecting the vision of the Rhône that had dominated the three previous decades, but their priorities for the "saved" river were not necessarily easy to reconcile.

Debates over Reserved Flow

The arguments over the volume of water to be allocated to the Rhône's original riverbed encapsulate these shifts in river management policy and development politics between the mid-1970s and 1986. Recall that dividing the river's flow and sending most of it through the diversion canal changed the ecology of the surrounding landscape by altering wider biological and hydrologic processes. In fact, the CNR's new term for the "former" or "dead" Rhône was now the descriptive and less politicized "short-circuited" Rhône *(Rhône court-circuité)*.[65] Given the intense environmental politics from the late 1960s on, it seems likely that CNR officials used this new language to present their proposed envirotechnical system for the upper river in a more neutral way, especially amid growing opposition to their agency's projects. Indeed, the amount of water to be allotted to the projects' reserved flows proved contentious even during preliminary discussions of the CNR's first two projects, Chautagne and Belley. The agency's officials disagreed with locals, environmentalists, fishers' associations, and the Ministry of the Environment over how much reserved flow was adequate.

In 1976, CNR officials planned to set aside 10 m^3/s for Belley's reserved flow. This recommendation prompted complaints from locals and town mayors, who, along with fishers and environmentalists, hoped the agency would raise the reserved flow to 50 m^3/s.[66] Officials of the Ministry of the Environment likewise declared that 10 m^3/s "is too weak with regard to the environment." Agency leaders defended their recommendation by comparing the reserved flows on the upper Rhône with those further downstream. As Gemaehling explained, the CNR's proposed reserved flow for Belley was approximately 6 percent of the river's average flow through the upper Rhône, in contrast to only 3–5 percent downstream of Lyon, including at Donzère-Mondragon. Although the river's total volume was much greater through the central and lower Rhône valley, a smaller share of water was "reserved" for the "dead" Rhône there. Proportionately, then, Belley's proposed reserved flow might be even twice that of the CNR's projects downstream. Such comparisons cast the CNR's recommendations for Chautagne and Belley's reserved flows in a favorable light and presumably were intended to garner support from supervisory state agency representatives, whose collaboration was ever more critical in a climate of mounting opposition.

As with the CNR's earlier projects on the central and lower Rhône, the economic implications of the agency's "technical" choices, including the projects' reserved flows, strongly shaped the design of its preferred envirotechnical system. France's economic recession during the 1970s heightened the agency's financial concerns, as did the marginal profitability of its projects along the upper Rhône. Augmenting the reserved flows appeared to threaten further the *rentabilité* of the CNR's projects. As Gemaehling put it, the project's "economical viability . . . will be compromised" by "increasing the reserved flow at Belley from 10 to 50 m^3/s as has been requested at several informational meetings [with the public]."

Given contemporary methods of calculating a project's economic viability, Gemaehling was right. Increasing the projects' reserved flows through the short-circuited Rhône would reduce the amount of water passing through the CNR's diversion canal and therefore the projects' hydroelectric turbines. The CNR's accounting procedures, which depended, in turn, on the EDF's methods of calculation, did not offer a way to quantify, let alone include, the ecological value of a former Rhône characterized by higher flows. It was even difficult to factor in the economic value if tourism, recreation, fishing, or other activities expanded thanks to these increased reserved flows. According to both agencies' equations, hydroelectrically productive water was an asset; environmentally productive water was a liability.[67]

Moreover, Gemaehling asserted that augmenting Chautagne and Belley's reserved flows would fail to produce "a notable improvement for the environment" but "dredging out the river bottom in order to create sufficient reservoirs" would. Reflecting the agency's persistent managerial view of the Rhône and perhaps exposing the insincerity of its recent environmentalist discourse, Gemaehling maintained that engineers and bulldozers could better sustain the ecology of the upper Rhône than more water could.[68]

The fact that most of Gemaehling's recommendations were ultimately disregarded over the next decade illustrates the extent of political and cultural change in France during the 1970s and early 1980s. Negotiations over the reserved flows on the upper Rhône serve as a useful envirotechnical index for these wider shifts. In short, the CNR's initial proposals incited increasing protest for each of its projects on the upper Rhône. But unlike many earlier criticisms of the CNR and its projects, these objections succeeded in modifying one key feature—volume of reserved flow—of the agency's envirotechnical system for this reach of the river.

In fact, although the CNR's administration publicly rejected a higher volume, the agency's engineers and scientists had already begun to study the potential financial implications of that outcome in internal reports. Gemaehling may have proclaimed that a higher reserved flow was environmentally futile, but the CNR's own experts conducted an in-house assessment of a modulated reserved flow that increased during the summer, a period of greater ecological sensitivity. According to this plan, the CNR's technical elites proposed reserved flows of 10 m³/s for three-fourths of the year, increasing to 20 m³/s at Chautagne and 30 m³/s at Belley during the summer. Subsequent considerations of even higher summer flows of 40 m³/s at Chautagne and 60 m³/s at Belley indicate that agency officials sensed the public's growing opposition over the issue and expected they would have to make additional concessions.[69]

This modulated reserved flow promised both to yield ecological benefits and to exact economic costs, as Gemaehling had feared. While environmental concerns characterized a widening array of groups, economic motives continued to drive the CNR's administration. Financial motives largely defined how CNR officials viewed a modulated reserved flow. For instance, they quickly translated higher reserved flows into the precise reduction in energy generation. Even more important, they lamented the loss of revenue, which might "seriously compromise the economic viability of the project[s]." They concluded that any reserved flow equal to or greater than 50 m³/s would make the projects unprofitable.[70]

One modification to the CNR's envirotechnical system offered a way to potentially reconcile the agency's economic interests with the growing environmental concerns of locals, activists, and the Ministry of the Environment. The agency's engineers proposed building a second hydroelectric plant on the reservoir dam at the head of the short-circuited Rhône in order to harness the energy (and thus financial) potential of the reserved flow.[71] The CNR might not be able to recover all of the "lost" economic potential of the increased reserved flow, but the second plant could produce enough energy to at least ensure that the agency's projects would remain *rentable*.

Although this solution appeared to offer a compromise between competing envirotechnical regimes, CNR officials continued to resist design modifications resulting in higher reserved flows. The agency's second EIS for Chautagne and Belley of March 1977 still proposed a year-round reserved flow of 10 m³/s. The CNR eventually made some concessions, yielding to pressure from locals and the Ministry of the Environment. By that June, the agency agreed to a modulated flow of 10 and 20 m³/s at Chautagne and 10 and 30 m³/s at Belley. But these increases did not satisfy many critics.[72] To the dismay of CNR administrators, the DUP inquiry commission sided with the agency's opponents the following spring. Although the commission granted Belley a favorable opinion, it mandated a minimum reserved flow of 30 m³/s for at least three-fourths of the year.[73] If this standard seemed high, the CNR soon faced even more exacting demands from the Direction de la Protection de la Nature, which called for 150 m³/s.[74] Much to the relief of the CNR's leaders, however, the minister of the environment promised to approve the projects' DUP if the CNR agreed to increase the reserved flow to 10 and 20 m³/s for Chautagne and 20 and 60 m³/s for Belley.[75] Perhaps the threat of such dramatic increases encouraged CNR officials to accept more modest changes.

These preliminary conflicts over the two projects' reserved flows reflected a new relationship among the CNR, the state, and the Rhône. The volumes of the reserved flows were proportionately greater than those of the CNR's projects downstream of Lyon. In addition, the modulated regime mimicked, to a limited extent, variable seasonal flows. Critics had, therefore, forced CNR officials to incorporate environmental concerns within their agency's envirotechnical regime and modify the design of its envirotechnical system. Even more significant changes were on the horizon.

Although the CNR's initial proposal had included Brégnier-Cordon, its separation from the Chautagne and Belley projects by the Ministry of the Environment helps to explain the different fate of its reserved flow. As with Chautagne and Belley, CNR officials initially proposed 10 m³/s year round but were widely criticized for this recommendation.[76] Internally, CNR engineers again analyzed the financial costs associated with higher reserved flows, apparently concerned their position was increasingly untenable. By December 1977, CNR officials agreed to "defend the interests of aquaculture [artificial propagation through fish breeding] and improve the environment" by raising Brégnier-Cordon's reserved flow to 20 m^3/s throughout the year.[77] However, when CNR engineers calculated the economic costs of even higher reserved flows in internal reports, they hinted that the agency might have to make additional concessions. Notably, they included volumes that far surpassed the earlier ranges considered at Chautagne and Belley. In two reports from 1978, they even studied the possibility of a year-round reserved flow of 100 m^3/s, ten times the CNR's initial recommendation.[78]

In light of this daunting option, the CNR's increase to 20 m^3/s at Brégnier-Cordon appeared modest. Moreover, it included a potentially advantageous caveat. The agency's officials argued that Brégnier-Cordon's reserved flow should be measured downstream of the confluence of the Rhône and Le Guiers, a large stream. By measuring the reserved flow here, they could subtract that stream's average flow, later estimated at 12 m^3/s, from the reserved flow. As a result, the CNR could boost Brégnier-Cordon's hydroelectric productivity by decreasing the amount of water flowing through the short-circuited Rhône to just 8 m^3/s.[79]

Because they could make this adjustment, to the advantage of their envirotechnical system, CNR officials seem to have been more amenable to raising Brégnier-Cordon's reserved flow. When the CNR submitted its EIS for the project in October 1978, agency experts proposed a reserved flow that fluctuated seasonally between 25 and 40 m^3/s. They had thus adopted a modulated regime and set the reserved flow even higher than at Chautagne and Belley; but they were careful to specify that these rates would be measured downstream of Le Guiers. Brégnier-Cordon's DUP made these figures official in April 1979.[80]

Yet, just as CNR officials had feared, the approval of the DUP no longer appeared to guarantee a project's realization or its completion according to preliminary technical specifications. Although Brégnier-Cordon had already received a DUP, both activists and state representatives continued to pressure the CNR to increase the project's reserved flow. Fishers' associations advocated 150 m^3/s and asserted that this volume of water was needed year round. State representatives also stepped up their pressure on the CNR; one boldly declared that "the environment wants 150 m^3/s."[81]

The agency's officials attempted not only to dismiss but also to undermine these demands. In April 1980, Gemaehling argued that high year-round flows did not actually make ecological sense because they did not mimic the Rhône's seasonal variations: they were too high too consistently for too much of the year. Gemaehling's concerns about the environmental integrity of the CNR's projects were certainly valid. His portrayal of the reserved flow as a near constant, however, failed to acknowledge that the Rhône's flow did vary and that, in turn, the reserved flow would inevitably also vary. A higher reserved flow would narrow the range of this variation, but its volume would actually be much closer to the river's average levels.[82] In any event, Gemaehling's reasoning may have been purely strategic, since he also emphasized the agency's financial motives, arguing that a reserved flow of 150 m^3/s would make Brégnier-Cordon economically untenable.[83]

Historically, the CNR's economic arguments had proved persuasive: a project's "technical" features could not undermine its economic feasibility. Yet by the early 1980s, Gemaehling's contention that a higher reserved flow would threaten the project's *rentabilité* no longer held water, so to speak. Although CNR officials claimed that the *rentabilité* of Brégnier-Cordon would be endangered if they were forced to accept a higher reserved flow, they faced a real dilemma. The agency would have to make concessions if it wanted to assure the project's continued viability *environmentally;* they were not going to receive approval for the project without these modifications, which in turn might reduce, if not ruin, the project's *economic* feasibility. Presumably recognizing this thorny situation, Gemaehling told the CNR's president in June 1980 that a modulated flow of 30 and 90 m^3/s—figures somewhere between the two extremes—might offer a way "to get out of this impasse."[84]

Although leaders of fishers' associations eventually backed down from their initial demands and admitted they would be satisfied with a year-round reserved flow of 70 m^3/s, the Ministry of the Environment continued to call for higher levels. Its new standard was 100 m^3/s.[85] To ensure the future of Brégnier-Cordon while reducing their agency's economic losses, CNR officials proposed building a second hydroelectric plant, which would "minimize the economic impact" of the higher reserved flow by recuperating some of the "lost energy."[86] Finally, after lengthy negotiations, these groups agreed on a three-season modulated regime: 80 m^3/s in November through March; 100 m^3/s in April, May, September, and October; and 150 m^3/s in June through August. Disputes over the site of measurement continued but were eventually resolved in the CNR's favor. Overall, the final figures at Brégnier-Cordon showed that CNR officials had agreed to a substantial increase in the reserved flow in less than a decade.[87] The "technical" design of the agency's third project thus demonstrated the growing importance and legitimacy of environmental concerns, which the CNR had to factor into its envirotechnical system even when they seemed to threaten its original envirotechnical regime.

If Brégnier-Cordon marked a turning point in the history of the development of the upper Rhône, then Sault-Brénaz denoted the transition between compromise and cancellation. The early part of the reserved flow debate at Sault-Brénaz echoed that of its predecessors. As with the CNR's first three projects, CNR officials agreed to increases from 10 to 20 m^3/s, but locals and environmentalists remained unsatisfied. At the end of 1981, activists plastered posters around the town of Sault-Brénaz, demanding that 150 m^3/s of water remain in the former Rhône. Their stipulation inspired a play-on-words headline in *Le Dauphiné libéré*: "*Débat sans réserve pour un débit réservé*" ("A no-holds-barred debate over a reserved flow").[88] Given the possible concessions at Brégnier-Cordon then still under debate, this demand no longer seemed extraordinary. After extensive discussion, in 1983, the parties tentatively agreed to a modulated regime of 20 and 60 m^3/s.[89] Just as at Brégnier-Cordon, the CNR planned a second hydroelectric plant to generate energy from its increased reserved flow.[90]

The debate over the reserved flow at Sault-Brénaz differed from its predecessors, however, in one important way. The nuclear development of the river, partly spurred by the energy crises, had altered the envirotechnical context of Sault-Brénaz's reserved flow. The fact that Bugey, a nuclear power plant from the 1960s, was located on this stretch of the Rhône's original riverbed meant that significant drops in the volume of the short-circuited Rhône could considerably increase the temperature of the remaining water. This could have severe consequences for the nuclear reactor, which needed a steady volume of water within a specific temperature range for its cooling processes. Members of the CNR's technical committee acknowledged that the thermal pollution of the Rhône downstream of Sault-Brénaz could even threaten the safe operation of Bugey. Locals expressed their own concerns about the compatibility of the CNR's proposed projects with "existing installations."[91] Consequently, in 1981, officials from the CNR and EDF began to study the river's thermic regime and analyzed the probable effects of Sault-Brénaz on the temperature of the Rhône.[92] In short, the CNR's proposed envirotechnical system for its new project appeared to threaten existing envirotechnical systems oriented to the production of nuclear power.

While state and local pressure had pushed the CNR's leaders to raise the reserved flows for its first three projects, apprehension about nuclear safety was the concern that ultimately spurred the increase at Sault-Brénaz. In 1983, CNR officials had already agreed to a modulated regime of 20 and 60 m³/s, but by the fall of 1986, they began to consider even higher flows due to the project's proximity to the nuclear facility. Bichet determined that Bugey required at least 150 m³/s of water in the short-circuited Rhône at all times, a minimum close to the average low-water flow of 180 m³/s in the upper Rhône. The CNR formalized this policy before the end of the year.[93] In less than a decade, then, CNR officials moved from a proposed reserved flow of 10 m³/s at Chautagne to 150 m³/s at Sault-Brénaz (though it must be allowed that the Sault-Brénaz concessions were due less to political pressure than to the upper Rhône's prior nuclearization). The reserved flow of Loyettes would prove even more striking: there would be none.

Unfortunately, the precise reasons for the cancellation of Loyettes have faded into the murk of archives not yet organized and the memories of incensed CNR administrators. Limited evidence from both agency and public archives, newspapers, and conversations with CNR employees does yield some tentative conclusions. Rather than opposing both projects, state officials, locals, and environmentalists focused their critiques on the Loyettes project. This decision may have been pragmatic, based on the assumption that it was unlikely that both projects would be canceled, and possibly reflecting a higher value placed on the ecology of Loyettes.

Envirotechnical regimes are not static, and their members and structures of power are historically contingent. As internal state reforms and growing opposition from various groups challenged the continued development of the Rhône, between the early 1970s and the mid-1980s, environmental issues complicated the state's development agenda from within and without. In particular, the creation of the Ministry of the Environment and passage of environmental laws meant the state came to have its own environmental agendas. As these changes upset existing power relations that had governed the state's environmental management practices for several decades, new and diverse groups began to view the Rhône in new ways, and no longer simply as a "mine" for energy and the means of national reconstruction and modernization. Environmentalists, scientists, intellectuals, outdoor enthusiasts, and locals forged new identities for the river and its future. These groups hoped to manage the Rhône according to new aims and to create new envirotechnical systems in ways that might forward their respective goals (even as those goals increasingly conflicted with one another). By the mid-1980s, outside critics and environmental constituencies within the state had forced the CNR to modify some elements of its envirotechnical system.

Over time, the coalition that had backed the CNR's projects for over three decades became so fractured it could no longer sustain the postwar era's regime for the Rhône's development. There simply was no longer any consensus over what "development" meant by the mid-1980s. Some constituencies continued to advocate projects like those of the postwar era. Others took a different approach to development, grounded in different understandings and interpretations of nonhuman nature and a different notion of the French nation. In their eyes, the Rhône needed to become a new envirotechnical system—one that would reconcile historic and more recent approaches to river management. By the late twentieth century, this hybrid of old and new, historic and recent envirotechnical systems, ultimately rendered the Rhône a light-green river.

CONCLUSION: LEGACIES OF THE RHÔNE

> Water is part of the common patrimony of the nation.
> —*Article 1 of the Water Law of January 3, 1992*

The vast archives of the CNR are filled with memos, reports, and other documents written by the agency's engineers and administrators over the past seventy-five years. Less visible but still present are the views of critics questioning the agency's activities. Also collected are the glossy brochures about the agency and its projects published by its public relations office. One from the 1990s particularly stands out: a single-page flyer that maps the Rhône's course through France. Remarkably, it depicts an almost exclusively natural river. An extensive key lists ten different kinds of sites along the Rhône, from natural reserves and ecological biotopes to hunting preserves and the ambiguous "other sites of ecological interest." Photographs of birds, wetlands, and other "natural" images of the valley are placed around the map. The flyer does picture lone fishermen in small boats, as well as one of the CNR's projects: the hydroelectric plant at Belley along the upper Rhône—notably, one of two examples of the agency's "integrated" approach to development, with trees and shrubs planted on top of the concrete structure. The title? "A New Goal: Uniting Technology and Nature" ("Une ambition nouvelle: Conjuguer Technique et Nature").[1]

While it is difficult to ascertain the intent of the flyer's creators, it does send several messages. By the late twentieth century, the CNR was framing its approach to the Rhône's management as a "uniting" of technology and nature. Whether or not this "new goal" actually describes the CNR's recent practices is a separate issue. But if we take seriously the rhetorical claim, the title clearly illustrates a conscious attempt by CNR leaders to strategically define—or rather, *redefine*—the perceived relationship between nature and technology on the Rhône.

For instance, it may be significant that *technique* is named first, even though conventional examples of technology are almost absent. They have been rendered nearly invisible in this representation of the Rhône, and the natural dominates both the textual and the visual narrative. By highlighting the CNR's environmental commitments, this narrative offers a selective portrayal of the agency's activities since 1934. Critics might claim that the greening of the agency and its envirotechnical regime that the flyer presents as voluntary occurred only after activists and state laws demanded changes, and after decades of large-scale development had already substantially transformed the Rhône.

The declaration of "A *New* Goal: Uniting Technology and Nature" also raises questions about their relationship in the past. The brochure's title makes a clean break between past and present, what historians and other scholars have called "rupture-talk." But was it a rupture? The landscape, and even CNR officials' own statements, appeared to belie this avowal of change. Agency officials had physically and discursively "united" technology and nature since the CNR's founding decades earlier—from physically building hydroelectric plants that incorporated the river's flow to constructing, both literally and metaphorically, the former Rhône and nuclear reactor hydrology. In addition, proponents of industrialized river management had repeatedly conflated nature and technology in order to justify development, especially in the postwar era. These examples demonstrate the material and cultural uniting of nature and technology long before this CNR flyer proclaimed it so. In other words, the CNR's rupture-talk was just that: rupture-*talk,* a rhetorical attempt to distinguish the present from the past.[2]

Furthermore, the idea of uniting technology and nature, whether or not it actually was new, assumed their initial separation. But the historical and contemporary landscapes of the Rhône indicated far more porous boundaries between both these concepts and their material form. Hydroelectric plants, nuclear reactors, irrigation networks, diversion canals, the "dead Rhône," even the river itself—whatever the river *was,* exactly—revealed earlier blendings of nature and technology. In short, *technique* and nature had been repeatedly brought together physically and discursively well before the late twentieth century.

Consider, for instance, the dramatic events that took place at the

Tricastin Nuclear Power Center along the central Rhône in the summer of 2008. On July 7, around ten o'clock in the evening, a storage tank retention basin at a radioactive effluent treatment plant overflowed. An estimated 6.25 cubic meters of uranium-contaminated effluent made its way, via the site's rainwater evacuation network, into the Gaffière and the Lauzon and eventually the Rhône. In some ways, this spill was not entirely abnormal: the facility regularly released treated effluent into Donzère-Mondragon's diversion canal. But that happened only after precise treatment and discharge procedures, which were bypassed in the July 7 accident.

On the following day, officials from the government's nuclear safety commission classified the accident as "level 1," the lowest of the commission's seven-tiered ranking of nuclear risk. However, on July 9, the Commission de Recherche et d'Information Indépendantes sur la Radioactivité (CRIIRAD), a French nongovernmental organization founded after Chernobyl, declared that "the spill surpassed by a factor of 27 (!) [sic] the maximum *annual* limit fixed by interministerial decree (article 18)." In fact, the spill had "surpassed by a factor of 161 (!) [sic] the maximum *monthly* limit fixed by [that] decree."[3]

These figures alone might have alarmed critics of the atomic age, not to mention those living near the enormous installation, but groundwater testing soon revealed even higher levels of radioactivity than expected. As a result, officials began investigating the entire site in an attempt to account for this disturbing discrepancy. On July 15, CRIIRAD announced that a huge mound of debris, four meters high and one hundred meters long, had been discovered near Donzère-Mondragon's diversion canal during the investigation. This mountain of approximately fifteen thousand cubic meters of soil contained more than 760 tons of radioactive waste from Pierrelatte, the now defunct uranium enrichment plant. In their report, CRIIRAD officials noted that the mound was highly susceptible to erosion and leaching into area groundwater; moreover, it was less than two hundred meters from the canal. The July 7 spill, the proximate cause of radioactive contamination of the central Rhône, ended up uncovering, then, a much older and potentially much more serious issue.[4]

These events took place precisely because large-scale "technological" systems and the Rhône had been "united" long before the CNR's pronouncement. In fact, the flows of radioactive effluent traced physical connections forged between the treatment plant and the remade Rhône

Conclusion: Legacies of the Rhône | 243

decades earlier. In contrast, the enormous mound containing—or rather, not containing—radioactive waste may not have been an official part of the nuclear facility. Nevertheless, in both these instances of pollution, the movement of radioactive elements between "technological" and "ecological" systems exposed their permeable boundaries.

Scholars who frame such events as the environmental impacts of technology, whether planned outcomes, unintended consequences, or "normal accidents" (to borrow Charles Perrow's phrase), risk neglecting the systemic connections forged, both purposefully and incidentally, between (in this case) nuclear technologies and hydrologic systems.[5] The remade Rhône was implicated in nuclear technologies; conversely, technologies of the atomic age became literally part of the hydrologic order of the central Rhône. The contamination at Tricastin in July 2008 illustrated the way the river had been drawn into complex, multiple, and competing envirotechnical systems—to the point that the Rhône itself had become one—long before the CNR's claim that it was newly uniting technology and nature.

This book has examined the dynamics of three inseparable confluences. Most literally, it has traced the French Rhône's shifting hydrology through time and space, focusing on the river's transformations since World War II. Environmental changes were, in fact, inextricably connected to political and cultural events in twentieth-century France, including reconstruction, modernization, decolonization, and European integration. French politics and culture thus both shaped and were shaped by the entangled processes of technological development and environmental management; for example, contemporaries made Donzère-Mondragon an icon of postwar *grandeur*, while ideas about nature underlay shifting conceptions of the French nation, whether after World War II or in the late twentieth century. This book has also explored the convergence and divergence of, as the CNR brochure put it, *technique et nature*, concepts often set in opposition to one another but which historical actors frequently merged, both physically and discursively, in the repeated remaking of the Rhône after 1945. Finally, to tell this multilayered story, I have drawn on central themes and methodological approaches from STS, environmental history, and the history of modern France. Confluences indeed.

As I hope the metaphor itself suggests, the confluence of these fields enriches each one while creating new interdisciplinary ground. For environmental historians, STS offers several valuable insights and analytic tools. Opening the "black box" of technology and exposing the contested process of technological development presents rich opportunities for investigating environmental historians' primary concern: the historical interactions between human and nonhuman nature. At the same time, a constructivist framework highlights the ways in which technologies are both artifacts and mediators of human-natural relations. Environmental historians might be well served, then, by focusing more attention on how technological artifacts and systems both shape and are shaped by interactions with the natural world. In short, a fine-grained analysis of technology and the process of technological development that uses tools from the history of technology to investigate the central questions of environmental history may extend the valuable insights environmental historians have to offer in fresh ways.

Yet, as the confluence metaphor again implies, disciplinary contributions are not one-sided, and perspectives from environmental history can enrich the social and historical studies of science and technology as well. The field of environmental history presupposes that nonhuman nature not only merits a place in human history but also helps account for that history. While the cultural turn has rightly refined environmental history over the past two decades, most environmental historians continue to insist that nature, apart from cultural and historical conceptions of it, shapes historical change. Although constructivism remains a point of tension between the fields, environmental historians have in fact helped expose some of the limits of a social constructivist framework, which a number of STS scholars have critiqued; many of the latter are moving instead toward a model of co-production or co-construction. In other words, insights from environmental history highlight the role of nonhuman nature in mediating scientific and technological change. The most persuasive arguments frame natural resources and ecological processes as neither unproblematic concepts nor deterministic forces but rather culturally shaped yet simultaneously partly autonomous, material constraints. These constraints limit but do not wholly determine the paths of either technology or society. Insights from environmental history, then, expose the ways objects of analysis within science and technology studies are at once sociotechnical *and* envirotechnical.

As the term itself implies, *envirotechnical analysis* seeks to build on the strengths and contributions of both fields. This book has focused on the dynamic envirotechnical character of France's Rhône River since World War II. While the envirotechnical dimensions of landscapes such as rivers, forests, and farms may be compelling, perhaps even self-evident, because they reveal visually the ways humans have managed natural resources and ecosystems through technological artifacts and systems, envirotechnical analysis does not need to be limited to such obvious cases. Rather, it can be extended to a whole host of (enviro)technologies—from the automobile and computer to irrigation networks and bioweapons—through all stages of their life cycles, including design, development, use, and disposal. It is precisely the fact that the "nature" of such diverse "technological" objects and systems is less readily visible that helps foster dichotomous thinking about nature and technology, whether by historical actors or analysts.[6]

In fact, technological artifacts and systems may end up obscuring human links to, even dependencies on, what is often framed as the natural world. Technological mediation might mask connections between seemingly disparate elements such as, say, twenty-first-century French citizens and local watersheds (via tap water), Provençal soil (wine), and African uranium (nuclear-generated electricity). For actors and analysts alike, the envirotechnical has been and remains too easily reduced to the technological, a move with important implications both politically and ecologically. Gabrielle Hecht has described how French advocates of nuclear power have reframed reactors' technopolitics as "only" technology, veiling the overt political dimensions of nuclear technologies—itself a profoundly political move.[7] Similarly, the reduction of envirotechnical artifacts and systems to "only" technologies obscures humans' profound interactions with nonhuman nature.[8] Within STS the "haunting ghost" of technological determinism and the field's understandable emphasis on sociopolitical processes both embedded within and constituted through technological development generally account for the less visible "nature" of technology.[9] Yet in the process, scholars have essentially black-boxed the environment. In short, illuminating the sociotechnical has ended up masking the envirotechnical.

These two approaches are, however, not mutually exclusive. As I argued in the Introduction, envirotechnical analysis cannot and should not ignore social processes—or, in a word, power. At the same time, sociotechnical analysis is enriched by incorporating nonhuman nature, both as a cultural construct and as a material constraint not wholly reducible to either cultural perceptions or human shaping. The concept of the envirotechnical obviously foregrounds environment and technology, but social processes, politics, and power are all embedded within the development and use of envirotechnical systems and are perhaps best encapsulated by the concept of envirotechnical regimes. An envirotechnical perspective, then, still leaves room for central issues within technology studies, including sociotechnical and technopolitical analyses. Foregrounding the envirotechnical simply, yet significantly, aims to reassert a place for nonhuman nature, both historically and conceptually, within technology studies while avoiding the realism or determinism of early work in the field.

Much of my discussion here has focused on seemingly abstract academic debates, but these issues have wider implications and potential applications. Take, for instance, recent debates about energy policy. Growing concern over global climate change has spurred a drive to reduce, if not replace, carbon-based fuels. Politicians, scientists, activists, and the media have analyzed numerous "renewable" energy sources, including biofuels, wind, and hydroelectricity. It is easy for advocates of so-called alternative energy to frame these sources as unproblematic, "natural" forms of energy that will solve global warming, sustainable energy development, and rising demands for energy worldwide simultaneously. Yet naturalizing alternative energy technologies is a political move, one with momentous consequences for the species and ecosystems affected by the development of these energy sources. Moreover, a "technofix" mentality essentially underlies these arguments—except that nature offers the purported fix.[10] In reality, *both* carbon-based fuels and alternative energy are envirotechnical systems, a point that has been largely overlooked in recent debates. In contrast, an envirotechnical perspective situates these various forms of energy squarely within both "ecological" and "technological" systems, thereby facilitating comparisons among these often difficult choices.[11]

While envirotechnical analysis therefore problematizes, for instance, ethanol as an easy solution to global climate change, this example also illustrates one of the pitfalls of hybridity as an analytic tool. Hybridity has become influential in many disciplines, including environmental history and STS, and it is obviously a premise of envirotechnical analysis itself; but hybridity—whether of environment and technology, technology and culture, or the like—risks flattening, even concealing, major differences *among* hybrids, even if one focuses on a single pair alone. Blurring conceptual boundaries may avoid certain problems while intro-

Conclusion: Legacies of the Rhône | 247

ducing others. Without reflective use, hybridity can lose some of its analytic utility: it tends to neglect specificities (in this case, comparisons and contrasts among envirotechnical systems) and mask change over time. More illuminating is an examination of specific envirotechnical systems in particular times and places. Such close analysis moves beyond the basic trope of hybridity to explore its particularities.

These analytic concerns about nature, technology, and their relationship are not, then, simply academic exercises. For one, they continue to have relevance in the Rhône valley today. Over the past two decades, the French government, the CNR, locals, activists, and other groups have sought to foster a new relationship between *technique et nature,* one that enhances changes begun in the 1970s and 1980s. In particular, improving environmental quality and undertaking ecological restoration have defined the recent management of the Rhône's hydrologic basin by the Bassin Rhône-Méditerranée-Corse (Rhône-Mediterreanean-Corsica Basin; RMC), one of six new "water agencies" *(Agences de l'Eau),* basically river basin-based water management agencies, created by the government in 1964. In 1991, the RMC's first "Plan d'action Rhône" outlined specific objectives such as eliminating pollution sources, increasing the reserved flow in sections of the Rhône, and restoring "high ecological quality over the entire river." These measures aimed to rehabilitate the "former Rhône," reestablish fish migration, and contribute to the overall "restoration" of the river.[12] They revealed not only the growing influence of ecological concerns in recent river management policy but also shifts in the power dynamics among envirotechnical regimes on the Rhône. In this respect, the Plan d'action Rhône, built on earlier decisions such as the cancellation of Loyettes, took them a step further, moving beyond abandonment of projects then under consideration to actual restoration, presumably to correct damages caused by earlier projects.

The path-breaking water law of 1992 solidified and even nationalized this approach. Its first article declared, "Water is part of the common patrimony of the nation." As such, "its protection, improvement, and the development of a useable resource, all the while respecting natural equilibriums, form part of the general interest."[13] This was a sweeping pronouncement, one that stressed the public and national value of water. Of course, it was unclear what "useable" meant or how protection, improvement, *and* development could be achieved while still "respecting natural equilibriums." Despite these ambiguities, this law, combined with the Plan d'action Rhône, helped lay the foundation for the river's restoration. By the early 1990s, then, the Rhône had become reinterpreted as essential patrimony, which echoed Minister of the Environment Crépeau's declaration at Loyettes almost a decade earlier. The future of water, the Rhône, and France remained connected, but they had been reforged in significant new ways.

With restoration serving as the new framework for the Rhône's management, state elites began to formulate plans to realize this overarching objective. In 1993, they developed the Charte d'Environnement du Rhône, drawing on the political authority that *charte* implied.[14] And in 1998, just one year after Dominique Voynet, newly appointed head of the newly merged Ministère de l'Aménagement du Territoire et de l'Environnement, cancelled the modernization of the Rhine-Rhône liaison, historically adversarial state ministries together asked the RMC to propose a ten-year program to restore the Rhône, allotting 10 million francs annually to pay for it. Restoring hydrologic and environmental quality was the program's primary goal. Representatives from the RMC hoped not only to prevent pollution and improve water quality along the entire French Rhône but also to identify stretches that might be restored to "a living and flowing river." In the end, RMC leaders decided to focus on the rehabilitation of the short-circuited Rhône and other fish-related issues, essentially using the river's fish as an indicator species for the health of the Rhône watershed.[15]

The CNR played a central role in the formulation and execution of this ten-year restoration program. As part of that program, government officials even declared the CNR the river's "environmental manager," something of an irony. After all, the CNR's administrators, experts, and workers had by then spent over a half century remaking the Rhône and had arguably already acted as the river's primary, though not sole, environmental manager for decades. Moreover, the goals and politics of restoration demanded that they modify the very envirotechnical systems their own predecessors had built. In one way or another, CNR officials have continued to play a central role in overseeing the Rhône through ever-shifting objectives, just as the atomic age, regional development, and agriculture altered the agency's envirotechnical system and regime in previous decades.

The CNR began to undertake projects based on these principles in the 1990s. The restoration of the short-circuited Rhône at Pierre-Bénite, just

Conclusion: Legacies of the Rhône

south of Lyon, was one of the agency's first projects under the new regime. There, CNR engineers attempted to "favor a return to the functioning of the Rhône system as close as possible to that which existed before development." In addition, they proposed raising the reserved flows of other projects, including Donzère-Mondragon, with the most promise for ecological renewal. The agency's officials also put forth their Plan Migrateurs Rhône-Méditerranée to help meet the agency's new mission to restore migratory fish populations.[16]

On closer examination, the CNR's pivotal role in the restoration of the Rhône is less surprising or ironic than one might suspect. On the one hand, we should not dismiss the shift in outlook or the growing importance of environmental concerns in the Rhône's management. Nor should we diminish the ecological repercussions of these changes. They matter, especially for the nonhuman species living in and near the Rhône. On the other hand, such restoration is not nearly as far-reaching as may first appear. The restoration of the Rhône is, to adopt Michael Bess's phrase, a "light-green" effort.[17] State officials, even from the RMC and Ministère de l'Aménagement du Territoire et de l'Environnnement, have not recommended that the CNR's dams or other large-scale projects be removed. Such a proposal would involve the radical remaking, even undoing, of earlier envirotechnical systems. Instead, "restoration" serves as a compromise between older and newer envirotechnical systems, ultimately seeking to reconcile technology and nature.

The reformist, rather than radical, character of restoration in the Rhône valley in the late twentieth century can be partly explained by the European Union's position on energy. The EU's requirements pushed France to privatize part of its energy sector. After decades of contentious relations with the EDF, the CNR has taken over electricity production and sales from its projects, effectively offering the central government a powerful technopolitical tool for complying with the European Union's neoliberal mandates. As a result, major modifications to the CNR's projects, including a significant reallocation of water to the short-circuited Rhône or perhaps even removal of several hydroelectric plants, would reduce the agency's ability to produce electricity and thus act as a domestic competitor to the EDF. Furthermore, growing concerns over global climate change and "peak oil," the resulting turn toward renewable energy, and revived justification for nuclear power make it even more unlikely that the state and private companies would demolish their extensive hydroelectric and atomic energy infrastructure in the Rhône valley or elsewhere in France. After all, both envirotechnical systems have allowed the country to stay at the cutting edge of international energy policy and technology development, particularly during the turbulent first decade of the twenty-first century.[18]

The limits of restoration are apparent in other ways—especially in other places. Restoration may have become increasingly influential and is therefore reshaping the Rhône once again, but the high modernism that underlay earlier approaches to the river's management has not entirely disappeared, even in western Europe. Witness the ambitious Rhône-Barcelona aqueduct proposal. In 1995, contemporaneous with both the proposed improvement of the Rhine-Rhône liaison and the plan to restore the Rhône, BRL experts and state engineers began to investigate directing some of the Rhône's flow to the parched northeastern coast of Spain. This project would potentially extend the BRL's water transfer network beyond the lower Rhône valley, Languedoc, and even France. The liaison and the aqueduct proposals share several key elements, chief among them an attempt to foster European integration through envirotechnical as well as technopolitical means.[19] Will this project share the fate of Loyettes and the liaison? Or will its proponents successfully lobby for its eventual materialization, especially as water receives growing attention in global politics and policy-making? A heat wave that struck Barcelona in July 2008 may only bolster the position of the aqueduct's advocates.

The high modernist legacy of the Rhône and its twentieth-century transformations is particularly alive and well in the so-called developing world. After all, if by the late 1980s there were no longer any projects to complete on the Rhône, except the fated modernization of the Rhine-Rhône liaison and later those related to the river's light-green restoration, the CNR would have to look elsewhere for work. And so it has done, marketing its expertise in river management to governments and international organizations around the world.[20] Advertising their "domains of competence," CNR officials have highlighted "more than 70 years of experience" with hydrologic studies, flood analysis, and physical modeling, along with other kinds of "expertise" related to water management. They emphasize four areas of special competency: hydroelectric development, management of navigable waterways, improvement and restoration of rivers and streams, and management and monitoring of river ecology. Over the past several decades, the CNR's engineers and technicians have served as "river consultants" for projects in over two dozen countries across Europe, Africa, the Middle East,

Asia, and South America, ranging from Bangladesh's Flood Action Plan 21 and improved navigation between the Nile and the Suez Canal to urban flood management in Bolivia and irrigation canals for India's controversial Narmada project. But the CNR is certainly not alone. The BRL has undertaken similar projects worldwide: it has affiliated offices in Madagascar, Algeria, and Ile de la Réunion, in addition to its headquarters in Nîmes, and it advertises activities in over eighty countries around the globe. In short, the knowledge, tools, and skills that experts acquired by managing and transforming the Rhône over the course of the twentieth century have thus flowed far beyond the river's watershed and even France itself.[21]

While critics within and outside the French state began to reject many of the assumptions that underlay the transformations of the Rhône between 1945 and the 1970s, the involvement of CNR and BRL experts in these recent projects suggests striking ideological and ultimately material continuities accompanying the agencies' expanding geographic focus. Within France, environmentalism and other critiques may have somewhat tempered the imperatives of high modernism, as epitomized by the turn toward "restoration" over the past two decades. But a quite different envirotechnical regime governs abroad. In fact, high modernism has actually had a remarkable second life in the waters of the postcolonial world. This is, then, perhaps the greatest paradox of "restoration," in that the global geographies of capitalism and technoscientific expertise restructured water management, and thus water itself, worldwide. For these reasons, a simple, progressive narrative arc about the Rhône and its remaking since 1945—from large-scale industrial development to ecological restoration—masks not only continuities within France but also the transnational legacies of the river's management.[22]

"A New Goal: Uniting Technology and Nature" may have been, then, a strategic representation of the CNR's management practices in the Rhône valley in the late twentieth century. Yet this construct, however problematic, sums up the history of the Rhône and its transformations since World War II. At the same time, it also describes how people have lived in the Rhône valley for centuries, if not millennia. The post-1945 transformations of the Rhône were therefore both a continuation of and a break from this much longer history of human interactions with the river. This declaration thus makes explicit the envirotechnical "landscape" that was, is, and undoubtedly will be the Rhône.

ABBREVIATIONS

AP1	L'aménagement du Rhône, Suivi de la liaison du Rhône au Rhin, Articles de presse, vol. 1 (1970–1975), BM-Lyon
AP2	L'aménagement du Rhône, Suivi de la liaison du Rhône au Rhin, Articles de presse, vol. 2 (1976–1977), BM-Lyon
AP3	L'aménagement du Rhône, Suivi de la liaison du Rhône au Rhin, Articles de presse, vol. 3 (1978–1983), BM-Lyon
AP4	L'aménagement du Rhône, Suivi de la liaison du Rhône au Rhin, Articles de presse, vol. 4 (1984–1987), BM-Lyon
AP5	L'aménagement du Rhône, Suivi de la liaison du Rhône au Rhin, Articles de presse, vol. 5 (1988–1992), BM-Lyon
BM-Lyon	Archives Régionales, Bibliothèque Municipale de Lyon
CAC	Centre des Archives Contemporaines (Fontainebleau)
CCE	Comité Central d'Entreprise
CE	Comité d'Etablissement
CEA	Commissariat à l'Energie Atomique
CGA	Confedération Générale de l'Agriculture
CNR	Archives, Compagnie Nationale du Rhône (Lyon)
DA	Direction Administrative
DE	Direction des Etudes (through 1972); Direction technique Equipement (starting in 1973)
DG	Direction Générale

Abbreviations

DT	Direction Technique
DX	Direction technique Exploitation
EDF	Electricité de France
FRAPNA	Archives de la Fédération Rhône-Alpes de Protection de la Nature (Lyon)
GEPAR	Le Groupe d'Etude des Perspectives Agricoles et Rurales Rhodaniennes
p.v.	procès-verbal, or minutes of meetings
SACTARD	Société Anonyme de Coordination des Travaux d'Aménagement du Rhône à Donzère
SOGREAH	Société Grenobloise d'Études et d'Applications Hydrauliques
TEX	Direction régionale des Travaux et de l'Exploitation des chutes (through 1972); Direction des Travaux chantiers et Exploitation (starting in 1973)
USSI	Société de Construction d'Usines de Séparation Isotopique

NOTES

Prologue

1. The following account is based on numerous primary and secondary sources that describe the Rhône and surrounding valley, either before World War II or between 1945 and the 1980s; please see note 12 for a complete list. I have been unable to locate a source documenting a specific tour of Plan officials through the Rhône valley after the war. However, Donzère-Mondragon, an enormous project completed between 1947 and 1952 and discussed in subsequent chapters, was an icon of the first Plan. In addition, its construction coincides exactly with the dates of the first Plan. It is extremely likely, then, that Plan officials were in the Rhône valley at this time. In addition, there is ample circumstantial evidence that American officials associated with the Marshall Plan visited the Rhône valley during the late 1940s and early 1950s, most likely accompanied by French officials and probably those from the (French) Plan. For instance, the U.S. House Committee on Foreign Affairs described the state of France, including its canals and waterways, in a 1945 report; the Rhône is prominently mentioned. See U.S. Congress, House Committee on Foreign Affairs, *Report to the President of the United States by Samuel I. Rosenman on Civilian Supplies for the Liberated Areas of Northwest Europe (Prepared Pursuant to Letter of Franklin D. Roosevelt Dated January 20, 1945)*, 79th Cong., 1st sess., 1945, 1–2, 134–135. In an oral history, David Bruce, chief of the Economic Cooperation Administration to France (May 1948–1949), also refers to the poor physical conditions of France after the war and states that he met with Jean Monnet "almost daily." See David K. E. Bruce, interview by Jerry N. Hess, Harry S. Truman Library, March 1, 1972, especially pp. 14–16 of the transcript, which is available online (www.trumanlibrary.org/oralhist/bruce.htm). Finally, U.S. government sources describe American bureaucrats going to France as well as French experts coming to the United States. See Economic Cooperation

Administration, *3 Years of the Marshall Plan* (Washington, DC: Government Printing Office, 1951). In fact, two Rhône projects, Génissiat and Donzère-Mondragon, are highlighted in this report. I thank Angie Boyce and Corinna Schlombs for their research assistance.

2. Tyler Stovall, *France since the Second World War* (New York: Longman, 2002), 29–31.

3. House Committee on Foreign Affairs, *Report to the President*, 135.

4. L.-L. Vallée, *Du Rhône et du Lac de Genève: Ou, des grands travaux à exécuter pour la navigation du Léman à la mer* (Paris: Librairie Scientifique-Industrielle de L. Mathias, 1843), 15.

5. Josette Ponce, interview by author, 1998.

6. Louis Bordeaux, "L'aménagement du Rhône: Étude d'économie politique" (Ph.D. diss., Université de Lyon, 1919), 35.

7. Jacques Bethemont, *Le thème de l'eau dans la vallée du Rhône: Essai sur la genèse d'un espace hydraulique* (Saint-Etienne: Imprimerie 'Le Feuillet Blanc,' 1972), 88; Mark Cioc, *The Rhine: An Eco-Biography, 1815–2000* (Seattle: University of Washington Press, 2002), 30.

8. Gilbert Tournier, *La vallée impériale: Couloir de l'Europe* (Lyon: Audin, 1977), 5; Marie-Françoise Souchon, *La Compagnie nationale d'aménagement de la région du Bas-Rhône-Languedoc* (Grenoble: Editions Cujas, 1968), 74.

9. House Committee on Foreign Affairs, *Report to the President*, 135. See also Chris Pearson, *Scarred Landscapes: War and Nature in Vichy France* (New York: Palgrave Macmillan, 2009).

10. Ari Kelman, *A River and Its City: The Nature of Landscape in New Orleans* (Berkeley: University of California Press, 2003).

11. Daniel Faucher, *L'homme et le Rhône* (Paris: Editions Gallimard, 1968), 82. For how the Rhône's flow compares to that of other rivers, see Jean Ritter, *Le Rhône* (Paris: Presses universitaires de France, 1973), 5; Cioc, *Rhine*, 31–32.

12. Charles Lenthéric, *La région du Bas-Rhône* (Paris: Librairie Hachette et Compagnie, 1881), 177; Charles Lenthéric, *Du Saint-Gothard à la mer, Le Rhône, Histoire d'un fleuve* (Paris: Libraire Plon, 1905), 316. Other sources from which this narrative draws but that are not cited specifically include: Daniel Bideau, *De Lyon à Valence au gré du Rhône* (Lyon: Elie Bellier, 1982); Albert Breittmayer, *Le Rhône: Sa navigation depuis les temps anciens jusqu'à nos jours* (Lyon: Henri Georg, Libraire-Editeur, 1904); M. Maurice Champion, *Recherches historiques sur les inondations du Rhône et de la Loire* (Paris: Typographie Panckoucke, 1856); Gabriel Faure, *The Banks of the Rhone from Lyons to Arles*, trans. Frank Kemp (Grenoble: J. Rey, 1922); Pierre George, *Les pays de la Saône et du Rhône* (Paris: Presses universitaires de France, 1946); Marie Mauron, *Au fil du Rhône: Des glaciers à la mer* (Paris: Horizons de France, 1957); Jean Michel, "Le problème du Rhône: L'aménagement intégral du fleuve. Son triple

point de vue: navigation, forces motrices, irrigations" (Ph.D. diss., Ecole des Sciences Politiques, 1932); Alain Pelosato and Bruno Floquet, *Le Rhône: Fleuve lumière* (Rennes: Editions Ouest-France, 1994); André Roussilhon, "L'utilisation du Rhône au point de vue de la navigation et de l'irrigation" (Ph.D. diss., Faculté de Droit, Université de Paris, 1914); Gilbert Tournier, *Rhône, dieu conquis* (Paris: Librarie Plon, 1952); Gilbert Tournier, *La vallée du Rhône: Son passé, son présent, son avenir. Exposition de documents historiques, projets et plans. Catalogue* (Sommières: Atelier Antoine Demontoy, 1954).

13. Compagnie Nationale du Rhône, *Le Plan Environnement: Programme d'action sur 10 ans* (Médiacité: May 1998), 5.

Introduction

The quotation from Hanotaux in the epigraph is from Daniel Bideau, *De Lyon à Valence au gré du Rhône* (Lyon: Elie Bellier, 1982), 8.

1. See, for instance, Don Mitchell, *Cultural Geography: A Critical Introduction* (Malden, MA: Blackwell, 2000). Naturecultures is from Donna J. Haraway, *The Companion Species Manifesto: Dogs, People, and Significant Otherness* (Chicago: Prickly Paradigm Press, 2003), 1; Donna J. Haraway, *When Species Meet* (Minneapolis: University of Minnesota Press, 2008), 16. See also Haraway, "A Cyborg Manifesto: Science, Technology, and Socialist-Feminism in the Late Twentieth Century," in *Simians, Cyborgs, and Women: The Reinvention of Nature* (New York: Routledge, 1991), 149–181. Natureculture is from Bruno Latour, *We Have Never Been Modern,* trans. Catherine Porter (Cambridge, MA: Harvard University Press, 1993), 7.

2. David E. Nye, *Technologies of Landscape: From Reaping to Recycling* (Amherst: University of Massachusetts Press, 1999), 10.

3. Amita Baviskar, *Waterscapes: The Cultural Politics of a Natural Resource* (New Delhi: Permanent Black, 2007).

4. Such terminology separates these histories, but they are, in fact, entwined.

5. On multipurpose development in the United States, see Karin Ellison, "The Making of a Multiple Purpose Dam: Engineering Culture, the U.S. Bureau Reclamation, and Grand Coulee Dam, 1917–1942" (Ph.D. diss., Massachusetts Institute of Technology, 2000).

6. As STS scholars have demonstrated, these categories, their relationships, and the sorting of objects and processes by actors into these groups are all problematic. I discuss these issues at more length below.

7. Most actors framed the river's transformation as *"l'aménagement du Rhône."* This translates roughly as "development" in English; the French verb,

aménager, is usually translated as "to develop" or "to equip." Like the English phrases, these terms connote improvement and are often associated with industrialization and modernization, making them apt examples of both sociotechnical systems and technopolitics, as political, economic, and social goals are all central to "development," whether in post-1945 France or the contemporary global South. At the same time, "development" was strongly (although not uniquely) tied to the Cold War and postcoloniality. For a discussion of the politics of postwar "development," see Heather Hoag and May-Britt Ohman, "Turning Water into Power: Debates over the Development of Tanzania's Rufiji Basin, 1945–1985," *Technology and Culture* 49 (2008): 624–651; Suzanne Moon, "Take-off or Self-sufficiency? Ideologies of Development in Indonesia, 1957–1961," *Technology and Culture* 39 (1998): 187–212. Historical sociologist Philip McMichael examines the history and politics of what he calls the postwar "development project" in *Development and Social Change: A Global Perspective,* 4th ed. (Thousand Oaks, CA: Pine Forge Press, 2007), especially chapters 2 and 3. For additional discussion of the politics of "projects," see Paul R. Greenough and Anna Lowenhaupt Tsing, eds., *Nature in the Global South: Environmental Projects in South and Southeast Asia* (Durham, NC: Duke University Press, 2003).

8. This phrase, "human and nonhuman nature," is prolific in environmental history, yet the shorthand is problematic because, as environmental historians have demonstrated, "nonhuman" nature is rarely entirely nonhuman. Somewhat ironically, then, the phrase ends up replicating the very dichotomy that most environmental historians seek to challenge: nature and culture.

9. "Naturalization" suggests both inevitability and naturalness, but my allusion here also hints at neutralization. In other words, these connotations are powerful strategies of neutralization; they help mask power—in and of itself a form of power. See also Timothy Mitchell, *Rule of Experts: Egypt, Techno-Politics, Modernity* (Berkeley: University of California Press, 2002), especially 50–53.

10. There is, however, no clear, deterministic relationship between "technological" "intentions" and "environmental" "outcomes." As many environmental historians and historians of technology have shown, there are often unintended consequences, both "social" and "environmental," associated with technological change. My point is that the contested process of technological development strongly shapes but does not wholly determine environmental change, a topic that is often the focal point of environmental history.

11. Important recent studies include Isabelle Backouche, *La trace du fleuve: La Seine et Paris (1750–1850)* (Paris: Editions de l'Ecole des hautes études en sciences sociales, 2000); David Blackbourn, *The Conquest of Nature: Water, Landscape, and the Making of Modern Germany* (New York: Norton, 2006);

Mark Cioc, *The Rhine: An Eco-Biography, 1815–2000* (Seattle: University of Washington Press, 2002); Matthew D. Evenden, *Fish versus Power: An Environmental History of the Fraser River* (New York: Cambridge University Press, 2004); Paul Josephson, *Industrialized Nature: Brute Force Technology and the Transformation of the Natural World* (Washington, DC: Island Press, 2002), especially chapter 1; Ari Kelman, *A River and Its City: The Nature of Landscape in New Orleans* (Berkeley: University of California Press, 2003); Christof Mauch and Thomas Zeller, eds., *Rivers in History: Perspectives on Waterways in Europe and North America* (Pittsburgh: University of Pittsburgh Press, 2008); Richard White, *The Organic Machine: The Remaking of the Columbia River* (New York: Hill and Wang, 1995); Donald Worster, *Rivers of Empire: Water, Aridity, and the Growth of the American West* (New York: Pantheon Books, 1985).

12. Maurice Agulhon, André Nouschi, and Ralph Schor, eds., *La France de 1940 à nos jours* (Paris: Nathan, 1995), 327–334; Robert Gildea, *France since 1945* (New York: Oxford University Press, 1997); Gabrielle Hecht, *The Radiance of France: Nuclear Power and National Identity after World War II* (Cambridge, MA: MIT Press, 1998), 1; Tyler Stovall, *France since the Second World War* (New York: Longman, 2002), 12–14.

13. Benedict Anderson, *Imagined Communities: Reflections on the Origin and Spread of Nationalism*, 2nd ed. (New York: Verso, 1991); Peter Sahlins, *Boundaries: The Making of France and Spain in the Pyrenees* (Berkeley: University of California Press, 1989).

14. Some critics of social construction have created a false opposition between the "social" and the "real," implying that social factors and processes are somehow not also real. However, extensive research on race and gender has persuasively demonstrated, for instance, how constructed categories and relations can be materialized and thus have profound consequences.

15. James C. Scott, *Seeing Like a State: How Certain Schemes to Improve the Human Condition Have Failed* (New Haven: Yale University Press, 1998).

16. Hecht, *Radiance of France*; Judith Schueler, *Materialising Identity: The Co-Construction of the Gotthard Railway and Swiss National Identity* (Amsterdam: Aksant, 2008).

17. Stephen Bocking argues that the space of northern Canada shaped the production of scientific knowledge there in "Science and Spaces in the Northern Environment," *Environmental History* 12 (2007): 867–894; he tends to equate, however, "space" and "environment." A number of scholars in technology studies and environmental history have troubled the binary of discourse and materiality in valuable ways. For instance, Hecht offers a productive framework for the relationship between technology and culture, and technology and politics; technologies are simultaneously cultural and political, while "culture" and

"politics" can take on specifically technological forms; see *Radiance of France*, introduction. In a recent essay, environmental historian Linda Nash maintains that "particular environments shape human intentions"; see "The Agency of Nature or the Nature of Agency?" *Environmental History* 10 (2005): 67–69. And, as Paul Sutter shows in his study of the Panama Canal's construction, physical conditions may reveal contradictions within actors' conceptualizations and thereby challenge them; see "Nature's Agents or Agents of Empire? Entomological Workers and Environmental Change during the Construction of the Panama Canal," *Isis* 98 (2007): 724–754. Integrating these insights highlights how material conditions and cultural frameworks are both dynamic and contested, by both other human groups and nonhuman nature.

18. Wiebe Bijker, "Dikes and Dams: Thick with Politics," *Isis* 98 (2007): 109–123. It is worth noting that various kinds of politics, whether "high," nationalist, gendered, and so on, can be materialized.

19. Chandra Mukerji, "The Great Forest Survey of 1669–1671: The Use of Archives for Political Reform," *Social Studies of Science* 37 (2007): 227–253; Blackbourn, *Conquest of Nature*; Karen M. O'Neill, *Rivers by Design: State Power and the Origins of U.S. Flood Control* (Durham, NC: Duke University Press, 2006).

20. Thomas Parke Hughes, "Technological Momentum in History: Hydrogenation in Germany 1898–1933," *Past and Present* 44 (1969): 106–132; Hecht, *Radiance of France*; Paul N. Edwards, "Infrastructure and Modernity: Force, Time, and Social Organization in the History of Sociotechnical Systems," in *Modernity and Technology*, ed. Thomas Misa, Philip Brey, and Andrew Feenberg (Cambridge, MA: MIT Press, 2003): 185–226; Bijker, "Dikes and Dams."

21. On explicit tensions between political and ecological borders, see Mark Fiege, "The Weedy West: Mobile Nature, Boundaries, and Common Space in the Montana Landscape," *Western Historical Quarterly* 36 (2005): 22–48; Robert M. Wilson, "Directing the Flow: Migratory Waterfowl, Scale, and Mobility in Western North America," *Environmental History* 7 (2002): 247–266; Robert M. Wilson, *Seeking Refuge: An Environmental History of the Pacific Flyway* (Seattle: University of Washington Press, 2010).

22. The concept of co-production can be extended, then, to several political geographies.

23. These arguments extend Hecht's critique of "technological style" because this concept takes for granted the close association between a national context and its technological approach, rather than examining how and why those links were forged; see *Radiance of France*, 12–14. Building also on Richard White, "The Nationalization of Nature," *Journal of American History* 86 (1999): 976–986, these arguments highlight how various political geographies were possible. Scholars should not, then, take historical actors' spatial scales or political frameworks, or even contemporary norms, at face value.

24. The literature on national identity in modern France is enormous. A key early work is the three volumes of Pierre Nora, ed., *Les lieux de mémoire* (Paris: Gallimard), published between 1984 and 1992.

25. Samer Alatout, "Towards a Bio-territorial Conception of Power: Territory, Population, and Environmental Narratives in Palestine and Israel," *Political Geography* 25 (2006): 601–621. Approaches to "environmentality" include Arun Agrawal, *Environmentality: Technologies of Government and the Making of Subjects* (Durham, NC: Duke University Press, 2005); Michael Goldman, "Constructing an Environmental State: Eco-Governmentality and Other Transnational Practices of a 'Green' World Bank," *Social Problems* 48 (2001): 499–523; Diana K. Davis, "Eco-Governance in French Algeria: Environmental History, Policy, and Colonial Administration," *Proceedings of the Western Society for French History* 32 (2004): 328–345. Several works in U.S. environmental history have emphasized the centrality of environmental issues to political ideology, state formation, the constitution of political power, and political processes more broadly; see Neil M. Maher, *Nature's New Deal: The Civilian Conservation Corps and the Roots of the American Environmental Movement* (New York: Oxford University Press, 2007); Sarah Phillips, *This Land, This Nation: Conservation, Rural America, and the New Deal* (New York: Cambridge University Press, 2007). A key early study linking political formation and water (management) was Donald Worster, *Rivers of Empires: Water, Aridity, and the Growth of the American West* (New York: Pantheon Books, 1985); Worster was influenced by Karl Wittfogel's "hydraulic thesis" in both intriguing and problematic ways.

26. Several of Donald Worster's essays particularly emphasize this argument: "Appendix: Doing Environmental History, in *The Ends of the Earth: Perspectives on Modern Environmental History*, ed. Donald Worster (New York: Cambridge University Press, 1988): 289–308; "A Transformation of the Earth: Toward an Agroecological Perspective in History," *Journal of American History* 76 (1990): 1087–1106; "A Long Cold View of History: How Ice, Worms, and Dirt Made Us What We Are Today," *American Scholar* 74 (2005): 57–66. Other scholars tend to combine materialist and discursive approaches, often complicating or even challenging this binary. See, for example, William Cronon, "Modes of Prophecy and Production: Placing Nature in History," *Journal of American History* 76 (1990): 1122–1131; William Cronon, "A Place for Stories: Nature, History, and Narrative," *Journal of American History* 78 (1992): 1347–1376; Arthur F. McEvoy, *The Fisherman's Problem: Ecology and Law in the California Fisheries, 1850–1980* (New York: Cambridge University Press, 1986); Arthur F. McEvoy, "Working Environments: An Ecological Approach to Industrial Health and Safety," *Technology and Culture* 36 (1995): S145–172; Ted Steinberg, "Down to Earth: Nature, Agency, and Power in History," *American Historical Review* 107 (2002): 798–820. For an insightful

analysis and critique of "agency" within (and beyond) environmental history, see Nash, "Agency of Nature or the Nature of Agency?" Sutter nicely frames the question of agency in terms of both historical actors and scholarly analysts in "Nature's Agents or Agents of Empire?"

27. See Chapters 3 and 7. See also Sara B. Pritchard, "The State of Nature: Redesigning the Rhône River, 1945–1955," paper presented at the annual meeting of the Society for the History of Technology, Santa Clara, CA, October 4–7, 2001; Andrew C. Isenberg, *Mining California: An Ecological History* (New York: Hill and Wang, 2005), especially pt. 1, "The Nature of Industry." To avoid the extremes of both technological and social determinism, many STS scholars speak of "shaping," "mediating," or "affording" possibilities rather than "determining" or "causing" outcomes. Most have focused, however, on "technological" or "material" affordances. In contrast, envirotechnical analysis seeks to highlight "natural" affordances without returning to environmental determinism. Incorporating insights from environmental history therefore suggests it is important to consider *materialities* (rather than materiality), including different kinds of "material" constraints. At the same time, envirotechnical analysis implies the blurry boundaries between these affordances.

28. William Cronon, "The Trouble with Wilderness: Or, Getting Back to the Wrong Nature," in *Uncommon Ground: Rethinking the Human Place in Nature,* ed. William Cronon (New York: Norton, 1995): 69–90; Mark David Spence, *Dispossessing Wilderness: Indian Removal and the Making of the National Parks* (New York: Oxford University Press, 1999).

29. For a historiographic review, see Jeffrey K. Stine and Joel A. Tarr, "At the Intersection of Histories: Technology and the Environment," *Technology and Culture* 39 (1998): 601–640. There are exceptions; for instance, Tarr and Martin Melosi have focused on urban environmental history. For several of Melosi's works, see *Garbage in the Cities: Refuse, Reform, and the Environment, 1880–1980* (College Station: Texas A&M University Press, 1981); "Cities, Technical Systems, and the Environment," *Environmental History Review* 14 (1990): 45–61; "The Place of the City in Environmental History," *Environmental History Review* 17 (1993): 1–23.

30. Scholars of European environmental history focused on such topics from the outset, undoubtedly reflecting Europe's own human and natural histories. For a nice summary, see Thomas Lekan and Thomas Zeller, "Scenery, Region, Power: Cultural Landscapes in Environmental History," in *Oxford Handbook of Environmental History,* ed. Andrew C. Isenberg (New York: Oxford University Press, forthcoming). See also Michael Bess, *The Light-Green Society: Ecology and Technological Modernity in France, 1960–2000* (Chicago: University of Chicago Press, 2003); Hecht, *Radiance of France,* 2nd ed., especially Michel Callon's foreword.

31. Many environmental historians still have more of a realist conception of nature than most STS scholars; I further discuss this point below.

32. As an example, Judith A. McGaw's *Most Wonderful Machine: Mechanization and Social Change in the Berkeshire Paper Industry, 1801–1885* (Princeton: Princeton University Press, 1987) examined changes in labor and gender relations following the industrialization of Massachusetts's paper industry. The first chapter describes the environmental conditions industrialists sought when selecting a mill site. The rest of the book details shifting social relations of industrial production. The book played an important role in integrating workers and gender within analysis of technological change and helped open up Hughes's systems theory, but it illustrates the way many historians of technology set up the relationship of "technology" and the "environment."

33. A few of the classic works that developed these approaches include the following. For systems theory, see Thomas Parke Hughes, *Networks of Power: Electrification in Western Society, 1880–1930* (Baltimore: Johns Hopkins University Press, 1983). For social construction of technology, see Trevor J. Pinch and Wiebe E. Bijker, "The Social Construction of Facts and Artifacts: Or How the Sociology of Science and the Sociology of Technology Might Benefit Each Other," in *The Social Construction of Technological Systems: New Directions in the Sociology and History of Technology*, ed. Wiebe Bijker, Thomas Hughes, and Trevor Pinch (Cambridge, MA: MIT Press, 1987): 17–50. For ANT, see Michel Callon, "Society in the Making: The Study of Technology as a Tool for Sociological Analysis," and John Law, "Technology and Heterogeneous Engineering: The Case of Portuguese Expansion," both in Bijker, Hughes, and Pinch, *Social Construction of Technological Systems*; Bruno Latour, "Give Me a Laboratory and I Will Raise the World," in *Science Observed: Perspectives on the Study of Science*, ed. M. Mulkay and K. Knorr-Cetina (London: Sage, 1993): 141–170; Bruno Latour, *Reassembling the Social: An Introduction to Actor-Network-Theory* (New York: Oxford University Press, 2005). Bijker critiques scholars for placing both arguments within the rubric of technological determinism; he argues that the idea of an autonomous technology is a theory of technology, while only the second idea, the thesis of technology determining society, is a true "technologically determinist" argument. See Wiebe E. Bijker, "Sociohistorical Technology Studies," in *Handbook of Science and Technology Studies*, ed. Sheila Jasanoff, Gerald E. Markle, James C. Petersen, and Trevor Pinch (Thousand Oaks, CA: Sage, 1995): 238.

34. ANT theorists argue that categories and the categorization of "nature," "society," and "technology" are precisely what is open to investigation. ANT's conception of "actants" serves as a way to include diverse agents without sorting them in advance. Hughes's systems theory has been refined and expanded since the publication of *Networks of Power*. For instance, most historians of

technology now include workers and users in ways Hughes did not originally envision. With the exception of scholars working at the intersection of environmental history and technology studies, few historians of technology have integrated "nature," which Hughes explicitly framed as external to the system, into it.

35. For a thorough discussion of the development of and precursors to envirotech, see Stine and Tarr, "At the Intersection of Histories." Envirotech is an official special interest group (SIG) within the Society for the History of Technology (SHOT); it also meets at the annual meetings of the American Society for Environmental History and the biennial conferences of the European Society for Environmental History. James C. Williams and I decided to found Envirotech at the August 1999 meeting of International Committee for the History of Technology in Belfort, France. The organization's listserv, originally based at Stanford University, began in January 2000. The group's first meeting (although the SIG was still unofficial at that point) was held at SHOT's conference in Munich in August 2000. Soon afterward, it received official SIG status.

36. William Boyd, "Making Meat: Science, Technology, and American Poultry Production," *Technology and Culture* 42 (2001): 631–664; Edmund Russell, "The Garden in the Machine: Toward an Evolutionary History of Technology," introduction to *Industrializing Organisms: Introducing Evolutionary History*, ed. Susan R. Schrepfer and Philip Scranton (New York: Routledge, 2004): 1–18, and the other essays in this volume; Scott Prudham, *Knock on Wood: Nature as Commodity in Douglas-Fir Country* (New York: Routledge, 2005); Ann Norton Greene, *Horses at Work: Harnessing Power in Industrial America* (Cambridge, MA: Harvard University Press, 2008); Timothy J. LeCain, *Mass Destruction: The Men and Giant Mines That Wired America and Scarred the Planet* (New Brunswick, NJ: Rutgers University Press, 2009); Robert Gardner, "Constructing a Technological Forest: Nature, Culture, and Tree-Planting in the Nebraska Sand Hills," *Environmental History* 14 (2009): 275–297. Of course, "landscape" has a long and rich history—and historiography, largely within the discipline of geography. For geographers, "envirotechnical landscape" is undoubtedly redundant, since landscape is by definition a blending of nature and culture. However, by calling a landscape envirotechnical, the phrase emphasizes its technological constitution and shaping. For hybrid landscapes more broadly, see Mark Fiege, *Irrigated Eden: The Making of an Agricultural Landscape in the American West* (Seattle: University of Washington Press, 1999); Richard White, "From Wilderness to Hybrid Landscapes: The Cultural Turn in Environmental History," *Historian* 66 (2004): 557–564.

37. Eric Kauffmann, "'Naturalizing the Nation': The Rise of Naturalistic Nationalism in the United States and Canada," *Comparative Studies in Society and History* 40 (1998): 666–695; Olivier Zimmer, "In Search of Natural Identity: Alpine Landscape and the Reconstruction of the French Nation,"

Comparative Studies in Society and History 40 (1998): 663–665; E. Elena Songster, "Cultivating the Nation in Fujian's Forests: Forest Policies and Afforestation Efforts in China, 1911–1937," *Environmental History* 8 (2003): 452–473; Sara B. Pritchard, "Reconstructing the Rhône: The Cultural Politics of Nature and Nation in Contemporary France, 1945–1997," *French Historical Studies* 27 (2004): 766–799; Lissa Roberts, "An Arcadian Apparatus: The Introduction of the Steam Engine into the Dutch Landscape," *Technology and Culture* 45 (2004): 251–276. See also several articles in the recent special issue, "New Directions in French Environmental History," of *French Historical Studies* 32 (2009), ed. Caroline Ford and Tamara L. Whited: Alice Ingold, "To Historicize or Naturalize Nature: Hydraulic Communities and Administrative States in Nineteenth-Century Europe"; Samuel Temple, "The Natures of Nation: Negotiating Modernity in the *Landes de Gascogne*"; Patrick Young, "A Tasteful Patrimony? Landscape Preservation and Tourism in the Sites and Monuments Campaign, 1900–1935."

38. These are my characterizations of a few emerging patterns in the literature thus far. See also Sara B. Pritchard, "'Le nouveau Rhône est né' (Donzère-Mondragon)," in *La Technologie au risque de l'histoire,* ed. Robert Belot, Michel Cotte, and Pierre Lamard (Belfort: l'Université de Technologie de Belfort-Montbéliard, 2000), 77–86.

39. Exposing the "nature" of diverse technologies (and conversely, the "technologies" of nature) is the heart of Schrepfer and Scranton, *Industrializing Organisms.*

40. "Mediators" of technology is a growing area of interest within the history of technology and STS, just as labor, gender, users, and other historically underrepresented groups have received greater attention since the 1980s. See, for instance, Carolyn M. Goldstein, "From Service to Sales: Home Economics in Light and Power, 1920–1940," *Technology and Culture* 38 (1997): 121–152. The topic of technological mediators parallels interest in the "translation" of science within the social and historical studies of science. Literature on mediation has focused, however, on human mediators of knowledge production and dissemination or the technological mediation of sociopolitical possibilities. My point here is to emphasize how technologies also mediate human-natural interactions. See also White, *Organic Machine,* especially 182.

41. McEvoy, "Working Environments," S152.

42. In many ways, then, envirotechnical analysis reflects the pendulum swing back from social constructivism and thus the growing interest in technological mediation and co-production.

43. Important exceptions include Kevin Dann and Gregg Mitman, "Exploring the Borders of Environmental History and the History of Ecology," *Journal of the History of Biology* 30 (1997): 291–302; Sutter, "Nature's Agents or Agents of Empire?" Ed Russell engages with some of these theoretical issues,

largely under the rubric of evolutionary history, in his introduction to *Industrializing Organisms* and "Evolutionary History: Prospectus for a New Field," *Environmental History* 8 (2003): 204–228. See also *The Illusory Boundary: Environment and Technology in History,* ed. Martin Reuss and Stephen H. Cutcliffe (Charlottesville: University of Virginia Press, 2010), introduction and afterword.

44. Cronon, *Uncommon Ground;* Michael E. Soulé and Gary Lease, eds., *Reinventing Nature? Responses to Postmodern Deconstruction* (Washington, DC: Island Press, 1995); Douglas R. Weiner, "A Death-Defying Attempt to Articulate a Coherent Definition of Environmental History," *Environmental History* 10 (2005): 404–420.

45. There are two key issues here. First, as Hecht argues, technologies such as nuclear reactors are not only politics (or only technologies). It is the specifically technological, material form of politics that matters. Second, materiality and material culture have garnered increasing attention, both empirically and theoretically, within STS. A focus on envirotechnical topics such as river development suggests, however, the importance of distinguishing *among* materialities.

46. See earlier citations regarding various approaches to nature's agency within environmental history, whether materialist, discursive, or a combination thereof. "Active player" is from McEvoy, "Working Environments," S149. Nature's "force" is from J. Donald Hughes, *What Is Environmental History?* (Cambridge: Polity, 2006). On the porous boundaries among actants, see Mitchell, *Rule of Experts,* especially chapter 1.

47. The difference between environmental history and STS here is especially sharp.

48. See, for instance, Helen Rozwadowski, *Fathoming the Ocean: The Discovery and Exploration of the Deep Sea* (Cambridge, MA: Harvard University Press, 2005); Scott Frickel, *Chemical Consequences: Environmental Mutagens, Scientist Activism, and the Rise of Genetic Toxicology* (New Brunswick, NJ: Rutgers University Press, 2004).

49. Mitchell, *Rule of Experts,* chapter 1.

50. On actors' tactical constructions of "nature" and "culture," see Chapter 4 here.

51. Contrast, for instance, two approaches to writing "oil histories": Alison Fleig Frank, *Oil Empire: Visions of Prosperity in Austrian Galicia* (Cambridge, MA: Harvard University Press, 2005); Myrna I. Santiago, *The Ecology of Oil: Environment, Labor, and the Mexican Revolution, 1900–1938* (New York: Cambridge University Press, 2006).

52. J. R. McNeill, "Observations on the Nature and Culture of Environmental History," *History and Theory* 42 (2003): 5–43; Worster, "Long Cold View of History."

53. For those who question the importance of the material, Pinch playfully suggests that they should "try walking into a wall." See Trevor Pinch, "On Making Infrastructure Visible: Putting the Non-Humans to Rights," *Cambridge Journal of Economics,* Advance Access published September 22, 2009, doi:10.1093/cje/bep044.

54. Sutter, "Nature's Agents or Agents of Empire?"

55. Sheila Jasanoff, ed., *States of Knowledge: The Co-Production of Science and the Social Order* (New York: Routledge, 2004), especially chapters 1 and 2.

56. Hecht, *Radiance of France,* 11.

57. This question emerged during one of our discussions at the NSF "Envirotech" workshop at the University of Maryland in 2006.

58. I am influenced by Latour and actor-network theory here, particularly the categorization of concepts and objects. However, ANT's mantra of "just follow the actors" does not adequately address the power dynamics of these networks. For one insightful critique, see Susan Leigh Star, "Power, Technology, and the Phenomenology of Conventions: On Being Allergic to Onions," in *A Sociology of Monsters: Essays on Power, Technology, and Domination,* ed. John Law (New York: Routledge, 1991): 26–56. In addition, it is worth noting that bringing analysts' categories to bear on historical research is not new. For instance, historians of gender bring that analytic lens to their studies, whether or not their actors did so.

59. This argument applies Hecht's charge to take seriously actors' ideas of technology, politics, and their relationship to nature, technology, and their interplay; see *Radiance of France,* 14–15. Hecht and Michael Allen develop a parallel point about actors' conceptions and uses of technological determinism in *Technologies of Power: Essays in Honor of Thomas Parke Hughes and Agatha Chipley Hughes* (Cambridge, MA: MIT Press, 2001), introduction.

60. In short, these arguments apply historical ontology to "nature" and "technology" specifically. For an excellent overview, see Michelle Murphy, *Sick Building Syndrome and the Problem of Uncertainty: Environmental Politics, Technoscience, and Women Workers* (Durham, NC: Duke University Press, 2006), introduction.

61. "Envirotechnical" builds on, but also is distinguished from, several related concepts. First, in *Human-Built World: How to Think about Technology and Culture* (Chicago: University of Chicago Press, 2004), Thomas Hughes introduces the term "ecotechnological environment," which he defines as "intersecting and overlapping natural and human-built environments" (153). Hughes discusses several examples that demonstrate the interpenetration of the "natural" and "technological" but does not situate his analysis within existing literature or develop the concept as an analytic tool. Hughes explains that "ecotechnological" focuses on "more sustainable" relations between nature and technology. I view "ecotechnological," at least as Hughes defines it, as a subset

of the envirotechnical, which are *any* connections between ecological and technological systems. Second, with respect to the term "envirotechnical," Peter Dear rightly prefers more grammatically correct terms such as "technoenvironment," "ecotechnology," or "ecotechnique." However, as I explain below, "technoenvironment" is problematic because it places technology before nonhuman nature. In addition, "ecotechnological" risks being politically freighted; in addition, my concerns about Hughes's "ecotechnological" resurface here. Finally, I am sympathetic with Hugh Gorman's concern that "technical" can imply neutrality or objectivity, a point against which historians of technology and STS scholars vehemently fight; this explains his preference for "envirotechnological." Despite these issues, "envirotechnical" does offer some linguistic simplicity.

62. Human impacts on the environment are obviously historically and culturally situated. By making this point, I do not intend to obscure important distinctions among regions, cultures, or historical eras, especially in terms of their spatial or temporal impact or its extent.

63. Nina Lerman, Ruth Oldenziel, and Arwen Mohun, eds., *Gender and Technology: A Reader* (Baltimore: Johns Hopkins University Press, 2003), 2. Similarly, Bijker emphasizes "at least three different layers of meaning" with respect to technology: physical artifacts, human activities, and knowledge; see Bijker, "Sociohistorical Technology Studies," 231. On the Anglo-American history of "technology," see Leo Marx, "The Idea of 'Technology' and Postmodern Pessimism," in *Does Technology Drive History?* ed. Merritt Roe Smith and Leo Marx (Cambridge, MA: MIT Press, 1994), 238–257.

64. For the purposes of defining envirotechnical systems here, these examples are "technologies." However, as the rest of the introduction maintains, I argue that they are in fact envirotechnical systems. At the same time, I am also interested in how contemporaries involved in the Rhône's remaking framed (or did not frame) these objects as "technology," "nature," or both.

65. Hughes, *Networks of Power*. "Systems" is one way to represent the links among diverse objects, ideas, and practices. Other related concepts in STS include networks and assemblages. I am using systems primarily because of its preeminence within the history of technology.

66. However, what is usually included within these categories is probably more limited than how I am using them. For instance, contrast popular definitions of technology with the much broader notion adopted by most scholars in technology studies.

67. Admittedly, the social, cultural, and political dimensions of technology are implicit; in contrast, "sociotechnical" or "technopolitical" makes them explicit. I am, however, conceptualizing the "technical" within "envirotechnical" systems and regimes very much according to now-common assumptions within

technology studies. One option would be to call them "enviro-socio-technical systems," but this is obviously cumbersome.

68. See my earlier discussion of "envirotechnical," its precursors, and its alternatives. In addition, this argument differs from wholly cultural constructivist positions, exemplified, for instance, by Simon Schama, *Landscape and Memory* (New York: Knopf, 1995), 61.

69. Of course, these "environmental" conditions were likely shaped by past actions by humans and nonhumans.

70. According to my definition, hydroelectric plants, nuclear reactors, and irrigation networks are all envirotechnical systems, as is the Rhône. Of course, the Rhône partly constitutes each of these systems, thereby highlighting the ways that multiple systems can intersect and shape one another.

71. I address the wider relevance and implications of envirotechnical systems in the Conclusion.

72. I use the adjective "natural" advisedly with several caveats. Technology is not "natural" in terms of its inevitability (e.g., autonomous technology). Technology is also not "natural," meaning untouched by humans. Rather, nonhuman nature, however mediated physically and discursively, partly constitutes technological artifacts and systems. At the same time, actors' invocations of technology as natural deserve analysis, since this claim performs strategic work for those groups.

73. Russell, "Garden in the Machine," 1.

74. Rachel Prentice reminded me that the modern biotech industry has certainly challenged this assumption.

75. Of course, as Marx pointed out, the production process also erases human labor involved in an object's creation. Marx was focused, however, on the commodity form. His critique is nonetheless useful here. Thanks go to Rachel Prentice for clarifying these points.

76. Again, language does not lend itself to a dynamic, interrelated conception of "nature" and "technology" here.

77. The geography of science and technology is a relatively new perspective within science studies. Two key works include Robert Kohler, *Landscapes and Labscapes: Exploring the Lab-Field Border in Biology* (Chicago: University of Chicago Press, 2002); David N. Livingstone, *Putting Science in Its Place: Geographies of Scientific Knowledge* (Chicago: University of Chicago Press, 2003).

78. Several works have influenced my thinking here, including McEvoy, "Working Environments"; Russell, "Garden in the Machine"; Russell, "Evolutionary History"; White, *Organic Machine;* White, "From Wilderness to Hybrid Landscapes."

79. This point is implicit in White, *Organic Machine,* and explicit in Russell's two essays.

80. A nice summary of Hegel and Marx's "first" and "second" nature is provided in William Cronon, *Nature's Metropolis: Chicago and the Great West* (New York: Norton, 1991), xix.

81. Admittedly, these arguments are in tension with current approaches in technology studies. Scholars have often pointed to (human) intentionality as a defining feature of historical agency. However, as Nash shows in her essay, human intentionality is far more complex and problematic than many historians and other scholars admit. In addition, by extending this argument that nature can be technological, humans are clearly not the only species who appropriate nonhuman nature as "technology" (consider bees, beavers, chimpanzees, and so on). At this point, I am setting aside the ways environmental historians have the potential to unsettle the implicit anthropocentrism of "technology" within technology studies. While scholars have widened "technology" and "actor" by considering workers, users, women, and so on, they have generally focused on the human and not considered the nonhuman. These arguments appear to reopen the old wound of technological determinism, but as I argue below, asserting that "nature" shapes or mediates technological change is not the same as claiming that nature determines the path of technological development.

82. At a conference several years ago, Thomas Hughes asked me, "Is the Rhône a technology?" Needless to say, the question stuck with me, and I hope this book provides an adequate response. At the same time, it is worth considering how, why, and when actors mobilized the idea of nature as technological, since it often served to defend development projects like those on the Rhône. In other words, the question can be posed both historically and analytically.

83. Hecht, *Radiance of France*, 17; Murphy, *Sick Building Syndrome*, 10.

84. I am using "politics" here in the broader Foucauldian sense.

1. Envisioning a New Rhône

1. Gabriel Faure, *The Banks of the Rhone from Lyons to Arles,* trans. Frank Kemp (Grenoble: J. Rey, 1922), 17.

2. Lapalud ("the marsh") is one example cited in Daniel Faucher, *L'homme et le Rhône* (Paris: Editions Gallimard, 1968), 164. The references to Aigues-Mortes and Les Brotteaux are based on my experiences living in the Rhône valley.

3. Giandou dates the beginning of this shift in the CNR's multipurpose objectives to 1962. See Alexandre Giandou, *La Compagnie nationale du Rhône (1933–1998): Histoire d'un partenaire régional de l'Etat* (Grenoble: Presses universitaires de Grenoble, 1999), 209. In contrast, Bethemont argues that hydroelectricity dominated until 1965, but he also refers to the 1962 *convention agricole* (revised in 1968) between the CNR and the state that altered the financial evaluation, reimbursement, and objectives of agriculture. See Jacques Bethemont, *Le*

thème de l'eau dans la vallée du Rhône: Essai sur la genèse d'un espace hydraulique (Saint-Étienne: Imprimerie 'Le Feuillet Blanc,' 1972), 202. The new politics of *l'aménagement du territoire* is also significant; see Chapter 5.

4. Josef Konvitz, *Cartography in France, 1660–1848: Science, Engineering, and Statecraft* (Chicago: University of Chicago Press, 1987), 105; Chandra Mukerji, *Territorial Ambitions and the Gardens of Versailles* (Cambridge: Cambridge University Press, 1997); Chandra Mukerji, "Cartography, Entrepreneurialism, and Power in the Age of Louis XIV: The Case of the Canal du Midi," in *Merchants and Marvels,* ed. Pamela Smith and Paula Findlen (New York: Routledge, 2002), 255–259; Chandra Mukerji, "Dominion, Demonstration, and Domination: Religious Doctrine, Territorial Politics, and French Plant Collection," in *Colonial Botany: Science, Commerce and Politics in the Early Modern World,* ed. Londa Schiebinger and Claudia Swann (Philadelphia: University of Pennsylvania Press, 2004), 19–33; Cecil O. Smith, "The Longest Run: Public Engineers and Planning in France," *American Historical Review* 95:3 (1990): 657–692.

5. Ludwik A. Teclaff, *The River Basin in History and Law* (The Hague: Nijhoff, 1967), 76; Ludwik A. Teclaff, *Abstraction and Use of Water: A Comparison of Legal Regimes* (New York: United Nations, 1972), 33; Bernard Barraqué, *Les politiques de l'eau en Europe* (Paris: Editions La Découverte, 1995), 142; Francisco Nunes Correia, ed., *Institutions for Water Resources Management in Europe,* vol. 1 (Rotterdam: A. A. Balkema, 1998), 85–86, 92; Tamara L. Whited, *Forests and Peasant Politics in Modern France* (New Haven: Yale University Press, 2000), 22.

6. On environmental management, both materially and symbolically, as a means of state-building in France, see Mukerji, *Territorial Ambitions;* Mukerji, "Cartography, Entrepreneurialism, and Power in the Age of Louis XIV"; Pierre Claude Reynard, "Moving Mountains: Local Reactions to Canal Building in Eighteenth-Century France," paper presented at the annual conference of the International Committee for the History of Technology, Prague, August 22–26, 2000.

7. Konvitz, *Cartography in France,* 123.

8. The "naturalization of the nation" builds on, but also revises, Richard White, "The Nationalization of Nature," *Journal of American History* 86 (1999): 976–986.

9. Teclaff, *Abstraction,* 33; Teclaff, *River Basin,* 90.

10. Teclaff, *River Basin,* 79, 197–198. For a study of prior appropriation doctrine at work in the U.S. West, the classic study is Donald Worster, *Rivers of Empire: Water, Aridity, and the Growth of the American West* (New York: Pantheon, 1985).

11. Teclaff, *Abstraction,* 34; Teclaff, *River Basin,* 88, 90–91; Pierre Claude Reynard, "Probing the Boundaries of Environmental Concerns: Reactions to

Hydraulic Public Works in Eighteenth-Century France," *Environment and History* 9 (2003), 251–273.

12. Correia, *Institutions*, 92.

13. For several articles on the history of water management technologies in the Netherlands, see the special issue, "Dutch Water Technologies," ed. Martin Reuss: *Technology and Culture* 43:3 (2002).

14. Correia, *Institutions*, 92.

15. On these watershed agencies, see Teclaff, *Abstraction;* Barraqué, *Les politiques de l'eau;* Francisco Nunes Correia, *Selected Issues in Water Resources Management in Europe,* vol. 2 (Rotterdam: A. A. Balkema, 1998); Correia, *Institutions;* M. Loriferne, *40 ans de politique de l'eau en France* (Paris: Economica, 1987).

16. Gilbert Tournier, *Rhône, dieu conquis* (Paris: Librarie Plon, 1952), 161.

17. Albert Breittmayer, *Le Rhône; Sa navigation depuis les temps anciens jusqu'à nos jours* (Lyon: Henri Georg, 1904), 51; Charles Lenthéric, *La région du Bas-Rhône* (Paris: Librairie Hachette et Compagnie, 1881), 51–52.

18. Various proposals are described in: Louis Barron, *Le Rhône* (Paris: Librairie Renouard, 1891); Louis Bordeaux, "L'aménagement du Rhône: Étude d'économie politique" (Ph.D. diss., Université de Lyon, 1919); Breittmayer, *Le Rhône;* Faucher, *L'homme et le Rhône;* Philippe Lamour, *Histoire des canaux du Rhône* (Paris: Editions Sodirep, 1961); Jean Michel, "Le problème du Rhône: L'aménagement intégral du fleuve. Son triple point de vue: navigation, forces motrices, irrigations" (Ph.D. diss., Ecole des Sciences Politiques, 1932); André Roussilhon, "L'utilisation du Rhône au point de vue de la navigation et de l'irrigation" (Ph.D. diss., Faculté de Droit, Université de Paris, 1914); Tournier, *Rhône, dieu conquis;* Henri Wohrer, "L'aménagement du Rhône et la loi du 27 mai 1921" (Ph.D. diss., Université d'Aix-Marseille, 1925).

19. For example, proposals for the lateral canal between Lyon and a major city in Provence reappeared at least four times between 1799 and the mid-nineteenth century, but never materialized. For an overview, see Bethemont, *Le thème de l'eau,* 181–182.

20. Alain Pelosato, *Le Rhône* (Paris: Presses universitaires de France, 1996), 15.

21. Tournier, *Rhône, dieu conquis,* 64, 71.

22. The flood of 1856 sparked investigation of earlier inundations. See, for instance, M. Maurice Champion, *Recherches historiques sur les inondations du Rhône et de la Loire* (Paris: Imp. de Panckoucke, 1856). For a summary of early flood control efforts, see Bethemont, *Le thème de l'eau,* especially chapter 1 of book 2.

23. Quoted in Whited, *Forests and Peasant Politics,* 58.

24. Pelosato, *Le Rhône,* 16.

25. The exception here was the high-chute dam at Génissiat, built between 1937 and 1947. Industrial development on the upper Rhône took off with the construction of two nuclear facilities, Bugey in the mid-1960s and Creys-Malville in the mid-1970s. See Chapter 7 for a discussion of how these projects reshaped the management of the upper Rhône, including mediating the design of other projects.

26. Pelosato, *Le Rhône,* 18–19; Alexandre Giandou, "Histoire d'un partenaire régional de l'Etat: La Compagnie nationale du Rhône (1933–1974)" (Ph.D. diss., Université Louis Lumière-Lyon II, 1997) (hereafter "Histoire d'un partenaire régional de l'Etat"), 23; Pierre Savey, *La CNR à 60 ans: Une brève histoire de la CNR à l'occasion de son anniversaire* (Lyon: CNR, 1993), 2; Tournier, *Rhône, dieu conquis,* 190.

27. Eugen Weber, *Peasants into Frenchmen: The Modernization of Rural France, 1870–1914* (Stanford: Stanford University Press, 1976).

28. Bethemont, *Le thème de l'eau,* 142.

29. Daniel Bideau, *De Lyon à Valence au gré du Rhône* (Lyon: Elie Bellier, 1982), 10, 17–18; Marie Mauron, *Au fil du Rhône: Des glaciers à la mer* (Paris: Horizons de France, 1957), 94; Martine Fournier and Pierre-Claude Tracol, *Hommage aux mariniers du Rhône* (Valence: Imp. réunies, 1980).

30. Reed G. Geiger, "Planning the French Canals: The 'Becquey Plan' of 1820–1822," *Journal of Economic History* 44:2 (1984): 329–339; Reed G. Geiger, *Planning the French Canals: Bureaucracy, Politics, and Enterprise under the Restoration* (Newark: University of Delaware Press, 1994). See also Weber, *Peasants into Frenchmen.*

31. For a discussion of the conflicts between transportation modes on the Rhône, see Savey, *La CNR.*

32. For a study of the Rhône's canals, see Bethemont, *Le thème de l'eau,* especially chapter 2 of book 2; Lamour, *Histoire des canaux du Rhône.* On provincial modernization during the Third Republic more broadly, see Weber, *Peasants into Frenchmen.*

33. Faucher, *L'homme et le Rhône,* 247; Alexandre Giandou, "Histoire de la Compagnie nationale du Rhône" (Projet de recherche, Université Lumière-Lyon II, 1991–1992) (hereafter "Histoire de la CNR"), 64.

34. Giandou, "Histoire d'un partenaire régional de l'Etat," 33. I address agriculture and the Rhône's transformation more extensively in the last section of Chapter 3.

35. Comité des travaux historiques et scientifiques, ed., *La ville et le fleuve: Colloque tenu dans le cadre du 112e Congrès national des sociétés savantes, Lyon, 21–25 avril 1987* (Paris: Editions du CTHS, 1989), 272.

36. Richard White describes parallel issues involving space, people, and hydroelectricity for the Columbia River in *The Organic Machine* (New York: Hill and Wang, 1995), 49–51.

37. Giandou, "Histoire d'un partenaire régional de l'Etat," 37.

38. Giandou, "Histoire de la CNR," 65–67. Jonage did not include, however, any measures for agriculture, which therefore differentiated it from later multipurpose projects.

39. Quoted in Bethemont, *Le thème de l'eau*, 185.

40. Giandou, "Histoire de la CNR," 68–69.

41. P. Salenc, *L'aménagement et la mise en valeur de la vallée du Rhône* (October 21, 1963), CNR, AR0086, box 48, Généralités, Fonds Salenc.

42. Giandou, *La Compagnie nationale du Rhône*, 27. He discusses this commission at greater length in "Histoire de la CNR," 70.

43. Quoted in Bethemont, *Le thème de l'eau*, 186–187.

44. Quoted in Giandou, "Histoire de la CNR," 71. These issues are also discussed in Bethemont, *Le thème de l'eau*, 186–187, and Giandou, *La Compagnie nationale du Rhône*, 26–27.

45. Giandou, *La Compagnie nationale du Rhône*, 29–31. An earlier but more extended discussion of these issues is found in Giandou, "Histoire d'un partenaire régional de l'Etat," 64–66, 69.

46. Giandou, *La Compagnie nationale du Rhône*, 29.

47. Wohrer, "L'aménagement du Rhône et la loi du 27 mai 1921," 71–72.

48. Giandou, "Histoire d'un partenaire régional de l'Etat," 48–50, 53–54; an abbreviated version is included in *La Compagnie nationale du Rhône*, 28. Giandou seems to underestimate, however, the crucial role of World War I.

49. Chapters 6 and 7 discuss the changing political climate that reshaped the economics of hydroelectricity during the 1960s and 1970s.

50. Bethemont, *Le thème de l'eau*, 224.

51. Giandou, "Histoire d'un partenaire régional de l'Etat," 59–60; Giandou, "Histoire de la CNR," 75.

52. The classic work here is Richard F. Kuisel, *Capitalism and the State in Modern France: Renovation and Economic Management in the Twentieth Century* (New York: Cambridge University Press, 1981).

53. Correia, *Institutions*, 95; Teclaff, *River Basin*, 92.

54. Paul R. Josephson, *Industrialized Nature: Brute Force Technology and the Transformation of the Natural World* (Washington, DC: Island Press, 2002).

55. Mark Cioc, *The Rhine: An Eco-Biography, 1815–2000* (Seattle: University of Washington Press, 2002); Edmund N. Todd, "Building a Hybrid Landscape to Purify the Ruhr Region, 1890–1935," *History of Technology* 22 (2000): 25–42.

56. Barraqué, *Les politiques de l'eau*, 273. See also Correia, *Institutions*, 93; Teclaff, *River Basin*, 4.

57. Karin Ellison, "The Making of a Multiple Purpose Dam: Engineering Culture, the U.S. Bureau Reclamation, and Grand Coulee Dam, 1917–1942" (Ph.D. diss., Massachusetts Institute of Technology, 2000).

58. On the environmental implications of the New Deal more broadly, see Neil M. Maher, *Nature's New Deal: Franklin Roosevelt's Civilian Conservation Corps and the Roots of the American Environmental Movement* (New York: Oxford University Press, 2007); Sarah T. Phillips, *This Land, This Nation: Conservation, Rural America, and the New Deal* (New York: Cambridge University Press, 2007); White, *Organic Machine*.

59. Giandou, *La Compagnie nationale du Rhône*, 36–46, 49; Giandou, "Histoire d'un partenaire régional de l'Etat," 77.

60. Bethemont, *Le thème de l'eau*, 192–193; Giandou, *La Compagnie nationale du Rhône*, 53; Giandou, "Histoire d'un partenaire régional de l'Etat," 101, 109, 122. In Chapter 3 I address the work of the CNR's technical committee in greater depth.

61. Giandou, "Histoire d'un partenaire régional de l'Etat," 99.

62. See Kuisel, *Capitalism and the State in Modern France*.

63. Giandou, *La Compagnie nationale du Rhône*, 52–53; Savey, *La CNR*, 4.

64. On Vichy's policies in La Crau, see Alexandre Giandou, "L'échec d'une colonisation agricole et ses conséquences: La Crau," *Ruralia*, 2000-06, http://ruralia.revues.org/document140.html (accessed June 15, 2009). For a broader discussion of environmental management under Vichy, see Chris Pearson, *Scarred Landscapes: War and Nature in Vichy France* (New York: Palgrave Macmillan, 2009).

65. Giandou, "Histoire d'un partenaire régional de l'Etat," 133, 371, 375, 378, 381; Giandou, "Histoire de la CNR," 86. Conflicts over groundwater, analyzed in Chapter 4, support Giandou's argument regarding tensions between the CNR and the Génie Rural.

66. Quoted in Giandou, "Histoire d'un partenaire régional de l'Etat," 440.

67. *Le problème d'aujourd'hui: La sécheresse et le charbon*, Les Actualités Françaises (December 7, 1945). Les Actualités Françaises released short newsreels, shown in cinemas throughout France from January 1945 to February 1967, all related to World War II, its consequences, and reconstruction. This film, along with several thousand others produced by Les Actualités Françaises, is available through the website of the Institut National de l'Audiovisuel: www.ina.fr (accessed August 12, 2008).

68. Maurice Agulhon, André Nouschi, and Ralph Schor, *La France de 1940 à nos jours* (Paris: Nathan, 1995), 106.

69. *Les travaux de barrage sur le Rhin et le Rhône*, Les Actualités Françaises (June 23, 1949), www.ina.fr (accessed August 12, 2008).

70. Maurice Larkin, *France since the Popular Front: Government and People, 1936–1986*, 2nd ed. (New York: Oxford University Press, 1997), 6.

71. Gabrielle Hecht, *The Radiance of France: Nuclear Power and National Identity in France after World War II* (Cambridge, MA: MIT Press, 1998).

I extend Hecht's discussion of the relationship between technology and national identity in Chapter 2.

72. La Documentation française, "L'équipement hydroélectrique de la France," *Notes, documentaires et études* 524, Série française 122 (1947): 496–587; Sécretariat d'état à la présidence du conseil et à l'information, *Le Rhône, son rôle economique (aménagement, navigation fluviale, énergie hydroélectrique, irrigation: Crau et Camargue), Notes documentaires et études, N. 309, 1946* (Paris: Direction de la Documentation, 1946); La Direction de la Documentation, ed., *Donzère-Mondragon et l'aménagement du Rhône* (Paris: La Documentation française, 1950); Georges Bonnefoy, "L'usine génératrice André Blondel," *Revue générale de l'électricité* 65 (1956): 5–33. For an example of the CNR's discourse on this point, see J. Bonnier, Note sur l'aménagement du tiers central du bas-Rhône et la construction de la chute de Donzère-Mondragon (October 1946), CNR, AR0103, Généralités.

73. La Direction de la Documentation, *Donzère-Mondragon*. This language is echoed in Roger Cédié, "Deux expériences d'économie mixte dans l'industrie électrique" (Ph.D. diss., Université de Paris, 1943), 44.

74. Sécretariat d'état à la présidence du conseil et à l'information, *Le Rhône*.

75. See Chapter 2 for a more extensive discussion of this theme and its political implications.

76. Sécretariat d'état à la présidence du conseil et à l'information, *Le Rhône*, 9. See also Résidence du conseil, *Le barrage de Génissiat et l'aménagement du Rhône, Notes documentaires et études, N. 839, 1948* (Paris: Direction de la Documentation, 1948).

77. Sécretariat d'état à la présidence du conseil et à l'information, *Le Rhône*.

78. *Les travaux de barrage.*

79. La Documentation française, "L'équipement hydroélectrique de la France."

80. Pierre Rousseau, *Glaciers et torrents, énergie et lumière* (Paris: Hachette, 1955).

81. La Documentation française, "L'équipement hydroélectrique de la France."

82. M. Giguet, Conversation de M. Giguet avec M. Calvat, Inspecteur général du Génie Rural le 4 novembre 1946 (November 4, 1946), CNR, AR0103, box 3, Donzère-Mondragon.

83. Résidence du Conseil, *Le barrage de Génissiat.*

84. Sécretariat d'état à la présidence du conseil et à l'information, *Le Rhône*; La Direction de la Documentation, *Donzère-Mondragon*.

85. By stating that the issue was eventually solved in this manner, I do not mean to imply that it was fixed forever as such. To the contrary, with the emer-

gence of the European Union, France's government has been forced to privatize part of its electricity sector. The CNR has become central to the process because it was a quasi-independent energy producer, unlike the nationalized Electricité de France.

86. Pelosato, *Le Rhône,* 24. For an extensive discussion of the CNR and the possibility of nationalization, see Giandou, *La Compagnie nationale du Rhône,* chapter 5.

87. For an explanation of this shift in the early 1960s, see Chapters 3 and 5.

88. The history of the Plan has received considerable attention by political scientists, economists, and historians. Certainly, the shifting role of the state in the French economy was controversial. I do not take up this question here. For secondary literature on this issue, see Kuisel, *Capitalism and the State in Modern France,* and Hecht, *Radiance of France.*

89. On engineering hierarchy within the CNR, see Giandou, *La Compagnie nationale du Rhône,* 59–64, 135. Secondary literature on French engineering is vast. For a few key works, see Ken Alder, "A Revolution to Measure: The Political Economy of the Metric System in France," in *The Values of Precision,* ed. Norton Wise (Princeton: Princeton University Press, 1995): 39–71; Ken Alder, *Engineering the Revolution: Arms and Enlightenment in France, 1763–1815* (Princeton: Princeton University Press, 1997); Geiger, "Planning the French Canals"; Hecht, *Radiance of France;* George G. Humphreys, *Taylorism in France, 1904–1920: The Impact of Scientific Management on Factory Relations and Society* (New York: Garland, 1986); Eda Kranakis, "Social Determinants of Engineering Practice," *Social Studies of Science* 19:1 (1989): 5–70; Eda Kranakis, *Constructing a Bridge: An Exploration of Engineering Culture, Design, and Research in Nineteenth-Century France and America* (Cambridge, MA: MIT Press, 1997); Richard F. Kuisel, "Technocrats and Public Economic Policy: From the Third to the Fourth Republic," *Journal of European Economic History* 2 (1973): 53–99; Theodore Porter, *Trust in Numbers: The Pursuit of Objectivity in Science and Public Life* (Princeton: Princeton University Press, 1995); Luc Rouban, *L'Etat et la science: La politique publique de la science et de la technologie* (Paris: ECNRS, 1988); Terry Shinn, "From 'Corps' to 'Profession': The Emergence and Definition of Industrial Engineering in Modern France," in *The Organization of Science and Technology in France, 1808–1914,* ed. Robert Fox and George Weisz (New York: Cambridge University Press, 1980): 183–203; Smith, "Longest Run"; Ezra N. Suleiman, *Politics, Power, and Bureaucracy in France: The Administrative Elite* (Princeton: Princeton University Press, 1974); Ezra Suleiman, *Elites in French Society: The Politics of Survival* (Princeton: Princeton University Press, 1978); Rosemary Wakeman, "La Ville en Vol: Toulouse and the Cultural Legacy of the Airplane," *French Historical Studies* 17 (1992): 769–790; Françoise Zonabend, *La presqu'île au nucléaire* (Paris: Odile Jacob, 1989).

90. Giandou, *La Compagnie nationale du Rhône*, 18–19, 65, 138.
91. The phrase is taken from the subtitle of Giandou's book.

2. Imagining the Nation's River

1. Paul Auriol to M. Bollaert (September 4, 1952), CNR, AR0103, box 175, Inauguration de Donzère-Mondragon, Conseils, Rapports avec la Présidence de la République et les Préfets locaux.

2. C. Barrière, note pour DA, Emissions à la Radio (October 3, 1952), CNR, AR0103, box 175, Inauguration de Donzère-Mondragon, Presse et Radio.

3. Emile Bollaert to M. Roger Ruchet, Ministre des Postes-Télégraphes-Téléphones (October 30, 1952), CNR, AR0103, box 175, Inauguration de Donzère-Mondragon, Timbre sur Donzère-Mondragon. Robert Cami designed and engraved this stamp, available to the public from October 6, 1956, to February 23, 1957. It was one of forty-one stamps issued in 1956. An image of and limited information about the stamp is available online: www.phil-ouest.com/Timbre.php?Nom_timbre:Donzere_Mondragon_1956 (accessed August 11, 2008).

4. My reconstruction of the inauguration is based on Jean Chignol, "Samedi, Donzère-Mondragon a été inauguré officiellement par le président de la République...et 2.000 CRS," *Les Allobroges*, October 27, 1952, 1; Jean Fangeat, "Devant les ambassadeurs de 45 nations, M. Vincent Auriol a mis en marche l'usine géant de Donzère-Mondragon," *Le Dauphiné libéré*, October 27, 1952, 1; "Invité par la Compagnie nationale du Rhône, M. Vincent Auriol accompagné de 53 ambassadeurs et diplomates étrangers inaugurera samedi prochain le Canal Donzère-Mondragon et l'Usine André Blondel," *Le Dauphiné libéré*, October 18, 1952.

5. Gabrielle Hecht, *The Radiance of France: Nuclear Power and National Identity after World War II* (Cambridge, MA: MIT Press, 1998).

6. Eric Kauffmann, "'Naturalizing the Nation': The Rise of Naturalistic Nationalism in the United States and Canada," *Comparative Studies in Society and History* 40 (1998): 666–695; Thomas Lekan, *Imagining the Nation in Nature: Landscape Preservation and German Identity, 1885–1945* (Cambridge, MA: Harvard University Press, 2004); see several of the essays in Thomas Lekan and Thomas Zeller, eds., *Germany's Nature: Cultural Landscapes and Environmental History* (New Brunswick, NJ: Rutgers University Press, 2005); Sara B. Pritchard, "Reconstructing the Rhône: The Cultural Politics of Nature and Nation in Contemporary France, 1945–1997," *French Historical Studies* 27 (2004): 766–799; Lissa Roberts, "An Arcadian Apparatus: The Introduction of the Steam Engine into the Dutch Landscape," *Technology and Culture* 45 (2004): 251–276; Samuel Temple, "The Natures of Nation: Negotiating Modernity in

the *Landes de Gascogne,*" and Patrick Young, "A Tasteful Patrimony? Landscape Preservation and Tourism in the Sites and Monuments Campaign, 1900–1935," both in *French Historical Studies* 32 (2009); Emily T. Yeh, "From Wasteland to Wetland? Nature and Nation in China's Tibet," *Environmental History* 14 (2009): 103–137; Olivier Zimmer, "In Search of Natural Identity: Alpine Landscape and the Reconstruction of the French Nation," *Comparative Studies in Society and History* 40 (1998): 663–665.

7. James C. Scott, *Seeing Like a State: How Certain Schemes to Improve the Human Condition Have Failed* (New Haven: Yale University Press, 1998), 4.

8. For a brief discussion of early infrastructure, see Chapter 1. For more extensive analysis, see Jacques Bethemont, *Le thème de l'eau dans la vallée du Rhône: Essai sur la genèse d'un espace hydraulique* (Saint-Étienne: Imprimerie 'Le Feuillet Blanc,' 1972); Alexandre Giandou, *La Compagnie nationale du Rhône (1933–1998): Histoire d'un partenaire régional de l'Etat* (Grenoble: Presses universitaires de Grenoble, 1999). Giandou's dissertation provides more historical detail than his subsequent book; see "Histoire d'un partenaire régional de l'Etat: La Compagnie nationale du Rhône (1933–1974)" (Ph.D. diss., Université Louis Lumière-Lyon II, 1997).

9. Quoted in Albert Breittmayer, *Le Rhône; Sa navigation depuis les temps anciens jusqu'à nos jours* (Lyon: Henri Georg, 1904), 11.

10. Henri Le Masson, "L'aménagement du Rhône de Genève à Donzère-Mondragon," *France Illustration,* June 19, 1948, 613; Marcel Carrière, "Découverte de la France, pays des merveilles, IX: Donzère, capitale du Rhône," *Détective, l'hébdomadaire des secrets du monde,* 1951. Both in CNR, Documentation, unarchived press clippings.

11. For the river as a "torrent," see Pierre de Latil, "Donzère-Mondragon, chantier n. 1 de France," *Science et vie,* February 1952, 100; Gilbert Tournier, *Rhône, dieu conquis* (Paris: Librarie Plon, 1952), 2. For the "undisciplined," "capricious," and "impetuous" river, see Marc Cluzea, "Rhône, merveilleuse mine d'énergie," *Le pèlerin,* February 16, 1958, 6; André Allix, "Le Rhône," *Réalisations industrielles,* June-July 1951, 65.

12. *Crin-Blanc: Cheval sauvage,* Films Montsouris, 1953. It won the Palme d'Or at the Cannes Film Festival.

13. Tournier, *Rhône, dieu conquis,* 18.

14. Albert Plécy, "Donzère-Mondragon, à la poursuite de 2 milliards de kilowatts," *Paris match,* November 1, 1952, 28; Cluzea, "Rhône"; Allix, "Le Rhône."

15. Plécy, "Donzère-Mondragon"; Chignol, "Samedi."

16. For early work on French anxiety about losing status as a "great power," see Stanley Hoffman, ed., *In Search of France* (Cambridge, MA: Harvard University Press, 1963). For studies that discuss the relationship between large-scale technology and politics in postwar France, see Hecht's *Radiance of France*

and several works by Michael Bess: "Ecology and Artifice: Shifting Perceptions of Nature and High Technology in Postwar France," *Technology and Culture* 36 (1995): 830–862; "Greening the Mainstream: Paradoxes of Antistatism and Anticonsumerism in the French Environmental Movement," *Environmental History* 5 (2000): 6–26; *The Light-Green Society: Ecology and Technological Modernity in France, 1960–2000* (Chicago: University of Chicago Press, 2003).

17. Consider the other major rivers in France: la Seine, la Loire, and la Saône. For literature on the gendering and control of nature, two classics are Carolyn Merchant, *The Death of Nature: Women, Ecology, and the Scientific Revolution* (San Francisco: Harper and Row, 1980); Londa Scheibinger, *Nature's Body: Gender in the Making of Modern Science* (Boston: Beacon, 1993).

18. Jean-François Virenque, *Nouveaux destins du Rhône: Navigation, électricité, irrigation* (Paris: Services d'Information de la mission spéciale en France de l'ECA, n.d.), 4, in CNR, Documentation, 09668. The Marshall Plan funded Economic Cooperation Administration (ECA) publications.

19. Office de Publicité Générale, *Donzère-Mondragon* (Paris: Editions Maurice André, 1953); Daniel Faucher, *L'homme et le Rhône* (Paris: Editions Gallimard, 1968). See also Chignol, "Samedi"; Paul-Louis Bret, "La formule du Rhône," *Rapports: France–Etats Unis,* November 1952; Carrière, "Découverte."

20. *L'or du Rhône,* Les Films Caravelle, 1950. This film, produced by Les Films Caravelle for the ECA (France) and the French government, won the 1951 Prix International at the Venice Biennale for the Section Technique et Travail. It was one of approximately three hundred films, contracted and produced by European filmmakers, that were paid for by the Marshall Plan. A copy of this film is in CNR, Documentation.

21. René Grosso, Pierre Nicolas, and Lucien Perret, "Donzère-Mondragon: Les chantiers (1947–1951)," *Bibliothèque de travail* 166 (1951). See also Allix, "Le Rhône"; Carrière, "Découverte"; de Latil, "Donzère-Mondragon"; Office de Publicité Générale, *Donzère-Mondragon.* For a few works that explore the relationship between gender and industrialization, see Maxine Berg, *The Age of Manufacturers, 1700–1820* (Oxford: Blackwell, 1985); Ruth Schwartz Cowan, *More Work for Mother: The Ironies of Household Technology from the Open Hearth to the Microwave* (New York: Basic Books, 1983); Nina E. Lerman, Ruth Oldenziel, and Arwen P. Mohun, eds., *Gender and Technology: A Reader* (Baltimore: Johns Hopkins University Press, 2003). One of the most sophisticated studies of gender and industrial technology is Laura Lee Downs, *Manufacturing Inequality: Gender Division in the French and British Metalworking Industries, 1914–1939* (Ithaca: Cornell University Press, 1995). On masculinity specifically, see Ruth Oldenziel, *Making Technology Masculine: Men, Women, and Modern Machines in America, 1870–1945* (Amsterdam: Amsterdam University Press, 1999).

22. Jacques Sabran, "Le plus puissant barrage d'Europe," *Le Dauphiné libéré,* April 3, 1952.

23. For visual evidence of the November 1951 flood, see *Inondations dans la vallée du Rhône,* Les Actualités Françaises, November 29, 1951, www.ina.fr (accessed August 12, 2008).

24. "Entre Montélimar, Bollène, et Orange, La France prend déjà son visage de demain," *Paris match,* n.d., in CNR, Documentation, unarchived press clippings.

25. Sara B. Pritchard, "Mining Land and Labor," *Environmental History* 10 (2005): 731–733.

26. Claudius Deriol, "Entre Donzère et Mondragon, Le Rhône va sauter une marche de vingt-deux mètres," *France Magazine,* June 25, 1950; Plécy, "Donzère-Mondragon"; J. P. Aymon, "Le Rhône dompté, III. Le chef éclusier de Donzère-Mondragon" (September 22, 1952), in CNR, AR0103, box 175, Inauguration de Donzère-Mondragon, Presse et Radio.

27. J. Labadie, "L'aménagement du Rhône de Donzère à Mondragon," *Science et vie* 381 (1949): 380–385; Cluzea, "Rhône"; Bret, "La formule."

28. Gilbert Tournier, *Le Rhône, fleuve dieu, vous parle* (Paris: Librairie Arthème Fayard, 1957).

29. "Entre Montélimar, Bollène, et Orange."

30. Bret, "La formule."

31. Tournier, *Rhône, dieu conquis,* ii. See also Cluzea, "Rhône"; "Dernier acte du chef-d'oeuvre de notre économie nationale, hier, Donzère-Mondragon a reçu l'approbation administrative," *Le progrès* [?], November 14, 1953, 15, in AD-Drôme, press clippings on the CNR.

32. For evidence of how other postwar technocrats articulated a discourse of love for their projects and expertise, see Hecht, *Radiance of France,* 34. On work, labor, and nature, the seminal essay is Richard White, "Are You an Environmentalist or Do You Work for a Living? Work and Nature," in *Uncommon Ground: Toward Reinventing Nature,* ed. William Cronon (New York: Norton, 1995): 171–185. See also Thomas C. Andrews, *Killing for Coal: America's Deadliest Labor War* (Cambridge, MA: Harvard University Press, 2008); Neil Maher, "A New Deal Body Politic: Landscape, Labor, and the Civilian Conservation Corps," *Environmental History* 7 (2002): 435–461. It is important to note, however, that an analytic connection between nature and labor is significantly different from the ways in which actors invoked (or obscured) this link.

33. De Latil, "Donzère-Mondragon"; Bret, "La formule," 3; Raoul Faure, "Le Rhône (II, Au fil de son histoire)," *Bibliothèque de travail* (1959), in CNR, Documentation, 09440; Groupement d'entreprises des travaux d'aménagement du Rhône à Donzère, *Donzère-Mondragon: Bulletin d'Information Technique*

et Professionelle, Numéro 3 (1947–1952), CNR. The CNR has not yet catalogued these bulletins.

34. Quoted in Daniel Bideau, *De Lyon à Valence au gré du Rhône* (Lyon: Elie Bellier, 1982), 8.

35. For another example of how history and historicism were invoked as strategies for justification, see Hecht, *Radiance of France*, chapter 1.

36. Quoted in Allix, "Le Rhône."

37. "The Rhône Valley," *Water Power* (1960): 214–223; Allix, "Le Rhône," 65.

38. Tournier, *Rhône, dieu conquis;* Virenque, *Nouveaux destins.*

39. Plécy, "Donzère-Mondragon," 3.

40. Carrière, "Découverte."

41. Jacques Sabran, "Le nouveau Rhône est né," *Le Dauphiné libéré,* October 24, 1952. See also J. P. Aymon, "Le Rhône sous le joug, I. Il y a quelque chose de changé" (September 19, 1952), in CNR, AR0103, box 175, Inauguration de Donzère-Mondragon, Presse et Radio; Deriol, "Entre Donzère et Mondragon."

42. Carrière, "Découverte"; Allix, "Le Rhône," 70, 117.

43. David Harvey, *The Condition of Postmodernity: An Enquiry into the Origins of Cultural Change* (Cambridge, MA: Blackwell, 1990), 16.

44. "Entre Montélimar, Bollène, et Orange"; Bret, "La formule"; Aymon, "Le Rhône dompté, III."

45. Tournier, *Rhône, dieu conquis,* 112.

46. Génissiat, technically the first project the CNR completed after World War II, was represented in similar ways at its inauguration. For instance, Minister of Industrial Production Robert Lacoste declared that "in Génissiat, we see an indisputable sign of France's rebirth." Quoted in *Le plus formidable barrage d'Europe occidentale,* Les Actualités Françaises, January 29, 1948, www.ina.fr (accessed August 12, 2008). Political and popular coverage of Donzère-Mondragon's inauguration usually emphasized the ways it had surpassed Génissiat.

47. Quoted in Fangeat, "Devant les ambassadeurs."

48. Quoted in Office de Publicité Générale, *Donzère-Mondragon.*

49. "Donzère-Mondragon," *Le Dauphiné libéré,* October 25, 1952.

50. Plécy, "Donzère-Mondragon"; Sabran, "Le plus puissant barrage."

51. Deriol, "Entre Donzère et Mondragon"; Plécy, "Donzère-Mondragon"; R. L. Lachat, "Alors que l'on s'apprête à inaugurer avec éclat le gigantesque ouvrage rappelons que ce sont 2 Grenoblois MM. Giroud et Perrin qui, en 1811, et pour la somme de 40 000 fr. lancèrent les premiers travaux de Donzère-Mondragon," *Le Dauphiné libéré,* October 23, 1952; Raymond Domergue, "Au cours d'une importante réunion tenue à Donzère, Les sinistrés du canal Donzère-Mondragon ont réaffirmé leur volonté d'obtenir la juste indemnisation

des dommages causés," newspaper unknown, April 12, 1952, in AD-Drôme, press clippings on the CNR; Le Masson, "L'aménagement du Rhône."

52. De Latil, "Donzère-Mondragon"; Carrière, "Découverte"; Plécy, "Donzère-Mondragon"; *Barrage de Donzère-Mondragon,* Les Actualités Françaises, October 5, 1950, www.ina.fr (accessed August 12, 2008).

53. De Latil, "Donzère-Mondragon," especially 100–101; Deriol, "Entre Donzère et Mondragon."

54. For discussion of Marcoule, a nuclear reactor located on the Rhône about twenty miles downstream of Donzère-Mondragon, see Gabrielle Hecht, "Peasants, Engineers, and Atomic Cathedrals: Narrating Modernization in Postwar France," *French Historical Studies* 20 (1997): 381–418.

55. Fangeat, "Devant les ambassadeurs"; de Latil, "Donzère-Mondragon."

56. Office de Publicité Générale, *Donzère-Mondragon.*

57. Miriam Levin, "Eiffel Tower Revisited," *French Review* 62 (1989): 1052–1064; Debora Silverman, *Art Nouveau in Fin-de-Siècle France: Politics, Psychology, and Style* (Berkeley: University of California Press, 1989).

58. Virenque, *Nouveaux destins,* 27.

59. For more on the idea of "technological modernity," see Bess, *Light-Green Society.*

60. *Barrage de Donzère-Mondragon.*

61. De Latil, "Donzère-Mondragon," 100–101.

62. Le Masson, "L'aménagement du Rhône."

63. Jean Durand, "Le 25 octobre 1952, le président Auriol inaugurait le complexe Donzère-Mondragon triomphe de la technique française," unknown newspaper, in CNR, Documentation, unarchived press clippings; Sabran, "Le plus puissant barrage." Regarding the project's world records, see Carrière, "Découverte"; Office de Publicité Générale, *Donzère-Mondragon;* Virenque, *Nouveaux destins,* 20; Deriol, "Entre Donzère et Mondragon."

64. René Grosso, Pierre Nicolas, Lucien Perret, "La peine des hommes à Donzère-Mondragon," *Bibliothèque de travail* 167 (1951): 13.

65. The literature on American-French relations after the war is extensive. Two classic studies are by Richard F. Kuisel: *Capitalism and the State in Modern France: Renovation and Economic Management in the Twentieth Century* (New York: Cambridge University Press, 1981), and especially *Seducing the French: The Dilemma of Americanization* (Berkeley: University of California Press, 1993).

66. Labadie, "L'aménagement du Rhône."

67. Claude Thomas, "L'équipement hydro-électrique en France, Du plus grand chantier au plus haut barrage," *France Illustration,* April 1, 1950, in CNR, Documentation, 09431.

68. Deriol, "Entre Donzère et Mondragon."

69. Cluzea, "Rhône."

70. One exception is the 1950 newsreel *Barrage de Donzère-Mondragon*, which does reference the Marshall Plan.

71. The exact figures for American financing of the CNR's projects are not entirely clear. Of the CNR's immediate postwar projects, 90 percent were funded by the Marshall Plan, according to U.S. Congress, Senate Hearings before Committee on Foreign Relations, 82nd Cong., 2nd sess., March/April 1952, 419–420. About 2 percent of the total Marshall Plan monies were devoted to the CNR's activities, according to Jean-Pierre Rioux, *La France de la Quatrième République 1. L'ardeur et la nécessité* (Paris: Editions du Seuil, 1980), 242.

72. Plécy, "Donzère-Mondragon," 17.

73. Plécy, "Donzère-Mondragon"; Durand, "Le 25 octobre 1952"; de Latil, "Donzère-Mondragon," 100.

74. "Donzère-Mondragon," *Le progrès,* October 27, 1952, 1.

75. Hecht, *Radiance of France.*

76. Chignol, "Samedi."

77. Aymon, "Le Rhône sous le joug, I."

78. Gilbert Tournier, *En descendant le cours du Rhône* (Lyon: Chambre de Commerce de Lyon, 1960).

79. "Dernier acte"; de Latil, "Donzère-Mondragon," 111.

80. Plécy, "Donzère-Mondragon"; *Barrage de Donzère-Mondragon.* James C. Scott would likely frame these changes in terms of state schemes of legibility and high modernism.

81. *L'or du Rhône.*

82. "Entre Montélimar, Bollène, et Orange."

83. Kuisel, *Capitalism and the State in Modern France;* Herrick Chapman, *State Capitalism and Working-Class Radicalism in the French Aircraft Industry* (Berkeley: University of California Press, 1991).

84. Fangeat, "Devant les ambassadeurs." Auriol did distinguish, however, between the "conception" and "execution" of the project, presumably contrasting engineers' contributions and those of workers.

85. "Donzère-Mondragon," *Le progrès,* October 27, 1952, 1.

86. The CNR's archives include numerous documents that refer to dozens of strikes held during the construction of Donzère-Mondragon, particularly in 1947 and 1948.

87. Tournier, *Rhône, dieu conquis,* 5–6.

88. Weber, *Peasants into Frenchmen.*

89. Sabran, "Le plus puissant barrage"; Carrière, "Découverte." For an account of the flooding of Tignes and local responses to the EDF's project, see Robert Frost, "The Flood of Progress: Technocrats and Peasants at Tignes (Savoy), 1946–1952," *French Historical Studies* 14 (1985): 117–140.

90. Local perspectives can also be gleaned from the Archives de l'Association des Amis du Vieux Donzère, Archives Municipales de Pierrelatte, and Archives Municipales de Saint-Paul-Trois-Châteaux.

91. In France, *écologiste* is roughly equivalent to the Anglo-American "environmentalist," but the term is loaded: *écologiste* and especially *écolo* are often used in a derogatory fashion to critique the environmental movement. In order to prevent confusion by using the literal translation, "ecologist," which might imply to American readers a scientist who studies ecology, I have used "environmentalist" (even though some French environmentalists did come from the scientific community). See Chapters 6 and 7 for discussion of environmentalists and their role in the Rhône's transformation since the late 1960s.

92. Plécy, "Donzère-Mondragon"; Deriol, "Entre Donzère et Mondragon"; Domergue, "Au cours d'une importante réunion." Chapter 3 discusses some of the environmental dimensions of the river's development, even if they were not framed as such.

93. Chignol, "Samedi."

3. Postwar Transformations

1. CNR, Procès-verbal de la réunion du comité technique du 20 décembre 1934 (Compte tenu des corrections décidées le 22 janvier 1935) (December 20, 1934), CNR, Comité n. 2, Procès-verbal de la réunion du comité technique du 22 janvier 1935 (April 15, 1935), and CNR, Comité technique n. 38 du 23 octobre 1941 (October 23, 1941), all in CNR, Comité technique, p.v. For retrospective accounts, see Jean Aubert, "La question du Rhône, Conférence du 16 avril 1943" (1943), in CNR, Documentation, 10039; Jean Aubert, "La canalisation du Bas-Rhône," in *Annales des Ponts et Chaussées* (1947), in CNR, Documentation, AE0953. Of note, documents in the CNR's archives tend to have long titles and folder names. Many include redundancies (such as acronyms for agencies followed by their full names) and inconsistencies (such as multiple date formats). Authors of these documents are not always specified; in these cases, I have indicated authorship by "CNR." In addition, complete first and last names are rare; first initials, if given, are more common. I have included all information "as is" to clarify historical provenance, especially because the CNR's archives are basically untapped.

2. CNR, Procès-verbal de la réunion du comité technique du 20 décembre 1934 (December 20, 1934). The Ponts et Chaussées engineers' concerns are raised in CNR, Comité n. 2, Procès-verbal de la réunion du comité technique du 22 janvier 1935 (April 15, 1935), 2. No source explains their shift in opinion, but it seems likely that the outbreak of World War II played a key role in helping transcend divergent views.

3. Lyon's Herriot port and Génissiat were both delayed by World War II. Donzère-Mondragon was the CNR's first project undertaken entirely after the war.

4. The project's name is explained in CNR, Comité technique n. 38 du 23 octobre 1941 (October 23, 1941). Aubert's quote is from "La question," 286. For an explanation of how the diversion approach would work, see Aubert, "La canalisation." On the diversion approach as the model of development and the unique status of Génissiat, see Sécretariat d'état à la présidence du conseil et à l'information, *Le Rhône, son rôle économique (aménagement, navigation fluviale, énergie hydroélectrique, irrigation: Crau et Camargue), Notes documentaires et études, N. 309, 1946* (Paris: Direction de la Documentation, 1946).

5. Aubert, "La canalisation," 473.

6. "Les réalisations françaises, Donzère-Mondragon, second Génissiat rhodanien," *Cahiers français d'information: La Documentation française* (December 1, 1948); J. Bonnier, Note sur l'aménagement du tiers central du Bas-Rhône et la construction de la chute de Donzère-Mondragon (October 1946), CNR, AR0103, Généralités; Sécretariat d'état à la présidence du conseil et à l'information, *Le Rhône*.

7. The quotations are from Pierre Delattre, *La chute de Donzère-Mondragon sur le Rhône* (Paris: Société des ingénieurs civils de France, 1952), in CNR, Documentation, AI0282; J. Bonnier, Note sur l'aménagement du tiers central du Bas-Rhône (October 1946); Cl. Gemaehling and P. Savey, "The Multipurpose Development of the River Rhône Valley," in *Water Power: IPC Electrical and Electronic Press* (November 1972), in CNR, Documentation, 09426; the reference to "new countries" is from both Aubert, "La canalisation," and Delattre, *La chute de Donzère-Mondragon*. On hydrologic and topographic challenges more broadly, especially for high-chute dams, see also "Les réalisations françaises, Donzère-Mondragon"; *Donzère-Mondragon* (Grenoble: La Houille Blanche, 1955); Georges Bonnefoy, "L'usine génératrice André Blondel," *Revue générale de l'électricité* 65 (1956): 5–33.

8. Delattre, *La chute de Donzère-Mondragon*. On the project's advantages, see La Direction de la Documentation, ed., *Donzère-Mondragon et l'aménagement du Rhône, La Documentation française illustrée* (Paris: La Documentation française, 1950). In referring to the "original riverbed," I am by no means suggesting that it was pristine or static; rather, this is how engineers referred to it.

9. Aubert, "La canalisation."

10. My references to the social, natural, and built environments of the central Rhône are not a reification of these categories. As STS scholars have shown, these concepts and the boundaries among them are historically and culturally contingent.

11. Marcel Bonnefoi, Compte-rendu de la réunion tenue le lundi 16 juin 1947 au siège de la "CGA" à Paris, entre: les représentants de la Compagnie nationale du Rhône et les représentants de la "CGA" (June 28, 1947), CNR, AR0103, box 3, Donzère-Mondragon (folder hereafter DM).

12. Jacques Bethemont, *Le thème de l'eau dans la vallée du Rhône: Essai sur la genèse d'un espace hydraulique* (Saint-Étienne: Imprimerie 'Le Feuillet Blanc,' 1972), 217–218.

13. The quote is from CNR, Comité technique n. 54 du 8 juillet 1946 (July 8, 1946), CNR, Comité technique, p.v. For early discussions of geologic integrity, see CNR, Comité technique n. 57 du 12 décembre 1946 (December 12, 1946), CNR, Comité technique, p.v.; Aubert, "La canalisation." Retrospective narratives particularly emphasize this point; see, for instance, Delattre, *La chute de Donzère-Mondragon; Donzère-Mondragon*, 16, 32.

14. To simplify, hydroelectric generation largely depends on two factors: the volume of water and the height difference (head) between the water's source (usually a reservoir) and the turbines.

15. CNR, Comité technique n. 57 du 12 décembre 1946 (December 12, 1946); CNR, Comité technique n. 60 du 7 juillet 1947 (July 7, 1947), CNR, Comité technique, p.v. For additional evidence of the CNR's preoccupation with its finances, including the economic dimensions of route selection, see Aubert, "La canalisation."

16. Delattre, *La chute de Donzère-Mondragon*.

17. These years were the basis of analysis in CNR, Aménagement de la chute de Donzère-Mondragon, Retenue à la cote 58,00 (January 25, 1944), CNR, AR0103, box 3, DM.

18. No detailed description of how they determined these flood averages remains. Most documents with information about sampling and "average" years simply include charts showing different flood intensities and their relative flows; they do not discuss the criteria for these categories. These flood averages are listed in CNR, Aménagement de la chute de Donzère-Mondragon, Retenue à la cote 58,00 (January 25, 1944); *Donzère-Mondragon*, 185. For examples of the average flows for other flood levels, see CNR, Chute de Donzère-Mondragon, Caractéristiques du barrage de retenue (August 23, 1943), CNR, AR0103, box 158, Rapports avec les Services du Contrôle, 1947–51; Bonnefoy, "L'usine."

19. The production of norms for other rivers has also had important consequences, both socially and ecologically. Probably the most famous example is the Colorado River. See Gary D. Weatherford and F. Lee Brown, eds., *New Courses for the Colorado River: Major Issues for the Next Century* (Albuquerque: University of New Mexico Press, 1986).

20. Delattre, *La chute de Donzère-Mondragon*.

21. "Les réalisations françaises, Donzère-Mondragon"; Delattre, *La chute de Donzère-Mondragon*. These debates suggest the importance of scale within

(but also among) envirotechnical systems. Each CNR project like Donzère-Mondragon might be considered its own system. However, each project was also a part of the CNR's entire river-based program, itself an envirotechnical system.

22. Bonnefoy, "L'usine."

23. Ponts et Chaussées's recommendation is described in "Les réalisations françaises, Donzère-Mondragon." The CNR's preference is outlined in Ingénieur en Chef du Service des nouvelles chutes, Chute de Donzère-Mondragon, Mémoire descriptif (April 20, 1942), CNR, AR0103, Chute de Donzère-Mondragon, Pièces générales et ouvrages hydroelectriques, 1, 5, 6.

24. R. Kirchner, Rhône, Aménagement de la chute de Donzère-Mondragon, la Cie nationale du Rhône, concessionnaire, Rapport de l'Ingénieur en Chef (October 23, 1942). A CNR document reports that the conventional low-water level is 53.05 NGF. With a reservoir level of 56.6, the height of the reservoir would be 3.55 meters higher than low-water level. The disparity between this conclusion and Kirchner's is therefore 0.7 meters, not an insignificant difference. Perhaps more important, it suggests the degree to which CNR and other engineers differed on these numbers. For the CNR's analysis, see CNR, Graphique de la tenue des eaux à l'échelle 170.500 donnant le nombre moyen annuel de jours pendant lesquels les eaux se sont tenues au-dessous d'une cote donnée, Moyenne 1906–1935 [1943?]. Both in CAC, 780467, box 24, Chute de Donzère-Mondragon, Prise en considération de l'avant projet, Depèche du 21 juillet 1943.

25. The figures for the 1856 flood appear less contentious. See Ingénieur en Chef du Service des nouvelles chutes, Chute de Donzère-Mondragon, Mémoire descriptif (April 20, 1942); CNR, Chute de Donzère-Mondragon, Note au sujet des nouvelles dispositions du barrage de garde (September 26, 1945), CNR, AR0103, box 4, Chute de Donzère-Mondragon, Dossier de résultats de l'enquête et des conférences et de propositions de la CNR; *Donzère-Mondragon*.

26. Again, demands for energy, largely spurred by postwar reconstruction, both material and symbolic, contributed to the prioritization of hydroelectric production. Agriculture and navigation did not disappear, but energy was the main concern after World War II. This certainly contrasts with the long-standing interest in navigation seen in Chapter 1. As we saw in that chapter, the emergence of hydroelectricity in the late nineteenth century altered the possible uses of the Rhône in important ways and contributed to the general shift from navigation to hydropower. Of course, as this and subsequent chapters show, the goals and power dynamics of river management would shift again.

27. Early CNR documents that refer to 58.00 NGF include CNR, Chute de Donzère-Mondragon, Caractéristiques du barrage de retenue (August 23, 1943); CNR, Chute de Donzère-Mondragon, Note au sujet des nouvelles dispositions du barrage de garde (September 26, 1945); Aubert, "La canalisation." On the

ensuing reservoir length, see La Direction de la Documentation, *Donzère-Mondragon;* Chambre de Commerce de la Drôme, *La Drôme* (Romans: J. A. Domergue, 1950). For a specific discussion of the energy consequences of raising the reservoir height, see CNR, Aménagement de la chute de Donzère-Mondragon, Retenue à la cote 58,00 (January 25, 1944). The standard was proposed as part of the project's preliminary design; see, for instance, CNR, Chute de Donzère-Mondragon, Résultats de l'enquête et des conférences (October 3, 1945), CNR, E17A3A5,6.

28. The CNR's terminology for the original riverbed does not seem to have been shared by the EDF for its hydroelectric projects. Robert Frost, personal communication, 1999.

29. In analyzing CNR documents from the 1940s to the present, I have noticed that "former" and "dead Rhône" have been used less frequently and often replaced by the more neutral phrase *Rhône court-circuité* ("short-circuited Rhône") since the 1970s. Although I have not compiled statistical data to document a clear trend, I assume that growing pressures from environmentalists and the Ministry of the Environment shaped the CNR's discourse in more recent years. See Chapter 7 for a discussion of the period since the early 1970s.

30. Again, the "original" river does not capture the Rhône's dynamism before the CNR's projects. As contemporaries pointed out, the "original" river was actually in constant flux. I am simply trying to emphasize here that the CNR's development program initiated a significant (literal) shift in the so-called original riverbed.

31. Aubert, "La canalisation."

32. Although the river's "hydrologic regime" was clearly mediated by experts' knowledge of it, this regime also presented material constraints regardless of the "accuracy" of their knowledge.

33. The CNR concluded that the "débit moyen" was 1,630 m^3/s in CNR, Aménagement de la chute de Donzère-Mondragon, Retenue à la cote 58,00 (January 25, 1944). It reported an average flow of 1,600 m^3/s in Delattre, *La chute de Donzère-Mondragon.* For the definition of "débit moyen," see *Donzère-Mondragon,* 129; Bonnefoy, "L'usine." Aubert outlined the minimum flow for five turbines in "La canalisation."

34. For 1,320 m^3/s, see P. Delattre to M. Haegelen, Inspecteur Général des Ponts et Chaussées, Chute de Donzère-Mondragon (October 19, 1943), CNR, AR0103, box 3, DM. For 1,530 m^3/s, see A. Mauchamp, Aménagement de la chute de Donzère-Mondragon, Durées comparatives des parcours du Rhône et de la dérivation par la batellerie (October 28, 1943), CNR, AR0103, box 49, Chute de Donzère-Mondragon, Comparaison des vitesses du courant et des temps de parcours de la batellerie dans le Rhône et dans le canal, Généralités. For 1,100 m^3/s, see CNR, Aménagement de la chute de Donzère-Mondragon, Retenue à la cote 58,00 (January 25, 1944). For 1,060 m^3/s, see Aubert, "La

canalisation." See also CNR, Chute de Donzère-Mondragon, Résultats de l'enquête et des conférences (October 3, 1945).

35. For the increase, see J. Bouvier to CNR (February 11, 1947), CNR, AR0103, box 26, Chute de Donzère-Mondragon, Marché SACTARD pour la construction du canal de dérivation, Préparation du marché. Several documents convey that the CNR hoped to increase the diversion canal's flow in order to boost (if not maximize) energy production; see, for instance, P. Delattre to M. Bourgin (January 3, 1950), CNR, AR0103, box 12, Chute de Donzère-Mondragon, Relations avec les Services du Contrôle concernant les travaux le long de la retenue; *Donzère-Mondragon,* 101; H. Babinet to M. le Préfet du Gard, Le Rhône, Dérivation du canal de Donzère-Mondragon, Conséquence de la réduction du débit du Rhône produite par la dérivation (October 15, 1955), CAC, 770764, box 19, Compagnie nationale du Rhône, Affaires diverses.

36. M. Bourgad, interview by author, 1999.

37. CNR, Comité technique n. 38 du 23 octobre 1941 (October 23, 1941). No document explains how CNR officials arrived at this figure.

38. Ministère de l'Agriculture et du Ravitaillement de Pampelonne, Direction Générale du Génie Rural et de l'Hydraulique agricole to M. le Ministre, Secretaire d'Etat à la Production Industrielle et aux Communications (November 15, 1943), CNR, AR0103, box 3, Chute de Donzère-Mondragon, Enquête, conférences, et correspondance, Services du Ministère de l'Agriculture (folder hereafter DM, Enquête, Min. de l'Ag.); Préfet de l'Ardèche to M. l'Inspecteur Général, Chef de la 6ème Circonscription Electrique (April 23, 1945), CNR, AR0103, box 4, Chute de Donzère-Mondragon, Dossier des résultats de l'enquête et des conférences et des propositions de la CNR, Départements du Gard, du Vaucluse, de la Drôme, et de l'Ardèche, Résultats de l'enquête (folder hereafter DM, Résultats de l'enquête, Départements).

39. Burdin, Rapport de l'Ingénieur en Chef du Génie Rural chargé de l'aménagement agricole des eaux (May 15, 1945), CNR, AR0103, box 4, DM, Résultats de l'enquête, Départements; CNR, Chute de Donzère-Mondragon, Résultats de l'enquête et conférences (August 10, 1945), CNR, AR0103, Box 4, Chute de Donzère-Mondragon, Enquête et conférences, Départements (Conseils généraux, communes).

40. These arguments revise James C. Scott, *Seeing Like a State: How Certain Schemes to Improve the Human Condition Have Failed* (New Haven: Yale University Press, 1999). I address issues related to shifts in fish populations and species composition, and how various constituencies viewed them, in more detail in "Envirotechnical Systems and Evolutionary Implications: Fish, Dams, and Development on France's Rhône River" (manuscript in preparation for *Technology and Culture*).

41. CNR, Aménagement de la chute de Donzère-Mondragon, Retenue à la cote 58,00 (January 25, 1944) (italics mine). The statement from the public in-

quiry comes from CNR, Chute de Donzère-Mondragon, Résultats de l'enquête et des conférences (October 3, 1945) (italics mine).

42. Marc Henry, Donzère-Mondragon, Conférences, Commission supérieur des sites, Réunion du 12 décembre 1951 (December 13, 1951), CNR, AR0103, box 3, DM.

43. E. Bollaert to M. Perchet, Direction de l'Architecture, Ministère de l'Education nationale (February 7, 1952), CNR, AR0103, box 3, Chute de Donzère-Mondragon, Enquête et conférences, Commissions des sites et monuments historiques; P. Delattre to M. Saulgeot, Directeur de l'Electricité (February 14, 1952), CNR, AR0103, box 5, Chute de Donzère-Mondragon, Rapports avec les Services du Contrôle au sujet de l'exécution.

44. Ministère de l'Industrie et du Commerce, "Décret du 7 décembre 1953 relatif à l'aménagement de la chute de Donzère-Mondragon, sur le Rhône," *Journal Officiel* (December 12, 1953), in CNR, AR0103, box 4, Chute de Donzère-Mondragon, Enquête et conférences, Décret de déclaration d'utilité publique, Convention, Cahier des charges (italics mine). On higher flows in the dead Rhône during floods, see J. Rostagni, note pour TEX, Conséquences de l'abaissement du niveau du Rhône entre Donzère et la restitution (February 3, 1953), CNR, AR0103, box 82, Chute de Donzère-Mondragon, Maintien de la nappe phréatique sur la rive droite du Rhône, Rapports avec l'administration.

45. A later part of this chapter addresses how the CNR managed the diversion canal during floods, especially as tensions emerged among the CNR, CEA, and EDF.

46. P. Delattre, note to M. Tournier (October 11, 1958), CNR, AR0106, Box 56, Chute de Donzère-Mondragon, Exploitation, Terrains industriels, CEA Commissariat à l'Energie Atomique, Centre nucléaire de Pierrelatte, Généralités et divers (folder hereafter DM, CEA, Pierrelatte, Généralités); Cl. Gemaehling, "Vers l'achèvement de l'aménagement du Rhône," *Techniques et Sciences Municipales* 3 (1974): 93–103. On the Mistral, see Gabrielle Hecht, *The Radiance of France: Nuclear Power and National Identity in France after World War II* (Cambridge, MA: MIT Press, 1998), 166.

47. USSI, Schéma de rejet des effluents (October 1, 1964), CNR, AR0106, box 51, Chute de Donzère-Mondragon, Exploitation, CEA Commissariat à l'Energie Atomique, Centre nucléaire de Pierrelatte, Relations avec le CEA, 1964 à 1967 inclus (folder hereafter DM, CEA, Pierrelatte, 1964–1967).

48. P. Delattre, note to M. Tournier (October 11, 1958).

49. I readily concede that the pre-CNR Rhône had already been reshaped by agricultural, industrial, urban, and other uses for centuries.

50. William Cronon, *Nature's Metropolis: Chicago and the Great West* (New York: Norton, 1991), xix.

51. R. Galley to M. l'Ingénieur en Chef du Génie Rural, Hydrologie du nouveau centre nucléaire du CEA à Pierrelatte (November 27, 1959), CNR,

AR0106, box 56, Chute de Donzère-Mondragon, Exploitation, CEA Commissariat à l'Energie Atomique (Centre nucléaire de Pierrelatte), Relations avec le CEA, 1959 à 1963 inclus. I have been unable to explain either the intense discussion of this issue or its sudden disappearance in 1964 other than the fact that the CEA's facility had become operational by that point. It is interesting to note, however, that the phrase "nuclear reactor hydrology" did not appear in discussions of later nuclear installations.

52. This definition is drawn from the *Oxford English Dictionary* and the *American Heritage Dictionary*.

53. French atomic testing has received the most attention here.

54. P. Delattre, note to M. Tournier (October 11, 1958). The initial figure is proposed in R. Galley to M. l'Ingénieur en Chef du Génie Rural, Hydrologie du nouveau centre nucléaire du CEA à Pierrelatte (November 27, 1959). His revised figure is outlined in R. Galley to CNR, Hydrologie du nouveau centre nucléaire du CEA de Pierrelatte (July 11, 1960), CNR, AR0106, box 56, Chute de Donzère-Mondragon, Exploitation, CEA Commissariat à l'Energie Atomique (Centre nucléaire de Pierrelatte), Relations avec diverses personnalités et administrations (Préfecture, EDF, SNCF, 6ème CE, Services Navigation, Génie Rural, etc. . . .) (folder hereafter DM, CEA, Pierrelatte, Relations).

55. The first increase is described in R. Galley to CNR, Hydrologie du nouveau centre nucléaire du CEA de Pierrelatte (June 16, 1961). The approval is from P. Bayard, note to DA, Donzère-Mondragon, Hydrologie du nouveau centre nucléaire de Pierrelatte (February 6, 1962), CNR, AR0106, box 56, DM, CEA, Pierrelatte, Généralités. The second increase is from C. Leduc to CNR, Hydrologie du centre nucléaire de Pierrelatte (October 7, 1964), CNR, AR0106, box 51, DM, CEA, Pierrelatte, 1964–1967. The convention is Avenant à la Convention CNR-CEA-EDF (March 16 and 18, 1964). Galley and the convention both in CNR, AR0106, box 56, DM, CEA, Pierrelatte, Relations.

56. R. Ponnelle, note to DA.GIA, Usine de séparation isotopique du Tricastin, Autorisation de prise d'eau (February 28, 1978). The official decree is Usine de séparation isotopique du Tricastin, Convention tripartite entre CNR-EURODIF-EDF (June 27, 1979). Both in CNR, AR0106, box 44, Chute de Donzère-Mondragon, Société de Construction d'Usines de Séparation Isotopique (USSI), Eurodif.

57. J. Garraud, Donzère-Mondragon, Projet d'implantation d'une usine thermique sur la commune de Pierrelatte, Compte rendu de la réunion tenue le mercredi 30 mai 1973 dans les bureaux du Siège Social de la CNR (June 5, 1973); P. de Gaujac, Site du Tricastin, Avant-projet d'implantation d'une centrale nucléaire sur le site du Tricastin, Compte rendu de la réunion REM-CNR le 30 mai 1973 à la CNR à Lyon (June 18, 1973). Both in CNR, AR0106, box 49, Généralités, 1973–1978.

58. On the CNR's reservations, see R. Galley to CNR, Hydrologie du nouveau centre nucléaire du CEA de Pierrelatte (July 11, 1960). For Audebrand's proposal, see G. Audebrand, Note pour TEX, Donzère-Mondragon, Hydrologie du centre nucléaire de Pierrelatte (September 14, 1960), CNR, AR0106, box 56, DM, CEA, Pierrelatte, Relations. See Chapter 4 for an extensive discussion of the CNR's counter-canal network and recharge program.

59. On the initial infrastructure, see R. Galley to CNR, Hydrologie du centre nucléaire de Pierrelatte (November 30, 1961). On the third pipe, see P. Bayard to Commissariat à l'Energie Atomique, Donzère-Mondragon, Hydrologie du centre nucléaire de Pierrelatte (November 16, 1960). Both in CNR, AR0106, box 56, DM, CEA, Pierrelatte, Relations. On the emergency policy, see C. Leduc to CNR, Hydrologie du centre nucléaire de Pierrelatte (October 7, 1964).

60. J. Forestier to CNR, Prise d'eau dans le canal de Donzère, Usine du Tricastin (May 8, 1974), CNR, AR0106, box 44 (no folder).

61. Several documents like Fioravante's indicate the Gaffière and Lauzon were drainage canals built by communities centuries ago. Recent maps and photographs show that both are extremely linear rivers, lending greater support to Fioravante's suggestion that they were actually managed, if not constructed, much earlier. Given their location, it seems likely that they were associated with the Canal de Pierrelatte. Note, however, that almost all recent documents refer to them as "rivers" or "streams" *(ruisseaux),* including those associated with a uranium spill in July 2008 (see the Conclusion). It appears that the human shaping of these systems has become so naturalized, culturally if not ecologically, over the past three centuries that state experts now represent them as rivers.

62. A copy of Fioravante's letter is attached to P. Delattre, Lettre de M. Fioravante au sujet de l'aménagement de la plaine de Pierrelatte, Transmis à M. Gres prié d'étudier cette question, après avoir pris contact avec M. Fioravante (August 26, 1959), CNR, AR0106, box 56, DM, CEA, Pierrelatte, Relations.

63. P. Delattre, Lettre de M. Fioravante au sujet de l'aménagement de la plaine de Pierrelatte (August 26, 1959). On the problems floods created for "drainage" networks, see CNR, Commissariat à l'Energie Atomique, Usine de Pierrelatte, Compte-rendu de l'entretien du 28 septembre 1959 (September 30, 1959). The Génie Rural study is Rollet, Rapport de l'Ingénieur d'arrondissement, Centre d'énergie atomique de Pierrelatte, Evacuation des eaux du site, Enquête hydraulique (August 31, 1961). All in CNR, AR0106, box 56, DM, CEA, Pierrelatte, Relations.

64. Galley's quote is from R. Galley to M. l'Ingénieur en Chef du Génie Rural, Hydrologie du nouveau centre nucléaire du CEA à Pierrelatte (November 27, 1959). The CNR's admission is quoted in Rollet, Rapport de l'Ingénieur d'Arrondissement, Centre d'énergie atomique de Pierrelatte, Evacuation des eaux du site, Enquête hydraulique (August 31, 1961). The resolution is described

in R. Galley to CNR, Hydrologie du nouveau centre nucléaire du CEA de Pierrelatte (July 11, 1960).

65. Cl. Gemaehling to M. l'Ingénieur en Chef, Usine de séparation isotopique du Tricastin (June 26, 1974), CNR, AR0106, box 44 (no folder); P. Bayard to SEDIM, Usine SFEC à Bollène, Rejets d'eaux usées dans le contre-canal (October 17, 1974), CNR, AR0106, box 94, Chute de Donzère-Mondragon, Déversements; E. Chauvet, Département de la Drôme, Commune de Saint-Paul-Trois-Chateaux, Centrale nucléaire du Tricastin, Demande de déclaration d'utilité publique (January 24, 1974), CNR, AR0106, box 49, Généralités, 1973–1978.

66. J. P. Hermet to CNR, Rejet des eaux, Nouvelle usine SFEC à Bollène (April 10, 1975), CNR, AR0106, box 94, unlabeled folder regarding SFEC.

67. On nuclear consumption of river-produced energy, see J. M. Delettrez, *Le Rhône de Genève à la Méditerranée* (France: B. Arthaud, 1974). Henry's concerns are outlined in M. Henry to Electricité de France, à l'attention de M. Boudrant, Usine et poste de Bollène, Alimentation du CEA (September 6, 1961), CNR, AR0106, box 56, DM, CEA, Pierrelatte, Relations.

68. Boudrant to CNR, A l'attention de Monsieur Henry, Usine & poste de Bollène, Alimentation du CEA à Pierrelatte (September 26, 1961), CNR, AR0106, box 56, DM, CEA, Pierrelatte, Relations.

69. Cl. Gemaehling, note to EU, Usine de Bollène, Alimentation du CEA à Pierrelatte (December 11, 1961), CNR, AR0106, box 56, DM, CEA, Pierrelatte, Généralités; M. Henry to M. Wyart, Direction de la Production et du Transport, Electricité de France, Donzère-Mondragon, Alimentation du CEA (April 20, 1962); M. Henry to M. le Directeur de la Production et du Transport, Electricité de France, Usine et poste de Bollène, Alimentation du CEA à Pierrelatte (December 14, 1961). Both of Henry's letters in CNR, AR0106, box 56, DM, CEA, Pierrelatte, Relations.

70. P. de Gaujac, Site du Tricastin, Avant projet d'implantation d'une centrale nucléaire sur le site du Tricastin, Compte rendu de la réunion REM-CNR le 30 mai 1973 à la CNR à Lyon (June 18, 1973).

71. P. Bayard to Electricité de France, A l'attention de M. Delattre, Aménagement de Donzère-Mondragon, Fermeture des barrages de garde en cas de crue (June 21, 1973), CNR, AR0106, box 49, Généralités, 1973–1978; CNR, Aménagement de Donzère-Mondragon, Modèle réduit de l'entrée de la dérivation, Consignes d'exploitation en temps de crues (August 26, 1976), CNR, AR0106, box 47, Donzère-Mondragon, Etudes au Labo de Gerland.

72. CNR, Aménagement de Donzère-Mondragon, Modèle de l'entrée de la dérivation, Evacuation de la crue millénaire et de la crue centenaire (September 3, 1976); CNR, Aménagement de Donzère-Mondragon, Modèle réduit de l'entrée de la dérivation, Programme des essais (September 14, 1976). Both in CNR, AR0106, box 47, Donzère-Mondragon, Etudes au Labo de Gerland.

73. The CNR was allowed to ignore its multipurpose mandate at Génissiat because it needed a pool of money to jumpstart its program. Consequently, it maximized hydroelectric production at that site by building a high-chute dam.

74. Leo Marx, *The Machine in the Garden: Technology and the Pastoral Ideal in America* (New York: Oxford University Press, 1964).

75. CNR, Compte rendu de la conférence du 26 avril 1947 pour la mise au point des travaux annexes du Génie Rural (April 26, 1947), CNR, AR0103, box 3, DM, Enquête, Min. de l'Ag.

76. A. David to M. le Ministre de l'Agriculture, Vaucluse et Drome, Le Rhône, Chute de Donzère-Mondragon (October 3, 1946), CNR, AR0103, box 3, DM, Enquête, Min. de l'Ag.

77. Chambre de Commerce d'Aubenas to M. Tournier, Directeur administratif de la CNR (July 31, 1952), CNR, AR0103, box 94, Chute de Donzère-Mondragon, Etudes et travaux le long du Rhône mort, Station de pompage de Bourg Saint Andéol, Irrigation des jardins.

78. J. Rostagni, note to M. le Directeur administratif, Entretien de M. Chazalon, Conseiller Général du Canton de St-Paul-Trois-Châteaux, M. Pommier, Maire de Saint-Paul-Trois-Châteaux, M. Blachon, Membre de l'Office Agricole de la Drôme (February 7, 1947), CNR, AR0103, box 3, DM, Enquête, Min. de l'Ag.

79. J. Salignon, note pour TEX, Donzère-Mondragon, Irrigation de la région de Pierrelatte, Influence des pompages (December 10, 1962), CNR, AR0106, box 23, Chute de Donzère-Mondragon, Exploitation, Irrigation.

80. The increase in irrigation water is from J. Rostagni, note to M. le Directeur Administratif, Entretien de M. Chazalon, Conseiller général du Canton de Saint-Paul-Trois-Châteaux, M. Pommier, Maire de Saint-Paul-Trois-Châteaux, M. Blachon, Membre de l'Office Agricole de la Drôme (February 7, 1947). David's assessment is from A. David to M. le Ministre de l'Agriculture, Vaucluse et Drôme, Le Rhône, Chute de Donzère-Mondragon (October 3, 1946).

81. The initial amount is stated in J. Bonnier, note to M. Henry, Chute de Donzère-Mondragon, Discussions avec le Service d'Aménagement Agricole des Eaux (November 9, 1946). The finalized figure is from M. Henry, note pour M. le Directeur Technique, Donzère-Mondragon, Reconstitution de la production agricole (November 15, 1946). Both in CNR, AR0103, box 3, DM, Enquête, Min. de l'Ag. It is worth noting that the 25 m^3/s does not include irrigation water diverted through the BRL's network; see Chapter 5.

82. On the conflicts between the CNR and the Génie Rural, see CNR, Compte-rendu de la conférence entre MM. David, Nourrit et Fioravanti, Ingénieurs en Chef du Génie Rural, MM. Tournier, Henry, Bonnier, de la CNR, et Pfahl, des Forces Hydrauliques (M. Haegelen a assisté au début de la conférence), le 18 novembre 1946 (November 19, 1946); CNR, Aide-mémoire de la

conversation de M. Calvert, Inspecteur Général du Génie Rural, et de MM. Giguet et Tournier, le mardi 3 décembre 1946, à Paris (December 4, 1946); A. David to M. le Directeur Général de la CNR, CNR, Chute de Donzère-Mondragon, Expropriation (August 11, 1947); CNR, Conseil d'administration du 16 décembre 1953, Autorisation de dépenses pour équipement rural à Donzère-Mondragon (December 16, 1953). On the CNR's narrow definition of reconstitution, see G. Tournier, note to DG, Donzère-Mondragon (March 31, 1950). All in CNR, AR0103, box 3, DM, Enquête, Min. de l'Ag.

83. The CNR's initial position is outlined in R. Giguet, note to M. le Directeur Technique et M. le Directeur Administratif, Donzère-Mondragon, Reconstitution de la production agricole (November 21, 1946). The CNR conveys the Ministry of Agriculture's stance in R. Giguet to M. Calvet (January 6, 1947). The Conseil d'Etat's decision is described in L. Saulgeot to M. le Ministre de l'Agriculture, Direction Générale du Génie Rural et de l'Hydraulique Agricole, 2ème Bureau, Compagnie nationale du Rhône, Chutes de Donzère-Mondragon et de Montélimar, Note du Conseil d'Etat en date du 27 janvier 1953 (February 23, 1953). All in CNR, AR0103, box 3, DM, Enquête, Min. de l'Ag.

84. *Remembrement* was pursued systematically after the 1950s, suggesting that projects like Donzère-Mondragon offered laboratories for the policy's development. See Michael Bess, *The Light-Green Society: Ecology and Technological Modernity in France, 1960–2000* (Chicago: University of Chicago Press, 2003), 41, 220.

85. Edmund Russell makes a parallel point in "The Garden in the Machine: Toward an Evolutionary History of Technology," introduction to *Industrializing Organisms: Introducing Evolutionary History,* ed. Susan R. Schrepfer and Philip Scranton (New York: Routledge, 2004), 1–18.

86. The first quote is from P. Salenc, L'aménagement et la mise en valeur de la vallée du Rhône (October 21, 1963), CNR, AR0086, Box 48, Généralités, Fonds Salenc (folder hereafter FS). Other documents that make a similar argument include: GEPAR [Le Groupe d'Etude des Perspectives Agricoles et Rurales Rhodaniennes], *Etude des perspectives agricoles et rurales rhodaniennes* (Lyon: Impression Super Plan, May 1972), in CNR, Documentation, L02984; CNR, La CNR et l'agriculture (March 17, 1973), CNR, AR0086, box 48, FS. Salenc's second quote is from Pierre Salenc, "Evolution des techniques de l'irrigation dans la vallée du Rhône en fonction des aménagements de la Compagnie nationale du Rhône," paper presented at the conference entitled Utilisation des ressources en eau d'un bassin dans le cadre de l'aménagement du territoire, Paris, 1970.

87. R. Sordoillet and J. Arrighi de Casanova, L'extension des irrigations dans le Département de Vaucluse (July 1958), CNR, AR0086, box 49; J. Fioravante to M. le Directeur de la CNR, Bollène, Irrigation de la région de

Pierrelatte (November 16, 1959), CNR, AR0106, box 23, Chute de Donzère-Mondragon, Exploitation, Irrigation; Le Préfet de la Drôme, M. le Chef du Service Régional de l'Aménagement des eaux, Conventions agricoles, Utilisation des crédits CNR (September 27, 1971), CNR, AR0086, box 48, FS.

88. Pierre Salenc, Les travaux d'hydraulique agricole en Ardèche (July 10, 1962); J. Fioravante, *Aménagement hydro-agricole Rhône Moyen* (Romans-sur-Isère: Ets J-A Domergue SA, May 1969), in CNR, Documentation, L01775; Schéma des études à entreprendre en vue de la mise en valeur agricole du moyen Rhône (November 19, 1965). Salenc's report and the 1965 study both in CNR, AR0086, box 48, FS.

89. GEPAR, Conventions agricoles—inventaire des actions envisageables (February 25, 1970), CNR, AR0086, box 48, FS. Salenc discusses the significance of Beauchastel in L'aménagement et la mise en valeur de la vallée du Rhône (October 21, 1963).

90. Salenc, "Evolution."

91. Schéma des études à entreprendre en vue de la mise en valeur agricole du moyen Rhône (November 19, 1965).

92. On the shift from energy to economic activity, see Perspectives de la Compagnie nationale du Rhône (March 1961), CNR, AR0086, box 45, Généralités, Fonds Salenc.

93. On *mise en valeur,* see P. Salenc to M. l'Ingénieur en Chef du Génie Rural chargé de l'aménagement agricole des eaux, Mise en valeur de la vallée du Rhône, Chute de Donzère-Mondragon, Région de Pierrelatte (October 26, 1962), CNR, AR0086, box 44, Donzère-Mondragon, Fonds Salenc; Salenc, L'aménagement et la mise en valeur de la vallée du Rhône (October 21, 1963).

94. Salenc, L'aménagement et la mise en valeur de la vallée du Rhône (October 21, 1963); J. Arrighi de Casanova, Travaux d'assainissement et d'irrigation de la basse vallée du Rhône (March 7, 1950), CNR, AR0086, box 49.

95. Sordoillet and Arrighi de Casanova, L'extension des irrigations dans le Département de Vaucluse (July 1958); Compagnie nationale du Rhône, *L'aménagement du Rhône et l'agriculture* (Lyon: Imprimerie Rivet, 1981).

96. Salenc, L'aménagement et la mise en valeur de la vallée du Rhône (October 21, 1963); CNR, Note et annexes (February 25, 1970), CNR, AR0086, box 20, Avantages économiques "non chiffrables" des investissements de la Compagnie nationale du Rhône, Renseignements demandés par M. L'Hermitte; GEPAR, Etude.

97. Fioravante, *Aménagement hydro-agricole Rhône Moyen.* On the desert metaphor, see Chapter 5.

98. The first quote is from Aménagement du Rhône par la Compagnie nationale du Rhône, Préparation de la réunion du sous-groupe "Agriculture," le 23 Avril 1965 (April 23, 1965), CNR, AR0086, box 46, Généralités, Fonds

Salenc; see also J. Fioravante to M. le Ministre de l'Agriculture, Direction Générale du Génie Rural et de l'Hydraulique Agricole (February 22, 1960). The second quote is from GEPAR, *Etude*. The Ministry of Agriculture's conclusion is from Ministère de l'Agriculture, Aménagement hydro-agricole de la Vallée du Rhône vu dans l'hypothèse d'une association des aménagements du programme de la Compagnie nationale du Rhône (May 1962). Fioravante and the Ministry of Agriculture documents are from CNR, AR0086, box 48, FS.

99. Sordoillet and Arrighi de Casanova, L'extension des irrigations dans le Département de Vaucluse (July 1958); Salenc, L'aménagement et la mise en valeur de la vallée du Rhône (October 21, 1963); Compte rendu de la réunion tenue le 2 février 1962 à la préfecture du Rhône à Lyon au sujet des problèmes de la mise en valeur de la vallée du Rhône par l'irrigation (February 2, 1962), CNR, AR0086, box 48, FS; Ministère de l'Agriculture, Aménagement hydro-agricole de la vallée du Rhône (May 1962).

100. On the working group's concerns, see R. Darves-Bornoz, Note à M. le Ministre de l'Agriculture, Financement des travaux de la CNR, Participation de l'agriculture (July 17, 1969). On the Ministry of Agriculture's complaints, see Jacques Duhamel to M. le Ministre délégué auprès du Premier Ministre, Chargé du Plan et de l'Aménagement du Territoire, Financement des travaux de la Compagnie nationale du Rhône, Participation du Ministère de l'Agriculture (July 30, 1969). Both in CNR, AR0086, box 48, FS. The 1972 GEPAR report is *Etude*.

101. P. Salenc, La mise en valeur agricole de la vallée du Rhône et la CNR, Préliminaires à une etude économique a posteriori des aménagements realizés (March 10, 1975), CNR, DX.GC 01.1012, box 009, Agriculture.

102. Arrighi de Casanova, Travaux d'assainissement et d'irrigation de la basse vallée du Rhône (March 7, 1950); Aménagements hydroagricoles de la vallée du Rhône (January 20, 1958), CNR, AR0086, box 48, FS.

103. G. Tournier to M. Nourrit, Ingénieur en Chef, Génie Rural et de l'Hydraulique Agricole (September 6, 1949), CNR, AR0103, box 3, DM, Enquête, Min. de l'Ag.

104. On the Canal de Pierrelatte's new flow, see Aménagement du Rhône par la Compagnie nationale du Rhône, Préparation de la réunion du sous-groupe "Agriculture" (April 23, 1965). On the CNR's changes to existing systems, see CNR, note to DE, Donzère-Mondragon, Irrigations (August 24, 1949); CNR to M. le Directeur Général de la CNR (July 8, 1953). Both in CNR, AR0103, box 3, DM, Enquête, Min. de l'Ag.

105. On raising flows, see Arrighi de Casanova, Travaux d'assainissement et d'irrigation de la basse vallée du Rhône (March 7, 1950). On improving existing canals, see CNR, Autorisation de dépenses pour l'équipement rural à Donzère-Mondragon (April 7, 1954), CNR, AR0103, box 3, DM, Enquête, Min. de l'Ag. On large-scale systems, see Salenc, "Evolution."

106. L. Nourrit to M. le Directeur Général de la CNR, Donzère-Mondragon, Projet d'une station de pompage devant permettre l'irrigation des terres situées à l'est du canal usinier (January 15, 1948); A. Blanc to M. le Directeur Général de la CNR (September 6, 1948); Ministère de l'Agriculture to M. le Directeur Général [of CNR] (November 3, 1950). All in CNR, AR0103, box 3, DM, Enquête, Min. de l'Ag.

107. Salenc, Les travaux d'hydraulique agricole en Ardèche (July 10, 1962).

108. Salenc, "Evolution"; J. Mathian, L'influence de l'aménagement du Rhône sur l'environnement au double point de vue humain et économique, Communication présentée au Comité français des grands barrages, Colloque du 9 juin 1971 (June 22, 1971), in CNR, Documentation, AI0213; GEPAR, *Etude*. See Chapter 5 for a discussion of this additional diversion of water.

109. Marceau Artaud to CNR (February 14, 1952); Coopérative agricole des usagers de la basse vallée de l'Arc to M. le Directeur de l'Agriculture (February 1952). Both in CNR, DA-GAM, box 40, Coopérative agricole des usagers de la basse vallée de l'Arc, Relations avec la Coopérative agricole des usagers de la basse vallée de l'Arc.

110. This point thus revises Scott, *Seeing Like a State*.

111. Convention (April 14, 1953), CNR, DA-GAM, box 40, Coopérative agricole des usagers de la basse Vallée de l'Arc, Contrat; CNR, La CNR et l'agriculture (March 17, 1973). The quote is from CNR, Aide-mémoire de la conversation de M. Calvert, Inspecteur Général du Génie Rural, et de MM. Giguet et Tournier, le mardi 3 décembre 1946, à Paris (December 4, 1946). Regarding subsidies, see CNR, Compte-rendu de la conférence entre MM. David, Nourrit et Fioravanti, Ingénieurs en Chef du Génie Rural, MM. Tournier, Henry, Bonnier, de la CNR, et Pfahl, des Forces Hydrauliques (November 19, 1946).

112. Sordoillet and Arrighi de Casanova, L'extension des irrigations dans le Département de Vaucluse (July 1958); F. Pelissier to M. le Ministre de l'Agriculture, Direction de l'aménagement rural et des structures, Service de l'hydraulique, Chute de Vallabrègues sur le Rhône, Convention agricole—article 6, Demande de participation financière de la CNR à l'implantation d'une station expérimentale de conservation de fruits et légumes par l'INVUFLEC (December 28, 1970), CNR, AR0086, box 3; Ministère de l'Agriculture, Aménagement hydro-agricole de la vallée du Rhône vu dans l'hypothèse d'une association des aménagements du programme de la Compagnie nationale du Rhône (May 1962); Salenc, Privas: 29 septembre 1960, Journée de la mécanisation (September 1960), CNR, AR0086, box 48, FS; Le Préfet de la Drôme to M. le Chef du Service Régional de l'Aménagement des eaux, Conventions agricoles, Utilisation des crédits CNR (September 27, 1971).

113. Michel Maisonneuve, *La conquête de l'eau: BRL, histoire de l'aménagement en Languedoc Rousillon* (Marseilles: Editions Editea, 1992), 65. The BRL is discussed at greater length in Chapter 5.

4. Local Responses

1. P. Delattre to M. Kirchner (March 14, 1949), CNR, AR0103, box 11, Chute de Donzère-Mondragon, Relations avec les Services du Contrôle, Maintien de la nappe et réalimentation en eau des propriétés (folder hereafter DM, Maintien); Charles Varennes to M. le Directeur de la CNR (December 29, 1952), CNR, AR0103, box 69, Chute de Donzère-Mondragon, Abaissements de la nappe dûs aux pompages et aux drainages nécessités pour les travaux, Réclamations des propriétaires, Mesures palliatives, Année 1953 (folder hereafter DM, Mesures palliatives, 1953) (underscored in the original).

2. Jacques Vollant to M. Salignon, Cie nationale du Rhône (December 8, 1956), CNR, AR0103, box 82, Chute de Donzère-Mondragon, Maintien de la nappe phréatique, Généralités, Divers (folder abbreviated hereafter DM, Maintien).

3. For an exemplary study of uncertainty in a socioenvironmental debate, see Michelle Murphy, *Sick Building Syndrome and the Problem of Uncertainty: Environmental Politics, Technoscience, and Women Workers* (Durham, NC: Duke University Press, 2006).

4. James C. Scott, *Seeing Like a State: How Certain Schemes to Improve the Human Condition Have Failed* (New Haven: Yale University Press, 1999).

5. For an overview of so-called controversy studies, see Brian Martin and Evelleen Richards, "Scientific Knowledge, Controversy, and Public Decision Making," in *Handbook of Science and Technology Studies,* ed. Sheila Jasanoff, Gerald E. Markle, James C. Petersen, and Trevor Pinch (Thousand Oaks, CA: Sage, 1995): 506–526. Two important controversy studies include Diane Vaughn, *The Challenger Launch Decision: Risky Technology, Culture, and Deviance at NASA* (Chicago: University of Chicago Press, 1997); Brian Wynne, "May the Sheep Safely Graze? A Reflexive View of the Expert-Lay Knowledge Divide," in *Risk, Environment, and Modernity: Towards a New Ecology,* ed. Scott M. Lash, Bronislaw Szerszynski, and Brian Wynne (Thousand Oaks, CA: Sage, 1996).

6. Maire de Bourg-Saint-Andéol to M. le Préfet de l'Ardèche à Privas (January 10, 1945); Commissaire-enquêteur d'Ardèche, Demande et avant-projet présentés par la CNR en vue d'obtenir l'autorisation comportant déclaration d'utilité publique d'aménager la chute de Donzère-Mondragon-sur-le-Rhône (January 15, 1945). Both in CNR, AR0103, box 4, Chute de Donzère-Mondragon, Dossier des résultats de l'enquête et des conférences et des propositions de la CNR, Départements du Gard, du Vaucluse, de la Drôme, et de l'Ardèche, Résultats de l'enquête.

7. The CNR's archives include locals' letters written to the agency beginning in 1949. This section is based on approximately one hundred of those letters. Uncited generalizations are based on the letters as a group; otherwise, I

have cited one or more letters exemplifying the pattern I have identified. It is also important to note several things about these letters. First, CNR employees oversee the agency's archive—from what documents are kept to how they are organized—including the letters on which much of this chapter is based. It is impossible for the historian to ascertain what documents might have been excluded, either then or now, purposefully or accidentally. Finally, the vast majority of locals' letters are typed. Based on the paper, typeface, and other evidence, it appears likely CNR secretaries made typed copies of all incoming letters. While this makes the historian's task much easier, it is impossible to know if the typist changed or excluded any statements from the original letters. At the same time, that so many letters have been kept suggests they have not been altered or destroyed. These issues are common "problems of the archive," and I sought to keep them in mind during my reading of the documentary evidence.

8. P. Delattre to M. Édouard Rastoin (September 30, 1952), CNR, AR0103, box 79, Chute de Donzère-Mondragon, Maintien de la nappe phréatique à l'aval de l'usine, Correspondance avec l'administration (folder hereafter DM, Maintien, Correspondance). The region's population is based on Georges Goldfard, Statistics of Communes near Donzère-Mondragon (1948); Georges Goldfard, Commune de Donzère, Saint-Paul-Trois-Châteaux, Bollène, Lapalud, La Motte, and Bourg-Saint-Andéol (1948). Both in CNR, AR0103, box 17, Energie, Chute de Donzère-Mondragon, Architecture générale et cités, Urbanisme. For a complaint from 1987, see Pierre Lambertin to M. le Directeur, CNR, Nappe rive gauche, Quartier Champredon, Bollène (April 28, 1987), CNR, AR0106, box 62, Donzère-Mondragon, Réalimentation en eau.

9. Locals who wrote letters to the CNR were unanimous in their view that the agency was responsible. It is more difficult to know the views of those who did not send letters to the agency. Press coverage did not mention locals' protests, but given France's postwar context, it was unlikely to have done so.

10. Each of these phrases has some slight variations in usage. Nonetheless, these three headings broadly capture the categorization of blame by locals.

11. Nouguier to M. le Secrétaire Général de la Préfecture (June 23, 1950), CNR, AR0103, box 69, Chute de Donzère-Mondragon, Abaissement de la nappe dû aux drainages nécessités pour les travaux, Réclamation des propriétaires, Mesures palliatives, Année 1950 (folder hereafter DM, Mesures palliatives, 1950); Henri Bastet to M. le Directeur de la CNR (November 11, 1952), CNR, AR0103, box 69, Chute de Donzère-Mondragon, Abaissement de la nappe dus aux drainages nécessités pour les travaux, Réclamation des propriétaires, Mesures palliatives, Année 1952 (folder hereafter DM, Mesures palliatives, 1952).

12. Fournier to M. le Directeur de la CNR (July 23, 1953), CNR, AR0103, box 69, DM, Mesures palliatives, 1953. Other exceptions include Joseph Roussin, Gilles Simon, and André Larmande to M. Rostagni, CNR (Decem-

ber 8, 1949), CNR, AR0103, box 67, Chute de Donzère-Mondragon, Abaissement de la nappe dû aux pompages et aux fracinages nécessités pour les travaux, Réclamations des propriètaires, Mesures, années 1944–1949 (folder hereafter DM, Réclamations, 1944–1949); Bouchet, Rapport sur les dommages causés dans la Commune de Bollène aux propriétés riveraines par le Canal de Donzère-Mondragon (1952), CNR, AR0103, box 69, DM, Mesures palliatives, 1952; Louis Argellier to M. le Directeur de la CNR (1953), CNR, AR0103, box 69, DM, Mesures palliatives, 1953.

13. Les propriétaires-exploitants, propriétaires, métayers et fermiers soussignés de la commune de La Garde-Adéhmar (Drôme) to M. le Ministre de l'Agriculture, Paris (November 25, 1949), CNR, AR0103, box 11, DM, Maintien.

14. Roussin, Simon, and Larmande to M. Rostagni, CNR (December 8, 1949).

15. Emile Lachaux to M. le Directeur de la CNR (January 18, 1953), CNR, AR0103, box 69, DM, Mesures palliatives, 1953. See also Joseph Rey, Léon Guion, Jean Feuillet, Louis Reynaud, and Charles Peyrard to M. Rostagni, Ingénieur CNR (December 7, 1949); Henri Granier to M. le Directeur (December 21, 1949). Both in CNR, AR0103, box 67, DM, Réclamations, 1944–1949.

16. Maire de La Garde-Adhémar to M. Rostagni (October 23, 1949), CNR, AR0103, box 67, DM, Réclamations, 1944–1949.

17. La Garde-Adhémar to M. Rostagni, Directeur des Travaux CNR (December 20, 1949), CNR, AR0103, box 67, DM, Réclamations, 1944–1949.

18. Argellier to M. le Directeur de la CNR (1953).

19. Varennes to M. le Directeur de la CNR (December 29, 1952).

20. A few departmental governments and administrative heads of state agencies did demand outside experts. See, for instance, Fontanel, Extrait du registre des arrêtés du Conseil de Préfecture (November 6, 1950), CNR, AR0103, box 69, DM, Mesures palliatives, 1950; Roger Rougier, Rapport d'expert (1952). The Services Agricoles de Vaucluse also conducted a study in 1952 mentioned in G. Tournier, Donzère-Mondragon, Abaissement de la nappe, Compte rendu de la réunion tenue à la Préfecture d'Avignon le 9-10-1952 (October 9, 1952). Both Rougier and Tournier in CNR, AR0103, box 69, DM, Mesures palliatives, 1952.

21. Préfecture de la Drôme to M. le Directeur (March 22, 1956), CNR, AR0103, box 82, Chute de Donzère-Mondragon, Maintien de la nappe phréatique, Généralités et divers.

22. Exceptions include André Broche to CNR (October 29, 1957); H. Salavert to M. le Directeur (September 22, 1957). Both in CNR, AR0103, box 87, Chute de Donzère-Mondragon, Maintien de la nappe phréatique sur la RD, Réclamations des propriétaires, Généralités (folder hereafter DM, Réclamations,

Généralités); Henri Jean Bouzigue to M. le Préfet du Gard (1953), CNR, AR0103, box 87, Chute de Donzère-Mondragon, Maintien de la nappe sur la RD, Réclamations des propriétaires, Généralités, de 1953 à 1961 (folder hereafter DM, Réclamations, 1953–1961); Auguste Sabatier and Louis Sabadel to M. le Directeur de la CNR (November 26, 1952), CNR, AR0103, box 69, DM, Mesures palliatives, 1952; Commission Départementale du Conseil Général du Gard, Voeu sur les mesures à prendre pour remédier à la situation résultant de la mise en service du canal de Donzère à Mondragon (February 24, 1955), CAC, 770764, box 19, Compagnie nationale du Rhône, Affaires diverses.

23. G. Dumas to CNR (August 5, 1954), CNR, AR0103, box 87, DM, Réclamations, 1953–1961; G. Jaume to M. le Préfet de la Drôme (July 24, 1950), CNR, AR0103, box 69, DM, Mesures palliatives, 1950; Jules Albin to M. l'Ingénieur en Chef de la CNR (November 2, 1949), CNR, AR0103, box 67, DM, Réclamations, 1944–1949.

24. The CNR's archives reference numerous strikes during Donzère-Mondragon's construction. On labor conflicts during the first half of the twentieth century more broadly, see Herrick Chapman, *State Capitalism and Working-Class Radicalism in the French Aircraft Industry* (Berkeley: University of California Press, 1990); Richard F. Kuisel, *Capitalism and the State in Modern France: Renovation and Economic Management in the Twentieth Century* (New York: Cambridge University Press, 1981).

25. L. Morand to M. le Directeur de la CNR (August 17, 1953), CNR, AR0103, box 69, DM, Mesures palliatives, 1953. Those that articulated threats include Rose Vincent to M. le Directeur de la CNR (September 23, 1952); R. Chabert to CNR (September 9, 1952); Gabriel Gillio to M. le Directeur de la CNR (September 23, 1952). All in CNR, AR0103, box 87, DM, Réclamations, 1953–1961.

26. In "May the Sheep Safely Graze?" Wynne argues, "Lack of overt public dissent or opposition towards expert systems is taken too easily for public trust" (50). Given locals' letters and other responses discussed below, it appears they expressed both overt and implicit concern.

27. For one example of eight residents signing a petition, see Efrem Richdounian et al. to M. Mathian (July 29, 1954), CNR, AR0103, box 136, Chute de Donzère-Mondragon, Abaissement de la nappe dû aux pompages et aux drainages nécessités par les travaux, Réclamations des propriétaires, Mesures palliatives, Année 1954 (folder hereafter DM, Mesures palliatives, 1954). Delattre mentions a petition with over six hundred signatures in Delattre to M. Édouard Rastoin (September 30, 1952).

28. For a discussion of the creation of a "groupement" to protest excess water around Pierrelatte, see Maire de Pierrelatte to M. l'Ingénieur en Chef de la 6ème Circonscription Electrique (November 9, 1953), CNR, AR0103, box 5, Chute de Donzère-Mondragon, Relations avec les Services du Contrôle,

Questions techniques, Récolement des travaux. In 1957, sixty-five individuals near Pierrelatte formed another organization. See G. Gres, CNR, Nappe de rive gauche du Rhône mort (November 4, 1957), CNR, AR0103, box 89, Chute de Donzère-Mondragon, Abaissement de la nappe, Réclamations des propriétaires, Années 1957–1958 (folder hereafter DM, Réclamations, 1957–1958); G. Audebrand, note pour TEX, Nappe de bordure du Rhône, Région de Pierrelatte (February 20, 1959), CNR, AR0103, box 81, Division Donzère-Mondragon, Maintien de la nappe phréatique le long du canal d'amenée, Généralités et divers (folder hereafter DM, Maintien, Généralités). See also Sinistrés du Canal de Donzère-Mondragon, announcement in *Le Dauphiné libéré,* October 13, 1953, in CNR, AR0103, box 69, DM, Mesures palliatives, 1953.

29. J. Rostagni, note to DA (February 20, 1950), CNR, AR0103, box 3, Donzère-Mondragon.

30. As is often the case with many historical records, it is difficult to know who authored this document, how widely distributed it was, how it was read, and what broader impact it may have had. For instance, letters from 1957 do not reference the poster. At the very least, however, the CNR official who included a copy of the poster in the agency's archive believed it was significant. In French, the text of the poster stated: "Visiteurs, Devant vous, La plaine du canton de Bollène dont les terrains ont été asséchés par les travaux du canal de Donzère-Mondragon. Situation. Avant les travaux: Terrain riche et frais; Eau abondante et pure; Nappe phréatique élevée et constanté; Batîments solides. Après les travaux: Terrain asséché; Eau non potable parfois contaminée; Nappe phréatique abaissée à un niveau non-bénéfique pour les cultures; Batîments lézardés parfois dangereux à habiter. Résultats: Rendements de la production agricole fortement diminués; Terrains lessives, steriles à bréve echéance; Ruine des agriculteurs en perspective. Conclusion: En ce qui concerne l'agriculture: N'accorder aucun crédit aux affirmations mensongères de la Compagnie nationale du Rhône. Allez dans les exploitations agricoles, goutez l'eau, régardez les batîments, et jugez par vous-même. Les Sinistrés du Canton de Bollène." Sinistrés de Bollène, Visiteurs, Devant vous, La plaine du canton de Bollène dont les terrains ont été assèché par les travaux du canal de Donzère-Mondragon (April 1957), CNR, AR0103, box 89, DM, Réclamations, 1957–1958.

31. Gabrielle Hecht, *The Radiance of France: Nuclear Power and National Identity after World War II* (Cambridge, MA: MIT Press, 1998), 29–30.

32. Again, we do not know the views of "silent" locals (i.e., those who did not send letters to the CNR). I cannot, therefore, claim that all locals were in agreement. Although my conclusions about the views of locals are limited to a small share of the actual population of the area, the number of letters and petitions are statistically significant, especially when compared to the region's population.

33. P. Delattre to M. le Préfet de Vaucluse (March 4, 1949); P. Delattre, note to DE and M. Gres, Donzère-Mondragon (March 5, 1949). Both in CNR, AR0103, box 67, DM, Réclamations, 1944–1949.

34. J. Rostagni to M. le Maire de La Garde-Adhémar (November 8, 1949), CNR, AR0103, box 67, DM, Réclamations, 1944–1949.

35. A 1955 map showing the influence of Donzère-Mondragon on the region's groundwater indicated that just west of the town center the area's water table had risen due to the feeder canal. CNR, Chute de Donzère-Mondragon, Influence des travaux sur le comportement de la nappe (1955), CNR, AR0106, box 70, Période de 1945 à 1955. I discuss this map at greater length later in the chapter. On CNR arguments about drought even in the 1980s, see P. Morand to M. le Maire, Saint-Paul-Trois-Châteaux (August 5, 1981), CNR, AR0106, box 46, Chute de Donzère-Mondragon, Nappe, Généralités, Correspondance.

36. See, for instance, J. Rostagni to M. Jules Albin (November 8, 1949), CNR, AR0103, box 67, DM, Réclamations, 1944–1949; CNR to M. Raynaud (October 6, 1953), CNR, AR0103, box 69, DM, Mesures palliatives, 1953; G. Gres to M. Bourrier, Chute de Donzère (October 21, 1953); M. Henry to M. Jean A. Bruchardt (July 16, 1955). Both in CNR, AR0103, box 82, DM, Maintien.

37. J. Rostagni, DA, Donzère-Mondragon, Abaissement de la nappe sur la RD du Rhône (July 6, 1953), CNR, AR0103, box 82, DM, Maintien.

38. CNR, DA, Assèchements des puits (July 7, 1950), CNR, AR0103, box 69, DM, Mesures palliatives, 1950.

39. J. Rostagni, DA, Abaissement de la nappe rive droite du Rhône (October 18, 1952), CNR, AR0103, box 82, DM, Maintien.

40. G. Tournier, DE, Abaissement de la nappe, Affaire Bresson (February 12, 1952), CNR, AR0103, box 69, DM, Mesures palliatives, 1952.

41. CNR, Donzère-Mondragon, Nappe rive droite, Conventions (March 10, 1959); CNR, Convention, Chute de Donzère-Mondragon (June 13, 1960). Both in CNR, AR0103, box 84, Chute de Donzère-Mondragon, Maintien de la nappe sur la RD du Rhône, Réclamations, Conventions avec les propriétaires.

42. P. Delattre to M. Pfhal, Ingénieur des Ponts et Chaussées (January 4, 1950), CNR, AR0103, box 11, DM, Maintien.

43. CNR to M. Fioravante, Abaissement de la nappe sur la rive droite du Rhône entre le barrage de Donzère et Pont St. Esprit (December 28, 1956), CNR, AR0103, box 82, Chute de Donzère-Mondragon, Maintien de la nappe phréatique sur la rive droite du Rhône, Rapports avec l'administration; P. Delattre to M. David, Chute de Donzère-Mondragon, Réalimentation de la nappe (May 23, 1950), CNR, AR0103, box 82, Chute de Donzère-Mondragon, Maintien de la nappe le long du canal d'amenée, Généralités, Divers (folder hereafter DM, Maintien de la nappe le long du canal d'amenée); P. Delattre to M. l'Ingénieur

en Chef de la 6ème Circonscription Electrique, Réalimentation de la nappe (October 30, 1952), CNR, AR0103, box 11, DM, Maintien.

44. P. Delattre to M. le Préfet de la Drôme (August 23, 1950), CNR, AR0103, box 69, DM, Mesures palliatives, 1950.

45. P. Delattre to M. E. Dalladier, Assemblée Nationale, Paris (October 15, 1952), CNR, AR0103, box 79, Chute de Donzère-Mondragon, Maintien de la nappe phréatique à l'aval de l'usine, Généralités et divers, 1947 à 1952 inclus, Correspondance avec l'administration (folder hereafter DM, Maintien, 1947–1952); Delattre to M. Édouard Rastoin (September 30, 1952).

46. This legal requirement is discussed in this chapter's final section.

47. J. Garraud, Donzère-Mondragon, Abaissement de la nappe de rive droite du Rhône, Compte-tenu de la réunion du 11 avril 1956 en Mairie de Bourg-Saint-Andéol (April 27, 1956), CNR, AR0103, box 82, DM, Maintien.

48. Rougier is quoted in A. Dagand, Compte rendu de la réunion d'expertise tenue en Mairie de Bollène le 18 mai 1954 à 10 heures, affaire "Groupement des Sinistrés de Bollène" (Abaissement de la nappe) (May 18, 1954), CNR, AR0103, box 79, Chute de Donzère-Mondragon, Maintien de la nappe phréatique à l'aval de l'usine, Généralités et divers, Année 1954/1958, plans, courriers (folder hereafter DM, Maintien, 1954/1958). For Dagand's follow-up, see A. Dagand, Compte rendu de la réunion d'expertise tenue à Bollène, le 22 juin 1954 à 10 heures (June 22, 1954), CNR, AR0103, box 79, Chute de Donzère-Mondragon, Maintien de la nappe phréatique à l'aval de l'usine, Généralités et divers. Both Rougier and Dagand allude to "zones" in the central Rhône, a reference to geologic studies Mathian carried out (discussed later in this chapter).

49. Brian Balogh's work on the American nuclear industry has suggested at least some technical experts within the industry contributed to its severe problems by failing to admit publicly the degree of uncertainty it discussed internally. See Brian Balogh, *Chain Reaction: Expert Debate and Public Participation in American Commercial Nuclear Power, 1945–1975* (Cambridge, MA: Cambridge University Press, 1991).

50. Société de Construction des Batignolles et al., Aménagement de la chute de Donzère-Mondragon, Avant-projet, Rapport sur les conditions hydrologiques des travaux du canal de Donzère-Mondragon (1946), CNR, AR0103, box 18, Aménagement de la chute de Donzère-Mondragon, Avant-projet, Exposé général des problèmes posés; B. Martino, Suite à ma note n. 191 du 3.12.47 de ma visite à Kembs et des observations dont elle a fait l'objet (December 15, 1947), CNR, AR0103, box 20, Chute de Donzère-Mondragon, Etudes et travaux éxécutés par le GETARD (ou SACTARD), Correspondance et notes, Généralités et divers.

51. J. Rostagni, note to DE, Etude de la nappe phréatique de la RG du Rhône entre Donzère et Lapalud (November 7, 1949), CNR, AR0103, box 67, DM, Réclamations, 1944–1949. For CNR documents conveying Rostagni's

results, see CNR, Tx, Abaissement de la nappe dans la région de Bollène, Réclamations et travaux exécutés (July 22, 1950), CNR, AR0103, box 79, DM, Maintien, Correspondance; J. Rostagni, DA, Abaissement de la nappe le long du canal d'amenée (September 27, 1950), CNR, AR0103, box 69, DM, Mesures palliatives, 1950; CNR, Aménagement de Donzère-Mondragon, Puits domestiques de la rive droite du Rhône, Approfondissement (April 6, 1951), CNR, AR0103, box 87, DM, Réclamations, 1953–1961; J. Mathian, DA, Assèchement de puits, Réclamation Barroux (March 28, 1952), CNR, AR0103, box 69, DM, Mesures palliatives, 1952.

52. Marc Henry, DA, Donzère-Mondragon, Abaissement de la nappe dans la région du canal d'amenée (October 10, 1950), CNR, AR0103, box 69, DM, Mesures palliatives, 1950 (emphasis in the original). Marc Henry to SACTARD, Chute de Donzère-Mondragon, Affaissements eventuels du sol par suite de la baisse du niveau de la nappe (April 30, 1952), CNR, AR0103, box 69, DM, Mesures palliatives, 1952.

53. G. Hatt, Résumé du rapport SACTARD du 8/6/48 relatif à l'abaissement de la nappe à l'aval de l'usine (May 31, 1948), CNR, AR0103, box 82, DM, Maintien de la nappe le long du canal d'amenée.

54. J. Mathian, Rapport, Conclusion (1952), CNR, AR0103, box 77, Chute de Donzère-Mondragon, Comportement de la nappe phréatique.

55. Hecht, *Radiance of France,* especially chapter 8; Balogh, *Chain Reaction.*

56. I found no document in the CNR's archives in which agency officials recommended or admitted lying.

57. Tournier, DMN, Réclamation des propriétaires de Donzère (September 16, 1950), CNR, AR0103, box 69, DM, Mesures palliatives, 1950; Rostagni, DA, Abaissement de la nappe le long du canal d'amenée (September 27, 1950).

58. CNR, Note pour DT, Abaissement de la nappe dû aux pompages de l'usine (February 14, 1949), CNR, AR0103, box 67, Chute de Donzère-Mondragon, Abaissements de la nappe dûs aux pompages et aux frainages nécessites pour les travaux, Réclamations des propriétaires, Mesures collectives années 1944–1949.

59. J. Rostagni, note pour TX, Abaissement de la nappe le long du canal d'amenée (August 17, 1950), CNR, AR0103, box 69, DM, Mesures palliatives, 1950. Other examples of descriptive language include P. Delattre, DE, Donzère-Mondragon, Abaissement de la nappe sur le rive droite du Rhône (February 9, 1951), CNR, AR0103, box 87, DM, Réclamations, Généralités; J. Mathian, DA, Abaissement de la nappe dans la région de la croisière (April 4, 1952), CNR, AR0103, box 69, DM, Mesures palliatives, 1952; Marc Henry, Donzère-Mondragon, Abaissement de la nappe (June 4, 1951); J. Mathian, note to M. Gemaehling, Donzère-Mondragon, Nappe phréatique voisine du Rhône court-circuité (March 17, 1956). Both in CNR, AR0103, box 82, DM, Maintien.

60. J. Mathian, DA, Réclamations diverses (November 8, 1952), CNR, AR0103, box 69, DM, Mesures palliatives, 1952; J. Mathian, DA, Réclamations pour fissurations des maisons (March 14, 1953), CNR, AR0103, box 69, DM, Mesures palliatives, 1953.

61. Bruno Latour, *We Have Never Been Modern,* trans. Catherine Porter (Cambridge, MA: Harvard University Press, 1993), 7.

62. J. Mathian, DE, Etude de la nappe (June 3, 1953), CNR, AR0103, box 82, DM, Maintien de la nappe le long du canal d'amenée.

63. CNR, DA, Abaissement de la nappe, Affaire "Groupement des Sinistrés de Bollène" (June 9, 1954), CNR, AR0103, box 136, DM, Mesures palliatives, 1954.

64. CNR, Chute de Donzère-Mondragon, Influence des travaux sur le comportement de la nappe (1955).

65. G. Gres, DA, Donzère-Mondragon, Abaissement de la nappe, Réclamation (February 21, 1958); G. Audebrand, note pour TEX, Assèchement de terrains, Réclamation Clariot (August 28, 1957). Both in CNR, AR0103, box 89, DM, Réclamations, 1957–1958.

66. For two examples of this at work in forestry, see Henry E. Lowood, "The Calculating Forester: Quantification, Cameral Science, and the Emergence of Scientific Forestry Management in Germany," in *The Quantifying Spirit in the Eighteenth Century,* ed. Tore Frangsmyr, J. L. Heilbron, and Robin E. Rider (Berkeley: University of California Press, 1990): 315–342; Karl Appuhn, "Inventing Nature: Forests, Forestry and State Power in Renaissance Venice," *Journal of Modern History* 72 (2000): 861–889.

67. The prevalence of the "improvement" argument is remarkable. I have included just a few examples here: Delattre to M. David, Chute de Donzère-Mondragon, Réalimentation de la nappe (May 23, 1950); Delattre to M. Édouard Rastoin (September 30, 1952); CNR, Aménagement de la chute de Donzère-Mondragon, Rétablissement de la nappe le long du canal de fuite, Notice (February 17, 1953), CNR, AR0103, box 11, DM, Maintien; CNR, Effet sur les cultures de l'alimentation artificielle de la nappe le long du canal de fuite de la chute de Donzère-Mondragon entre Bollène et Mondragon (October 4, 1954), CNR, AR0103, box 79, DM, Maintien, 1954/1958.

68. M. Henry, DE, Donzère-Mondragon, Puits domestiques rive gauche, Zone nord (February 27, 1952), CNR, AR0103, box 69, DM, Mesures palliatives, 1952; CNR to M. l'Ingénieur en Chef de la 6ème Circonscription Electrique, Situation de la nappe (September 5, 1952), CNR, AR0103, box 11, DM, Maintien.

69. CNR, Donzère-Mondragon, Nappe rive droite, Conventions (March 10, 1959); CNR, Convention, Chute de Donzère-Mondragon (June 13, 1960).

70. Cahier des charges spéciales pour l'aménagement de Donzère-Mondragon, December 7, 1953. It is strange enough that the *cahier* was ap-

proved after construction had been completed, but that the laws frame groundwater issues in terms of "if" they occur is also noteworthy. Of course, in December 1953, the causality of groundwater change was still being contested.

71. M. Henry, note pour DG, Donzère-Mondragon, Réglage de la nappe souterraine (May 22, 1950), CNR, AR0103, box 82, DM, Maintien de la nappe le long du canal d'amenée.

72. The abandoned siphon proposal is mentioned in M. Bouvet to CNR (October 6, 1947), CNR, AR0103, box 81, DM, Maintien, Généralités. The wall is from Martino, Suite à ma note n. 191 du 3.12.47 (December 15, 1947). The expansion of irrigation networks is from J. Bonnier, Chute de Donzère-Mondragon, Réalimentation de la nappe sur la rive droite du canal de fuite (March 13, 1950), CNR, AR0103, box 81, Maintien de la nappe phréatique sur la rive droite du canal de fuite, Correspondance avec l'administration (folder hereafter DM, Maintien, Correspondance).

73. Bouvet to CNR (October 6, 1947).

74. Pierre Claude Reynard, "Charting Environmental Concerns: Reactions to Hydraulic Public Works in Eighteenth-Century France," *Environment and History* 9 (2003): 251–273, especially 254.

75. M. Henry, Memo: Chute de Donzère-Mondragon, Questions agricoles (September 2, 1949), CNR, AR0103, box 3, Chute de Donzère-Mondragon, Enquête, conférences, et correspondance, Services du Ministère de l'Agriculture.

76. Bonnier, Chute de Donzère-Mondragon, Réalimentation de la nappe sur la rive droite du canal de fuite (March 13, 1950).

77. A. David to M. le Directeur des Etudes, CNR, Chute de Donzère-Mondragon, Abaissement de la nappe phréatique (May 9, 1950), CNR, AR0103, box 81, DM, Maintien, Généralités; M. Henry, Chute de Donzère-Mondragon, Réalimentation de la nappe, Compte rendu de l'entretien du 5 juin MM. Henry et Bonnier avec M. Nourrit, Ingénieur en Chef du Génie Rural à Avignon et de son adjoint M. Daure (June 8, 1950), CNR, AR0103, box 82, DM, Maintien de la nappe le long du canal d'amenée.

78. Bourgin is quoted in M. Henry, Entretien de M. Henry avec MM. Bourgin & Pfahl (12 Juillet 1950) (July 19, 1950), CNR, AR0103, box 5, Chute de Donzère-Mondragon, Relations avec les Services du Contrôle et l'Administration Supérieure.

79. A. David to M. le Directeur Général de la CNR, Chute de Donzère-Mondragon, Réalimentation de la nappe (October 3, 1950), CNR, AR0103, box 79, DM, Maintien, Correspondance.

80. P. Delattre to M. David (November 2, 1950), CNR, AR0103, box 81, DM, Maintien, Correspondance.

81. On problems with the test wells, see CNR, note to DE, Réalimentation de la nappe (January 12, 1951), CNR, AR0103, box 81, DM, Maintien,

Correspondance. On unexpected flow rates, see CNR, Aménagement de la chute de Donzère-Mondragon, Rétablissement de la nappe le long du canal de fuite, Notice (February 17, 1953). On the "central hole," see J. Mathian, DE, Etude de la nappe (May 29, 1952), CNR, AR0103, box 79, Chute de Donzère-Mondragon, Maintien de la nappe phréatique à l'aval de l'usine, Généralités et divers, Année 1953, plans, courriers, réalimentation de la nappe à l'aval de l'usine.

82. J. Bonnier, DE, Chute de Donzère-Mondragon, Puits domestiques entre le Rhône et le canal de fuite (June 9, 1952), CNR, AR0103, box 69, DM, Mesures palliatives, 1952.

83. CNR, note to DE, Réalimentation de la nappe (January 12, 1951).

84. G. Gres to M. Sentenac, Nappe de Donzère-Mondragon (August 6, 1952), CNR, AR0103, box 79, DM, Maintien, Correspondance.

85. Tournier, Donzère-Mondragon, Abaissement de la nappe, Compte rendu de la réunion tenue à la Préfecture d'Avignon le 9-10-1952 (October 9, 1952).

86. Marc Henry, Chute de Donzère-Mondragon, Maintien de la nappe phréatique, Conférences avec différents services (December 4, 1952); P. Delattre to M. l'Ingénieur en Chef de la 6ème Circonscription Electrique (January 6, 1953). Both in CNR, AR0103, box 11, DM, Maintien; Marc Henry, Donzère-Mondragon, Nappes le long du canal de fuite (November 16, 1953), CNR, AR0103, box 79, DM, Maintien, 1954/1958.

87. Solétanche, CNR, Chute de Donzère-Mondragon, Canal de fuite, Nappe rive droite, Etude par analogie électrique, Compte-rendu des essais (May 15, 1953), CNR, AR0103, box 78, CNR, Chute de Donzère-Mondragon, Canal de fuite, Nappe rive droite, Etude par analogie électrique, Compte-rendu des essais; Solétanche, Chute de Donzère-Mondragon, Canal de fuite, Constitution d'un écran étanché en bordure du canal (June 24, 1953), CNR, AR0103, box 78, CNR, Chute de Donzère-Mondragon, Canal de fuite, Constitution d'un écran étanché en bordure du canal.

88. CNR, Année 1958, Nappe phréatique (1958), CNR, AR0106, box 73, Année 1958; CNR, Donzère-Mondragon, Nappe phréatique de la rive droite, Observations de 1959 (March 8, 1960), CNR, AR0103, box 82, Chute de Donzère-Mondragon, Maintien de la nappe phréatique, Généralités.

89. CNR, Mémoire, Année 1961, Nappe phréatique (1961), CNR, AR0106, box 72, Année 1961.

90. P. Bayard to M. le Directeur Général, Chute de Donzère-Mondragon, Réalimentation de la nappe, Rive gauche (January 31, 1968); CNR, Compte rendu de la réunion d'information tenue à la salle des conférences à Bollène le 19 juin 1969 (June 23, 1969). Both in CNR, AR0106, box 62, Donzère-Mondragon, Réalimentation en eau.

91. Richard White, *The Organic Machine* (New York: Hill and Wang, 1995); Richard White, "'Are You an Environmentalist or Do You Work for a

Living?' Work and Nature," in *Uncommon Ground: Toward Reinventing Nature,* ed. William Cronon (New York: Norton, 1995): 171–185.

92. Alexandre Giandou, *La Compagnie nationale du Rhône (1933–1998): Histoire d'un partenaire régional de l'Etat* (Grenoble: Presses universitaires de Grenoble, 1999), 197.

93. My ideas on modeling have been influenced by Latour and other actor-network theorists, and by Matthew W. Klingle, "Plying Atomic Waters: Lauren Donaldson and the 'Fern Lake Concept' of Fisheries Management," *Journal of the History of Biology* 31 (1998): 1–32.

94. This controversy raises some fascinating questions about the politics of symmetry within STS. The classic work here is David Bloor, *Knowledge and Social Imagery* (Chicago: University of Chicago Press, 1976), especially chapter 1. The principle of symmetry has "elevated" fields usually excluded from modern science. However, the groundwater controversy opens up several important issues. The honesty and integrity of intellectual positions is generally assumed in controversy studies. Yet when the stakes are big and knowledge claims are deeply politicized (as they often are), those positions may not be entirely forthcoming. Furthermore, the principle of symmetry treats the ideas of diverse groups equally; however, they do not usually have equal power. Without adequate attention to these differential power dynamics, analysts' symmetry risks flattening (perhaps even obscuring) these uneven power relations. I thank Suman Seth for pushing my thinking on these points.

5. Rethinking the Nation

1. Antoine Pinay, "Une alliance d'avenir à l'échelle de l'Europe: Le grand delta," *Delta, Revue d'action régionale* (1968): 7–13.

2. J. F. Gravier, *Paris et le désert français* (Paris: Flammarion, 1947).

3. Modeling, whether geographic or scientific, produces landscape in a double sense: it produces material forms discursively, and it often ends up reshaping that landscape physically. On the power of models and their implications for landscapes in the Rhône valley, see Chapter 4. See also Henry E. Lowood, "The Calculating Forester: Quantification, Cameral Science, and the Emergence of Scientific Forestry Management in Germany," in *The Quantifying Spirit in the Eighteenth Century,* ed. Tore Frangsmyr, J. L. Heilbron, and Robin E. Rider (Berkeley: University of California Press, 1990): 315–342; Matthew W. Klingle, "Plying Atomic Waters: Lauren Donaldson and the "Fern Lake Concept" of Fisheries Management," *Journal of the History of Biology* 31 (1998): 1–32; Karl Appuhn, "Inventing Nature: Forests, Forestry and State Power in Renaissance Venice," *Journal of Modern History* 72 (2000): 861–889.

4. Economist W. W. Rostow, among others, contributed to the formulation of modernization theory. On the Cold War and "development" more broadly,

see Suzanne M. Moon, "Take-off or Self-sufficiency? Ideologies of Development in Indonesia, 1957–1961," *Technology and Culture* 39 (1998): 187–212, especially 190–194.

5. Michael Hechter, *Internal Colonialism: The Celtic Fringe in British National Development* (Berkeley: University of California Press, 1975).

6. Diana K. Davis, *Resurrecting the Granary of Rome: Environmental History and French Colonial Expansion in North Africa* (Athens: Ohio University Press, 2007). The concepts of metropole and periphery are central to Wallerstein's world systems theory. They have been rightly criticized and complicated by subsequent scholars in a variety of disciplines. Nonetheless, they do capture the idea of uneven political and economic relations between different parts of the globe.

7. Quoted in "Pour un plan national d'aménagement du territoire" (1950), reproduced in Christel Alvergne and Pierre Musso, *Les grands textes de l'aménagement du territoire et de la décentralisation* (Paris: La Documentation française, 2003). For analysis of Claudius-Petit's ideas, see Benoît Pouvreau, "La politique d'aménagement du territoire d'Eugène Claudius-Petit," *Vingtième siècle, Revue d'histoire* 79 (2003): 43–52.

8. Cecelia Applegate, "A Europe of Regions: Reflections on the Historiography of Sub-National Places in Modern Times," *American Historical Review* 104 (1999): 1157–1182.

9. Quoted in Gravier, *Paris et le désert français*, 1.

10. Vivien A. Schmidt, *Democratizing France: The Political and Administrative History of Decentralization* (New York: Cambridge University Press, 1990), 4–5, 12–15.

11. Pierre Deyon, *Paris et ses provinces: Le défi de la décentralisation, 1770–1992* (Paris: Armand Colin, 1992).

12. Peter Alexis Gourevitch, *Paris and the Provinces: The Politics of Local Government Reform in France* (Berkeley: University of California Press, 1980), 32, 162.

13. Paul Claval, *Histoire de la géographie française de 1870 à nos jours* (Paris: Nathan, 1998), 223, 444.

14. After receiving his Ph.D., Gravier served as a Vichy bureaucrat, joining the general secretary of youth in 1941 and the Alexis Carrel Foundation two years later. After the war, he served in the Ministry of Reconstruction until 1949. From 1950, Gravier worked as a member of the Commissariat Général au Plan for fifteen years. He was also a member of the Conseil Economique et Social for regional economic development, another supervisory agency where he could implement his ideas, during 1959–1964.

15. Gourevitch, *Paris*, 32, 162.

16. Quoted in Deyon, *Paris*, 119.

17. Michel Michel, *L'aménagement régional en France: Du territoire aux territoires* (Paris: Masson, 1994).

18. On health issues, see David Barnes, *The Making of a Social Disease: Tuberculosis in Nineteenth-Century France* (Berkeley: University of California Press, 1995). Regarding crime and criminality, see Robert Nye, *Crime, Madness, and Politics in Modern France* (Princeton: Princeton University Press, 1984); Ann-Louise Shapiro, *Breaking the Codes: Female Criminality in Fin-de-Siècle Paris* (Stanford: Stanford University Press, 1996). On Parisian sewers as part of urban modernization, see Donald Reid, *Paris Sewers and Sewermen: Realities and Representations* (Cambridge, MA: Harvard University Press, 1991). For an overview of Haussmannization, see Jeremy Popkin, *A History of Modern France*, 3rd ed. (Englewood Cliffs, NJ: Prentice-Hall, 1994), 128–130.

19. A classic, if somewhat dated, study of French agriculture in the twentieth century is Michel Gervais, Marcel Jollivet, and Yves Tavernier, *Histoire de la France rurale: La fin de la France paysanne depuis 1914* (Paris: Editions du Seuil, 1977). On Vichy's environmental dimensions, see Chris Pearson, *Scarred Landscapes: War and Nature in Vichy France* (New York: Palgrave Macmillan, 2009). On Vichy's policies in La Crau, see Alexandre Giandou, "L'échec d'une colonisation agricole et ses conséquences: La Crau," *Ruralia*, 2000-06, http://ruralia.revues.org/document140.html (accessed June 15, 2009). On the regime's policies in the Camargue, see Chris Pearson, "A 'Watery Desert' in Vichy France: The Environmental History of the Camargue Wetlands, 1940–1944," *French Historical Studies* 32 (2009): 479–509.

20. Jean-Robert Pitte, *Philippe Lamour: Père de l'aménagement du territoire* (Paris: Fayard, 2002), 226.

21. J. F. Gravier, *L'aménagement du territoire et l'avenir des régions françaises* (Villeneuve-Saint-Georges: Imp. l'Union Typographique, 1964).

22. Niles M. Hansen, *French Regional Planning* (Bloomington: Indiana University Press, 1968), especially chapter 10. See also Michael Bess, *The Light-Green Society: Ecology and Technological Modernity in France, 1960–2000* (Chicago: University of Chicago Press, 2003), 50–52.

23. Lamour's biography and the early institutional history of what became the BRL is culled from Pitte, *Philippe Lamour*. See also Michel Maisonneuve, *La conquête de l'eau: BRL, histoire de l'aménagement en Languedoc-Roussillon* (Marseilles: Editions Editea, 1992), especially 48–49.

24. Quoted in Pitte, *Philippe Lamour*, 160.

25. Maisonneuve, *La conquête de l'eau*, 7.

26. Ibid., 53.

27. Quoted in Pitte, *Philippe Lamour*, 176. This figure is listed in old francs, the currency before de Gaulle's change in monetary policy implemented in January 1960.

28. Maisonneuve, *La conquête de l'eau,* 68–69. Locals near Donzère-Mondragon did not appear to become Poujadistes, but those living farther south and in the eastern Languedoc did (see Chapter 4).

29. Maisonneuve, *La conquête de l'eau,* 65.

30. Décret n. 55-254 du 3 février 1955 relatif à l'irrigation, à la mise en valeur et à la reconversion de la région du Bas-Rhône et du Languedoc, *Journal Officiel* (February 15, 1955), in AD-Hérault, A62i.

31. Maisonneuve, *La conquête de l'eau.*

32. Claude Delmas, *L'aménagement du territoire* (Paris: Presses universitaires de France, 1962), 78; Société du Canal de Provence et d'Aménagement de la Région Provençale, Demande de concession, Modification au mémoire explicatif général du dossier de demande de concession (January 1962), CAC, 19840241, box 8, Aménagement hydraulique et agricole du bassin de la Durance, Demande de concession générale et de déclaration d'utilité publique présentée par la "Société du Canal de Provence et d'Aménagement de la Région Provençal." See also AD-Hérault, EAF0, for documents referencing Marseille's growing water needs.

33. R. Perronnet, *La société d'économie mixte d'aménagement régional: L'exemple de la Compagnie nationale d'aménagement de la région du Bas-Rhône et du Languedoc* (Montpellier: Paysan du Midi, 1962), 4. This list comes from several sources in the AD-Hérault, including Perronnet.

34. Maisonneuve, *La conquête de l'eau,* 65–66, 72–73.

35. Numerous documents in the AD-Hérault provide detailed analysis of various *casiers* or subsections thereof. For example, folder AC12b-34 includes a technical study of "Casier 1, Secteur D1, 4e Zone." This is only one example of dozens. On the colonial use of *casiers,* see David Biggs, "Managing a Rebel Landscape: Conservation, Pioneers, and the Revolutionary Past in the U Minh Forest, Vietnam," *Environmental History* 10 (2005): 448–476, especially 462; David Biggs, "Breaking from the Colonial Mold: Water Engineering and the Failure of Nation-building in the Plain of Reeds, Vietnam," *Technology and Culture* 49 (2008): 599–623, especially 607.

36. The BRL's western zone is discussed in AD-Hérault, A49, A60, A61, A61C2, A62, A63, A64, A65, and A66. On the Lauragais Audois specifically, see AC91, AC92a, DC91a, DC91a2, DC92, DC94a, and DC94j. This list of relevant documentation is illustrative, not exhaustive.

37. SOGREAH, Basses-plaines situées à l'est du Département de l'Hérault, Avant-projet sommaire d'irrigation, Zone basse, Réseau de distribution par canaux, Note explicative et estimative (1961), AD-Hérault, AC32c1, Aménagement de la région du Bas-Rhône-Languedoc, Casier n. 3 (Vidourle-la Gardiole), Partie au sud du canal principal du Vidourle à l'aérodrome de Fréjorgues, Avant-projet général.

38. Maisonneuve, *La conquête de l'eau.*

39. On coastal tourism in the Languedoc-Roussillon, see Bess, *Light-Green Society*, 223. For an example of urban water networks in Sète, see AD-Hérault, AC08h.

40. See Chapter 2, as well as Robert Frost, "The Flood of Progress: Technocrats and Peasants at Tignes (Savoy), 1946–1952," *French Historical Studies* 14 (1985): 117–140.

41. Countless BRL hydrologic and hydrogeologic studies are included in the archives of the AD-Hérault; see, for instance, A41, A42, A43, A44a, A44b, A44c, A44d, A12, A13, and A14. For analysis of historic water usage, see B23. For detailed analysis of the BRL's water distribution network in one site, see AC32a-13 and AC32a-14. For extensive analysis of the possibilities of sprinkler irrigation, see DC67-02b, DC67-02, DC67-04, DC67-05, DC67-07, DC67-09, DC67-11, and DC67-12. For an early calculation of possible rate structures, see A010.

42. For instance, the Bureau des grands aménagements régionaux is mentioned in AD-Hérault, EA00.

43. Lamour was a prolific author in the mid-1960s. His books include *L'aménagement du territoire: Principes, éléments directeurs, méthodes et moyens* (Paris: Les Editions de l'épargne, 1964), *La réforme du régime foncier* (Paris: Les Editions de l'épargne, 1964), and *60 millions de Français* (Paris: Buchet, Chastel, 1967). All three studies were published within a few years of DATAR's founding.

44. Maisonneuve, *La conquête de l'eau*, 8.

45. Quoted in ibid., 116; see also 66.

46. Important early environmental histories of French North Africa include Davis, *Resurrecting the Granary of Rome*; Caroline Ford, "Reforestation, Landscape Conservation, and Anxieties of Empire in French Colonial Algeria," *American Historical Review* 113 (2008): 341–362; Will D. Swearingen, *Moroccan Mirages: Agrarian Dreams and Deceptions, 1912–1986* (Princeton: Princeton University Press, 1987). On the tight relationship between Algeria and "France," see Todd Shepherd, *The Invention of Decolonization: The Algerian War and the Remaking of France* (Ithaca: Cornell University Press, 2006).

47. Jacques Soustelle, *L'avenir du "Grand Delta" et l'Europe* (Paris: IMP, n.d.), in CNR, Documentation, AE 0753, 3. See also Abel Thomas, *Sillon Rhodanien: Axe-Rhin Méditerranée: Rapport du Commissaire à l'aménagement du territoire* (Paris: R. Lacer, 1960), 155–156; P. Bayard, "Le vrai visage de l'aménagement de Pierre-Bénite sur le Rhône," *L'Ingénieur* 73 (1967): 25.

48. Ian Thompson, *The Lower Rhône and Marseille* (Oxford: Oxford University Press, 1975), 46.

49. On the relationship between human bodies and environments in the United States, see Conevery Bolton Valencius, *The Health of the Country: How*

American Settlers Understood Themselves and Their Land (New York: Basic Books, 2002); Linda Nash, *Inescapable Ecologies: A History of Environment, Disease, and Knowledge* (Berkeley: University of California Press, 2006).

50. Soustelle, *L'avenir,* 2; "Le Grand Delta: un rendez-vous avec l'Europe ... des journées internationales 1972 à l'horizon 1980," *La vie française* 1452 bis (1973): 5, 7; Thompson, *Lower Rhône,* 46.

51. Pierre Terrin, "Le Grand Delta," *Transports* 147 (1962): 391–394.

52. One example is René Mayer, "Le Rhône, Principal axe de transport français," *Delta: Revue d'action régionale* 19 (1968): 28–35, in CNR, Documentation, AE 0167. Mayer held a number of important ministerial positions (1944–1952). He also served briefly as President of the Council (1953), the equivalent of the Fifth Republic's Prime Minister, and was later the head of the European Coal and Steel Community (1955–1957).

53. "Le Grand Delta," 7.

54. Soustelle, *L'avenir,* 6.

55. Ibid., 4; Mayer, "Le Rhône," 28–29; Terrin, "Le Grand Delta," 391–392; "Le Grand Delta," 10–11. On "*la Californie européenne,*" see Ministère de l'Equipement et du Logement et du Ministère des Transports, "Equipement et développement du sillon rhodanien et du sud-est français," *Equipement, Logement, Transports* (1969): 9.

56. Soustelle, *L'avenir,* 4; Mayer, "Le Rhône," 28–29.

57. Terrin, "Le Grand Delta"; Mayer, "Le Rhône," 28–29.

58. Ministère de l'Equipement et al., "Equipement," 16.

59. "Le Grand Delta," 5, 7, 9–10.

60. Soustelle, *L'avenir,* 2.

61. Ibid., 11; "Le Grand Delta," 5, 7.

62. Hechter, *Internal Colonialism.* Valerie Kuletz developed these ideas with respect to "nuclear colonialism" in *The Tainted Desert: Environmental and Social Ruin in the American West* (New York: Routledge, 1998).

63. Pinay, "Une alliance," 7, 9, 12–13.

64. Thompson, *Lower Rhône,* 46.

65. Terrin, "Le Grand Delta"; "Le Grand Delta," 6.

66. Ministère de l'Equipement et al., "Equipement," 3, 39.

67. Soustelle, *L'avenir,* 11; "Le Grand Delta," 25–26; Thompson, *Lower Rhône,* 46.

68. "Le Grand Delta," 5, 26.

69. Ministère de l'Equipement et al., "Equipement," 6.

70. Soustelle, *L'avenir,* 3–4, 6.

71. Lamour, *L'aménagement du territoire,* 6–7.

72. "Le Grand Delta," 24.

73. Ministère de l'Equipement et al., "Equipement," 4, 24.

74. Lamour, *L'aménagement du territoire,* 34; Pinay, "Une alliance," 7–13.

75. Jean-Michel Hoerner, *Géopolitique des territoires: De l'espace approprié à la suprématie des Etats-Nations* (Perpignan: Presses universitaires de Perpignan, 1996).
76. Soustelle, *L'avenir,* 17, 19.
77. Mayer, "Le Rhône," 34–35.
78. Pinay, "Une alliance," 9–12.
79. Soustelle, *L'avenir,* 17, 19.
80. Pinay, "Une alliance," 9–12.
81. Gabrielle Hecht discusses how French technologists after 1945 celebrated both techniques of knowledge production in *The Radiance of France: Nuclear Power and National Identity after World War II* (Cambridge, MA: MIT Press, 1998); for systems thinking, see 36–39, 102–105; for *la prospective,* see 44–47. For an overview of systems thinking, its implementation, and its implications, see Agatha C. Hughes and Thomas P. Hughes, eds., *Systems, Experts, and Computers: The Systems Approach in Management and Engineering, World War II and After* (Cambridge, MA: MIT Press, 2000).
82. Soustelle, *L'avenir,* 2; Pinay, "Une alliance," 10–11.
83. Olivier Guichard, *Aménager la France: Inventaire de l'avenir* (Paris: Robert Laffont, 1965); Pinay, "Une alliance," 13.
84. "Le Grand Delta," 27.
85. Ministère de l'Equipement et al., "Equipement," 6.
86. Pinay, "Une alliance," 9.
87. Soustelle, *L'avenir,* 14.
88. Ibid., 11–13, 19; Pinay, "Une alliance."
89. Richard White, "The Nationalization of Nature," *Journal of American History* 86 (1999): 976–986.
90. Jacques Bethemont, "Le Rhône: Entre nation et région," *Revue de géographie de Lyon* 72 (1997): 67–75.

6. Rethinking the Rhône

1. Fernand Braudel, *The Identity of France,* trans. Siân Reynolds, 2 vols. (New York: Harper and Row, 1988), chapter 7.
2. Given the cancellation of these two projects, it is worth noting that STS scholars have emphasized the importance of investigating "failures" in addition to "successes" in science, technology, and engineering, an intellectual commitment rooted in two concerns: first, Bloor's principle of symmetry, in which "successful" and "failed" knowledge claims are treated symmetrically, and second, that researching and writing about only successes fosters a sense of inevitability, even teleology. Furthermore, the success/failure binary evades a central question: *Whose* success? Environmental historians might add: What does "success" and "failure" mean for nonhumans?

3. Daniel Faucher, "'Du Rhône sauvage au Rhône discipliné,'" *Association universitaire d'études drômoises* 3–4 (1975): 48–52; Marcel Weckel, "L'aménagement du Rhône, exposé par M. l'Ingénieur général, lors du 796ᵉ déjeuner-conférence de la Société de Géographie Commerciale le 12 décembre 1973," *Revue économique française* 2 (1974): 43–58.

4. André Castelnau, "A Brens, près de Belley, on a remis le Rhône dans son lit... pour le faire travailler! (455 millions de kWh par an)," *Le progrès,* March 21, 1982, in BM-Lyon, AP3; Jacqueline Beaujeu-Garnier, "Rhône, couloir du futur," newspaper unknown (n.d.), in BM-Lyon, AP1; "Econotes, Chautagne et Belley: Les barrages pour 900.000 habitants," *Objectif Savoie-Léman,* June 18, 1977, in BM-Lyon, AP2.

5. For all of the sources associated with the BM-Lyon's Articles de presse, it is worth noting that library staff clipped, compiled, and organized articles from local, regional, and national newspapers and popular magazines. These volumes of thematically arranged press clippings greatly facilitated my research because I was able to gain access to dozens of popular articles in one centralized location, even though newspaper titles, dates, page numbers, and so on were frequently missing.

6. "L'aménagement du fleuve," *Bulletin du PCM* 10 (1976): 22–26; Faucher, "'Du Rhône sauvage'"; Beaujeu-Garnier, "Rhône," 49.

7. Castelnau, "A Brens."

8. "Aménagement du Rhône, un confluent tout neuf," *Métropole,* October 18, 1978. See also Jean Osterman, "Le Rhône dompté," *La vie française,* March 17, 1980; "Point final de l'équipement du Bas-Rhône, Le barrage de Vaugris inauguré par Raymond Barre," *Lyon Matin,* October 4, 1980. All in BM-Lyon, AP3.

9. Gilbert Tournier, Michel Laferrère, and François Tornas, *Lyonnais: Beaujolais, Forez, Vivarais* (Paris: Librairie Larousse, 1973); Castelnau, "A Brens."

10. A. Hiely, "Dans la vallée du Rhône, industrie et environnement veulent aller de pair," newspaper unknown (n.d.), in BM-Lyon, AP1; CNR, "Publi-information. La CNR: Un aménageur fluvial au service des collectivités," *Vie publique,* October 1991, in BM-Lyon, AP5.

11. CNR, "Publi-information."

12. Beaujeu-Garnier, "Rhône."

13. J. M. Delettrez, *Le Rhône de Genève à la Méditerranée* (France: B. Arthaud, 1974). See also Tournier et al., *Lyonnais;* Gilbert Tournier, *La vallée impériale: Couloir de l'Europe* (Lyon: Audin, 1977).

14. François Grosrichard, "L'inauguration du barrage de Vaugris, 'La liaison Rhin-Rhône se réalisera si elle constitue vraiment une grande ambition collective,'" *Le Monde,* October 5–6, 1980, in BM-Lyon, AP3.

15. Daniel Faucher, *L'homme et le Rhône* (Paris: Editions Gallimard, 1968), 369.

16. On the depoliticization of development, obviously in quite different contexts, see also James Ferguson, *The Anti-politics Machine: "Development," Depoliticization, and Bureaucratic Power in Lesotho* (Minneapolis: University of Minnesota Press, 1994); Timothy Mitchell, *Rule of Experts: Egypt, Techno-Politics, Modernity* (Berkeley: University of California Press, 2002), especially chapter 1.

17. See Chapter 7 and the Conclusion for more on this point.

18. Beaujeu-Garnier, "Rhône"; Delettrez, *Le Rhône*.

19. Delettrez, *Le Rhône*.

20. Quoted in "L'aménagement du fleuve," *Bulletin du PCM* 10 (1976): 22–26.

21. Cl. Gemaehling, "Vers l'achèvement de l'aménagement du Rhône," *Techniques et sciences municipales* 3 (1974): 93–103.

22. Grosrichard, "L'inauguration."

23. In my survey of newspaper, magazine, and other popular print media from the 1970s through 1990s, I have found no reference to Notre Dame or other icons discussed in Chapter 2.

24. Vincent Simon, "Le Rhône, fleuve-dieu, fleuve-béton," *Métropole: Magazine d'informations de Lyon et Rhône-Alpes,* June-July 1979, 17–21.

25. Much of the following background on the political history of the Fifth Republic is based on Jeremy D. Popkin, *A History of Modern France,* 3rd ed. (Upper Saddle River, NJ: Pearson, 2006), 285–320.

26. "M. Paul Ribeyre: 'Cette grande idée peut constituer pour la jeunesse le symbole des ambitions nouvelles de la France,'" *Le progrès,* November 25, 1975; "Giscard: 'La France au rendez-vous de l'Europe,'" *Le quotidien de Paris,* November 25, 1975; Tournier, *La vallée impériale;* "Seurre: La liaison Rhône-Rhin-Danube grande entreprise européenne,"*Le Monde,* November 6, 1973. For other examples of the "Europeanization" of the project, see Philippe Lamour, "Rhin-Rhône: Une liaison pour l'Occident," *Le Monde,* July 11, 1975; Roland Guinier du Vignard, "Rhin-Rhône: Un trait d'union qui divise," *Le quotidien de Paris,* November 25, 1975. All except Tournier in BM-Lyon, AP1.

27. "Europeanization" both builds off of and revises White's "nationalization of nature." In other words, the Rhône's history emphasizes both national and transnational frames. See Richard White, "The Nationalization of Nature," *Journal of American History* 86 (1999): 976–986.

28. D. Largeron, "Les lônes remplacées par des zones," *Le progrès,* October 3, 1980; Bruno Fournier, "Inauguration: L'usine hydraulique de Chautagne, premier maillon de l'aménagement du haut Rhône," *Le Dauphiné libéré,* February 13, 1981. Both in BM-Lyon, AP3; Simon, "Le Rhône."

29. There are both similarities and differences among "Western" environmentalisms, let alone between "Western" and "non-Western" movements. Of course, all of these terms are problematic. For an overview, see Ramachandra Guha, *Environmentalism: A Global History* (New York: Longman, 1999).

30. Michael Bess, "Greening the Mainstream: Paradoxes of Antistatism and Anticonsumerism in the French Environmental Movement," *Environmental History* 5 (2000): 6–26; Anna Bramwell, *Ecology in the Twentieth Century: A History* (New Haven, CT: Yale University Press, 1989); Philip G. Cerny, ed., *Social Movements and Protest in France* (New York: St. Martin's Press, 1982); Guillaume Sainteny, *L'introuvable écologisme français?* (Paris: PUF, 2000); Guillaume Sainteny, *Les Verts*, 3rd ed. (Paris: PUF, 1997); Guillaume Sainteny, "L'écologisme en Allemagne et en France: Deux modes différents de construction d'un nouvel acteur politique," working paper from *Barcelona Institut de Ciències Polítiques i Socials* 78 (1993).

31. Not surprisingly, many of FRAPNA's documents are duplicated in CNR and CAC archives or public statements issued in the media.

32. CNR, Chute de Belley, Demande d'autorisation de travaux avec déclaration d'utilité publique, Dossier d'enquête (March 10, 1977), CNR, AR0121, box 6, Chute de Belley, Dossier d'enquête. I address the EIS law and its role in mediating the transformation of the upper Rhône in more detail in Chapter 7.

33. For a study of "scientist-activists" and a discussion of the concept of "scientist-activists" as a distinct social movement, see Scott Frickel, *Chemical Consequences: Environmental Mutagens, Scientist Activism, and the Rise of Genetic Toxicology* (New Brunswick, NJ: Rutgers University Press, 2004). For analysis of an individual scientist that raises some related issues, see Michael Egan, *Barry Commoner and the Science of Survival: The Remaking of American Environmentalism* (Cambridge, MA: MIT Press, 2007).

34. Monique Coulet (FRAPNA cofounder), interview by author, June 2003.

35. Jean-Paul Bravard (professor, Université de Lyon-2), interview by author, November 1998.

36. M.J.G., "Nos lecteurs nous écrivent, Le Haut-Rhône," *Le progrès*, August 6, 1976, in BM-Lyon, AP2.

37. "Environnement, La CNR contestée," *Le progrès*, January 24, 1982, in BM-Lyon, AP3.

38. Bernard Clavel, "Témoignage, J'ai vu mourir le Rhône, je ne veux pas voir mourir le Doubs," *Le Monde*, June 22, 1978; "Les automobilistes pêcheurs sportifs: Un grand fleuve est mort," *Le progrès soir*, December 24, 1981. Both in BM-Lyon, AP3.

39. "Killed" is from Simon, "Le Rhône." "Assassinated" is from J. L. and M. Mandrin Soulié, "Le Rhône assassiné," *La gueule ouverte: Combat non-violent, Hebdomadaire d'écologie politique et de désobéissance civil* 215

(June 21, 1978): 2–3. "Massacred" is from Guy Jourdan, "Aménagement du Haut-Rhône, la FRAPNA pas d'accord," *Le progrès,* January 22, 1982, in BM-Lyon, AP3.

40. "Ecologie: Construction du barrage de Loyettes, la CNR rassure les opposants . . . ," *Le progrès,* January 16, 1982, in BM-Lyon, AP3; Jourdan, "Aménagement."

41. Carolyn Merchant, *The Death of Nature: Women, Ecology, and the Scientific Revolution* (San Francisco: Harper and Row, 1980).

42. Beaujeu-Garnier, "Rhône," 65.

43. A vast literature on American environmental history has deconstructed "wilderness" to demonstrate how it is a managed and transformed landscape. For a variety of essays that explore this theme, see William Cronon, ed., *Uncommon Ground: Toward Reinventing Nature* (New York: Norton, 1995).

44. Simon, "Le Rhône."

45. Robert Fritsch, "Un paysage à sauvegarder: La vallée du Rhône jurassien de Bellegarde à Lagnieu," *Bulletin de la société d'histoire naturelle de la Savoie* 185 (1977): 7; "Défense du fleuve Rhône et de la rivière d'Ain: Création d'un comité de coordination," *Le progrès,* December 26, 1981; Guy Jourdan, "Remous autour du projet de barrage de Loyettes," *Le progrès,* January 14, 1982; A. G., "Aménagements du Rhône, La CNR devra remonter le courant," *Voix de l'Ain,* January 15, 1982. All but Fritsch in BM-Lyon, AP3.

46. "Une pétition pour la sauvegarde du Haut-Rhône et de la rivière-d'Ain," *Le Dauphiné libéré,* February 24, 1982, in BM-Lyon, AP3.

47. Such a movement illustrates the concept of co-construction or co-production within STS. See, for instance, Bruno Latour, "Give Me a Laboratory and I Will Raise the World," in *Science Observed: Perspectives on the Social Study of Science,* ed. Karin Knorr-Cetina and Michael Mulkay (London and Beverly Hills, CA: Sage, 1983), 141–170; Bruno Latour, "Technology Is Society Made Durable," in *A Sociology of Monsters: Essays on Power, Technology, and Domination,* ed. John Law (London: Routledge, 1991), 103–131.

48. Françoise Holtz-Bonneau, "Des barrages à contre-sens," *Ecologie,* 344 (March 15–April 15, 1982): 19–20; R. Michel, "Derniers aménagements hydrauliques du haut Rhône," *Le progrès,* February 26, 1982, in BM-Lyon, AP3. See also "Philippe Lebreton: 'L'Ain c'est ma patrie,'" *Le Dauphiné libéré,* November 16, 1982, in BM-Lyon, AP3.

49. Philippe Lebreton, "Non aux barrages de Loyettes et de Sault Brenaz," *Le courrier de la nature* 79 (May-June 1982): 31–33. On Lebreton, see Michael Bess, *The Light-Green Society: Ecology and Technological Modernity in France, 1960–2000* (Chicago: University of Chicago Press, 2003), 120–121.

50. For an extensive discussion of states' simplification schemes, see James C. Scott, *Seeing Like a State: How Certain Schemes to Improve the Human Condition Have Failed* (New Haven, CT: Yale University Press, 1998). Ted Porter's work has

illustrated the power of numbers in masking the politics of decision-making. See Theodore Porter, *Trust in Numbers: The Pursuit of Objectivity in Science and Public Life* (Princeton, NJ: Princeton University Press, 1995).

51. "Aménagement, Débat sans réserve pour un débit réservé," *Le Dauphiné libéré,* January 10, 1982, in BM-Lyon, AP3; Guinier du Vignaud, "Rhin-Rhône: Un trait d'union qui divise"; M.-R. G., "Economie et écologie, Les questions auxquelles il faut réfléchir pour que la liasion Rhône-Rhin nous apporte un maximum d'avantage et un minimum de nuisances," *Le progrès,* January 31, 1975, in BM-Lyon, AP1.

52. Jourdan, "Aménagement."

53. F. Bueb, "La bataille du canal à grand garabit, Le canal Rhin-Rhône: Tous les coups sont permis, Le rapport du professeur Linder qui concluait à une 'catastrophe écologique' avait été réécrit,'" *Libération,* May 9, 1977; "Le canal à grand garabit contesté par les scientifiques," *Ecologie hebdo,* November 18, 1977. Both in BM-Lyon, AP2.

54. C. H., "Aménagement du Rhône: Peu de conséquences sur la qualité des eaux du fleuve," *Le Journal,* September 2, 1983, in BM-Lyon, AP3.

55. "L'aménagement du haut Rhône, une commission sénatoriale a étudié sur le terrain la future chute de Belley," *Le progrès,* March 4, 1975, in BM-Lyon, AP1; Mairie de Givors Coordination des associations de la vallée du Rhône, "Centrales nucléaires: pour ou contre?" *Stop Pollution/Vallée du Rhône* 1 (1975): 1; M.J.G., "Nos lecteurs nous écrivent."

56. Gilles Morel, "L'aménagement du Rhône: De très nombreux et très graves problèmes humains," in BM-Lyon, AP2. Nationalism is invoked in Jourdan, "Aménagement." The related notion of patrimony is from Fritsch, "Un paysage." The "subjugated Rhône" is from André Castelnau, "Equipement, Le nouveau visage du Rhône canalisé," *Le progrès,* April 14, 1982, in BM-Lyon, AP3. On unbalanced development, see "Pour Condrieu, L'aménagement du Rhône pose des problèmes," *Le Journal,* September 21, 1978, in BM-Lyon, AP3; "Le développement du Rhône Moyen: À la recherche d'une stratégie," *Liaisons Rhône-Alpes* 24 (March 1975): 43–51.

57. Paul R. Josephson, *Industrialized Nature: Brute Force Technology and the Transformation of the Natural World* (Washington, DC: Island Press, 2002), 24.

58. Alain Gilbert, "Les chantiers de la CNR à Bregnier et à Belley, Bétonner, exploiter . . . et respecter le Rhône," *Voix de l'Ain,* December 26, 1980. See also Morel, "L'aménagement du Rhône." On the potential costs to the nation, see Mairie de Givors Coordination des Association de la Vallée du Rhône, "Centrales nucléaires: pour ou contre?" 1; "Comurhex: La main dans le fût," *Hebdo-Pays* 19 (July-August 1979): 5. On the EDF-like CNR, see Holtz-Bonneau, "Des barrages à contre-sens." FRAPNA's complaints: Jourdan, "Aménagement."

59. Richard Kuisel, *Seducing the French: The Dilemma of Americanization* (Berkeley: University of California Press, 1993).

60. Bess, *Light-Green Society*; Gabrielle Hecht, *The Radiance of France: Nuclear Power and National Identify in France after World War II*, 2nd ed. (Cambridge, MA: MIT Press, 2009), especially Michel Callon's foreword; Bruno Latour, *Politics of Nature: How to Bring the Sciences into Democracy* (Cambridge, MA: Harvard University Press, 2004).

61. The environmental justice literature is vast. It tends to be divided into two camps: one focuses on inequalities of race and class within individual national contexts; the other centers on global inequalities, especially in terms of modern industrial capitalism. A major contribution of environmental justice movements, whether national or transnational, has been to highlight the ways "environmental" problems affect humans, although rarely equally.

7. A New Modern

1. Michel Crépeau to CNR (?) (March 10, 1982), CNR, AR0126, box 2, Chute de Sault-Brénaz, Enquête d'utilité publique, Enquête (folder hereafter SB, Enquête).

2. The agency has gone through numerous name changes; originally, it was called the Ministère de la Protection de la Nature et de l'Environnement. Maurice Larkin, *France since the Popular Front: Government and People, 1936–1986*, 2nd ed. (New York: Oxford University Press, 1997), 332; Michael Bess, *The Light-Green Society: Ecology and Technological Modernity in France, 1960–2000* (Chicago: University of Chicago Press, 2003), 83–84.

3. CNR, Chute de Belley, Demande d'autorisation de travaux avec déclaration d'utilité publique, Dossier d'enquête (March 10, 1977), CNR, AR0121, box 6, Chute de Belley, Dossier d'enquête. On the strengths and weaknesses of the EIS law, see Bess, *Light-Green Society*, 198–201.

4. Michael Bess, "Greening the Mainstream: Paradoxes of Antistatism and Anticonsumerism in the French Environmental Movement," *Environmental History* 5 (2000): 6–26. I borrow the phrase "greening of the French state" from Florian Charvolin, "L'invention de l'environnement en France (1960–1971): Les pratiques documentaires d'agrégation à l'origine du Ministère de la protection de la nature et de l'environnement" (Ph.D. diss., Université Pierre Mendès-France de Grenoble and Ecole Nationale Supérieure des Mines de Paris, Centre de Sociologie de l'Innovation, 1993).

5. Robert Gildea, *France since 1945* (New York: Oxford University Press, 1996), 94, 101, 204; Brendan Prendiville, *Environmental Politics in France* (Boulder: Westview Press, 1994), 12. However, as Gabrielle Hecht has pointed out, "nuclear" is actually a deeply problematic term; see her "Nuclear Ontol-

ogies," *Constellations* 13 (2006): 320–330, and "A Cosmogram for Nuclear Things," *Isis* 98 (2007): 100–108.

6. Gildea, *France since 1945*, 192; Prendiville, *Environmental Politics in France*, 14. See also Alain Touraine, *La prophétie anti-nucléaire* (Paris: Seuil, 1980).

7. "CNR: Crise de l'energie et aménagement du Haut-Rhône," *Bref Rhône Alpes* 365 (December 5, 1973), in BM-Lyon, AP1.

8. Gabrielle Hecht, *The Radiance of France: Nuclear Power and National Identity in France after World War II* (Cambridge, MA: MIT Press, 1998), 109.

9. On quantification in French engineering and its relationship to politics and expertise, see Theodore Porter, *Trust in Numbers: The Pursuit of Objectivity in Science and Public Life* (Princeton, NJ: Princeton University Press, 1995).

10. CNR, L'aménagement énergétique du Haut-Rhône, Sous-dossier A, Description de l'aménagement, Notice (February 12, 1974); CNR, Comité technique, Procès-verbal n. 150 des réunions des 9 et 16 avril 1974 (April 16, 1974). Both in CNR, Dossier for Comité technique n. 150, 9 et 16 avril 1974.

11. F. Collomb and A. Billiemaz to Monsieur le Premier Ministre, Raymond Barre (April 27, 1978), CAC, 770764, box 23, Dossier I, Chautagne-Belley interventions (folder hereafter Dossier I, C-B).

12. The "declaration of public utility" process for the two projects was initiated May 1976. See Collomb and Billiemaz to Monsieur le Premier Ministre (April 27, 1978).

13. This chronology is based on CNR, Comité technique, Procès-verbal n. 153 de la réunion du 18 février 1976 (February 18, 1976), CNR, Dossier for Comité technique n. 153 du 18 février 1976; Max Moulins to M. le Ministre de l'Industrie (April 21, 1978), CNR, AR0121, box 10, Demande de mise à l'enquête, Arrêté préfectoral.

14. The DUP process is described in CNR, La chute de Sault-Brénaz, Note de présentation, Compagnie nationale du Rhône (1980), CNR, AR0126, box 4, Chute de Sault-Brénaz, Enquête d'utilité publique, Note de présentation + photos. Certainly, France has had inquiry procedures over the public costs and benefits of projects since at least the eighteenth century. See, for example, Pierre Claude Reynard, "Public Order and Privilege: Eighteenth-Century French Roots of Environmental Regulation," *Technology and Culture* 43 (2002): 1–28.

15. P. Savey, Extrait du p.v. du Conseil d'administration du 18 janvier 1979 n. 431 (January 18, 1979), CNR, AR0121, box 10, Rapport d'activités au Conseil d'administration, Chautagne, Belley, Brégnier-Cordon, Années 1978 à 1981.

16. CNR, Chute de Belley, Demande d'autorisation de travaux avec déclaration d'utilité publique, Dossier d'enquête (March 10, 1977).

17. M. Legrand, note pour Monsieur le Ministre, Compagnie nationale du Rhône, Aménagement du Haut-Rhône, Chutes de Chautagne et de Belley (June 27, 1978), CAC, 770764, box 23, Dossier I, C-B.

18. *Pour un environnement "à la française": Textes et déclarations de M. Valéry Giscard d'Estaing, Président de la République* (Paris: La Documentation française, 1977).

19. For the longer history of *chartes* and their relationship to political regimes, political legitimacy, and history explored within the context of the archives and libraries of France, see Lara Jennifer Moore, *Restoring Order: The Ecole des Chartes and the Organization of Archives and Libraries in France, 1820–1870* (Duluth: Litwin Books, 2008).

20. Max Moulins to M. Doublet, Paris (December 21, 1976), CNR, AR0121, box 2, Informations sur l'administration.

21. J-F Carrez to M. le Ministre de l'Industrie et de la Recherche, Chutes de Chautagne, Belley, Brégnier-Cordon sur le Rhône, Demande d'autorisation de travaux avec DUP, CNR, pétitionnaire (November 4, 1976), CAC, 770764, box 23, dossier I, Beige folder (folder hereafter Dossier I, Beige); Ministre de la Qualité de la Vie, Note technique sur les aménagements CNR de Chautagne et Belley sur le Haut Rhône (March 21, 1977), CAC, 770764, box 23, dossier I. The refusal is discussed in M. Legrand to M. le Directeur Général (December 2, 1976), CNR, AR0121, box 6, Impact sur l'environnement.

22. Moulins to M. le Ministre de l'Industrie (April 21, 1978).

23. This point therefore revises James C. Scott, *Seeing Like a State: How Certain Schemes to Improve the Human Condition Have Failed* (New Haven: Yale University Press, 1999).

24. D. Cardot to M. l'Ingénieur en Chef de la Circonscription Electrique Sud-Est, CNR, Chutes de Chautagne, Belley et Brégnier-Cordon, Demande d'autorisation d'exécuter les travaux avec déclaration d'utilité publique (August 10, 1976), CNR, AR0123, box 5, Chute de Brégnier-Cordon, Mise à l'enquête (folder hereafter BC, Mise); CNR, Comité technique, Séance n. 153 du 18 février 1976, Annexe n. II au p.v. n. 153, Aménagement énergétique du Haut-Rhône, Chutes de Chautagne, Belley et Brégnier-Cordon, Résumé de l'exposé présenté par M. Savey (April 5, 1976), CNR, Dossier for Comité technique n. 153 du 18 février 1976.

25. The vocabulary of "the state" and "external" groups admittedly reifies artificial divides between the two.

26. For documents expressing support from municipal councils, chambers of commerce, and political representatives, see CNR, Comité technique, Procès-verbal n. 150 des réunions des 9 et 16 avril 1974 (April 16, 1974); Chambre de Commerce et d'Industrie de Vienne, Extrait du registre des délibérations (June 25, 1979), CAC, 820752, box 22, Aménagement du Haut-Rhône, Chute de Brégnier-Cordon, II Résultats de l'enquête et des conférences, Chute de Brégnier-Cordon, DUP, Dossier C, Registre d'enquête déposé à la Préfecture de l'Ain et pièces y annexés (folder hereafter BC, II, Résultats, Dossier C). Hunters hoped the CNR's projects would improve habitat for certain species: Paul Havet

to M. le Directeur des Etudes et Travaux, CNR (November 29, 1978), CNR, AR0123, box 14, Relations sur l'administration et les mairies de BC et des Avenières.

27. P. Savey to Comité de défense de l'environnement Belley-Est, Dépôt d'ordures projeté par la municipalité de Belley (May 7, 1974), CNR, AR0121, box 26, Lutte contre la pollution, protection nature: Comptes rendus de réunion (folder hereafter Lutte)

28. Sources describing FRAPNA activities include J. Commerot and R. Javellas to M. le Ministre de l'Industrie, Chutes hydroliques [sic] et conséquences (April 4, 1977); Cl. Gemaehling to Service de l'Industrie et des Mines Rhône-Alpes, Aménagement du Haut-Rhône, Avis Fédération de Protection de la nature (May 16, 1977). Both in CAC, 770764, box 23, Dossier I, C-B; R. Buisson to M. le Président de la Commission d'enquête portant sur l'aménagement de la chute de Brégnier Cordon sur le Rhône (June 26, 1979); J. Monteau, Letter to Commission (July 11, 1979). Both in CAC, 820752, box 22, BC, II, Résultats, Dossier C. L. Charlot, A. Chabert, and J. Mercier, Aménagement de la chute de Brégnier-Cordon sur le Rhône, Enquête, préalable à la déclaration d'utilité publique du projet d'aménagement de la chute de Brégnier-Cordon, Avis de la Commission d'Enquête (August 31, 1979), CNR, AR0123, box 5, Chute de Brégnier-Cordon, Résultats de l'enquête (folder hereafter BC, Résultats); FRAPNA, Voilà ce qu'il faut faire ... (1981), CNR, AR0126, box 2, SB, Enquête; Association pour l'environnement de la vallée du Rhône amont, Assemblée générale, 10 mars 1984, Brégnier-Cordon, Compte-rendu (March 10, 1984), CNR, AR0123, box 14, Impact sur l'environnement (folder hereafter Impact); H. Maneint, Compte rendu interne de la réunion annuelle sur l'environnement du 26 février 1987 (March 4, 1987), CNR, AR0097, box 1, Brégnier-Cordon, Lutte contre la pollution, Etudes écologiques.

29. The activities of other environmental groups are discussed in C. Pernin to M. L. Charlot, Ingénieur Divisionnaire des T.P honoraire, Commissaire Enquêteur (June 29, 1979); Jean-Pierre Raffin to M. le Commissaire Enquêteur (July 6, 1979). Both in CAC, 820752, box 22, BC, II, Résultats, Dossier C. Association pour l'environnement de la vallée du Rhône amont, Assemblé générale, 10 mars 1984, Brégnier-Cordon, Compte-rendu (March 10, 1984). R. Bichet, note pour DT.E., Aménagement des chutes de Loyettes et Sault-Brénaz, Enquête d'utilité publique, Réunions d'information (December 15, 1981); R. Bichet, DT.E, Aménagements de Loyettes et de Sault-Brénaz, Enquête d'utilité publique, Associations d'écologistes (December 24, 1981); FRAPNA to M. Michel Crépeau (1982). All three in CNR, AR0126, box 2, SB, Enquête.

30. P. Arnaud, Y. Martin, and L. Cottin, Chute de Brégnier-Cordon, DUP, Dossier E, Rapport des ingénieurs (March 5, 1980), CAC, 820752, box 22, Aménagement du Haut-Rhône, Chute de Brégnier-Cordon, II Résultats de

l'enquête et des conférences, Chute de Brégnier-Cordon, DUP, Dossier E, Pièces préparées par les ingénieurs.

31. For fishers' concerns, see Gemaehling to Service de l'Industrie et des Mines Rhône-Alpes, Aménagement du Haut-Rhône, Avis Fédération de Protection de la nature (May 16, 1977); Bichet, DT.E, Aménagements de Loyettes et de Sault-Brénaz, Enquête l'utilité publique, Associations l'écologistes (December 24, 1981); Association pour l'environnement de la vallée du Rhône amont, Assemblé générale, 10 Mars 1984, Brégnier-Cordon, Compte-rendu (March 10, 1984); Maneint, Compte rendu interne de la réunion annuelle sur l'environnement du 26 février 1987 (March 4, 1987). Maneint also expressed apprehension about agricultural issues. For other farming interests, see Bichet, ibid.

32. FRAPNA, Pétition pour la sauvegarde du Haut-Rhône et de la rivière d'Ain (1982), CNR, AR0126, box 2, SB, Enquête. Of note, duplicate copies of many letters from FRAPNA members or the organization as a whole are included in the CNR's and CAC's archives.

33. A. Lalegerie, Groupe de travail Haut-Rhône, Réunion du 22 février 1977 (February 22, 1977), CNR, AR0121, box 6, Impact sur l'environnement; Roger Ninin, Procès-verbal de la réunion de la Commission Départementale des Sites (July 9, 1979), CNR, AR0123, box 5, BC, Résultats; Maneint, Compte rendu interne de la réunion annuelle sur l'environnement du 26 février 1987 (March 4, 1987).

34. René Rousseau, Réunion départementale de l'environnement (December 2, 1981), CNR, AR0121, box 26, Lutte.

35. Roger Ninin, Projet d'aménagement hydro-électrique du Haut-Rhône par la CNR, Réunion d'information du 18 mars 1976 (March 18, 1976), CNR, AR0121, box 10, Enquête hydraulique. See also A. Blin, Extrait du régistre des délibérations du Comité syndical (May 26, 1977), CNR, AR0121, box 34, Lutte contre la pollution du Lac du Bourget, Rejet des eaux usées du Lac du Bourget; Paul Chatelain to M. Savey (August 13, 1979), CNR, AR0123, box 5, BC, Résultats; R. Bichet, note pour DT, Section d'études des chutes de Loyettes et Sault-Brénaz, Enquête d'utilité publique, Position de la municipalité de Sault-Brénaz (December 1, 1981), CNR, AR0126, box 2, SB, Enquête.

36. Jean-François Noblet to M. le Président de la Commission d'enquête portant sur l'aménagement de la chute de Brégnier Cordon sur le Rhône (June 26, 1979); L. Berteaux to M. le Président (July 7, 1979); Yvonne Survoi, letter (July 8, 1979); Association pour la connaissance et la protection de l'environnement Nature et Vie Sociale to M. le Président, Commission enquête du barrage de Brégnier-Cordon (July 10, 1979). All in CAC, 820752, box 22, BC, II, Résultats, Dossier C.

37. Noblet to M. le Président de la Commission d'enquête portant sur l'aménagement de la chute de Brégnier Cordon sur le Rhône (June 26, 1979);

Association pour la connaissance et la protection de l'environnement Nature et Vie Sociale to M. le Président, Commission d'enquête du barrage de Brégnier-Cordon (July 10, 1979); G. Maurel, Rapport sur les incidences sur l'environnement de l'aménagement de la chute de Brégnier-Cordon sur le Haut-Rhône, Séance du 28 février 1980 (February 28, 1980), CNR, AR0123, box 14, Impact; Odile Gaschignard to CNR (January 10, 1981), CNR, AR0126, box 2, Chute de Sault-Brénaz, Enquête d'utilité publique, Résultats de l'enquête; P. Arnaud to M. le Directeur du Gaz, de l'Electricité et du Charbon (June 1, 1977), CAC, 770764, box 23, Dossier I, C-B; Procès-verbal de la réunion de la Commission Départementale des Sites (July 9, 1979), CAC, 820752, box 22, BC, II, Résultats, Dossier C.

38. An example of anti-CNR perspectives is discussed in CNR, Aménagement de Belley, Projet-variante proposé par le Comité de Défense du Rhône Savoyard-Bugiste (September 22, 1977), CNR, AR0121, box 10, Contre projet du Comité de Défense du Rhône Savoyard Bugiste. Examples of constituencies that favored the projects but still issued reservations include Albert Buschini to M. le Directeur de la SNR [sic], Aménagement du haut Rhône (suggestion) (August 3, 1976), CNR, AR0121, box 2, Informations sur l'administration; Jacques Pallama to M. Charlot, Commissaire Enquêteur (July 10, 1979), CAC, 820752, box 22, BC, II, Résultats, Dossier C; Olivier Philip to M. le Directeur Interdépartemental de l'Industrie (April 4, 1980), CNR, AR0123, box 5, BC, Résultats.

39. Early complaints about the reserved flow are referenced in Georges Peyronne to M. l'Ingénieur en Chef du Service de la Navigation de Lyon (May 12, 1980); R. Bichet, note pour DT.E, Aménagement de la chute de Sault-Brénaz, Enquête d'utilité publique, Commune de Sault-Brénaz (December 18, 1981). Both in CNR, AR0126, box 2, SB, Enquête.

40. René Million to M. Max Moulins (May 12, 1976), CNR, AR0121, box 55, Pêche; Gemaehling to Service de l'Industrie et des Mines Rhône-Alpes, Aménagement du Haut-Rhône, Avis Fédération de Protection de la nature (May 16, 1977); Jean Commerot to M. le Président de la Commission d'enquète [sic] du barrage de Brégnier Cordon (July 4, 1979), CAC, 820752, box 22, BC, II, Résultats, Dossier C.

41. Buschini to M. le Directeur de la SNR [sic], Aménagement du haut Rhône (suggestion) (August 3, 1976); Procès-verbal de la réunion de la Commission Départementale des Sites (July 9, 1979); Pallama to M. Charlot, Commissaire Enquêteur (July 10, 1979); CNR, Réunion du 24 Juin 1980 avec les représentants des pêcheurs (July 1, 1980), CNR, AR0121, box 26, Organismes régionaux de l'aménagement du territoire, Correspondance INRA.

42. On the Rhône's beauty, see Million to M. Max Moulins (May 12, 1976); Paul Dufournet to M. le Ministre de l'Equipement (July 5, 1977), CNR, AR0121, box 10, Contre projet du Comité de Défense du Rhône Savoyard Bugiste. On the river's remaining wildness, see Roger Mathieu to M. le Président, Commission

d'Enquête [sic] (1979), CAC, 820752, box 22, BC, II, Résultats, Dossier C; Anne Line Majo to M. Le Président de la République, Barrage de Brégnier-Cordon sur le Rhône (July 16, 1980), CNR, AR0123, box 14, Lutte contre la pollution, Protection de la nature.

43. Robert Hainard to M. le Président de la Commission d'Enquête au barrage de Brégnier Cordon (July 2, 1979), CAC, 820752, box 22, BC, II, Résultats, Dossier C. On Hainard, see Bess, *Light-Green Society,* 120.

44. The first quotation is from CNR, Selon communication téléphonique de M. Bichet le 21 décembre 1981 (December 21, 1981), CNR, AR0126, box 2, SB, Enquête. The remaining quotations are from A. Buisson, District de Yenne, Assemblée générale du 15 décembre 1978 (December 15, 1978), CNR, AR0121, box 55, Chute de Belley, Assainissement de Yenne, Comptes-rendus de réunions; Ministère de l'Agriculture, Projet de réserve naturelle de Brégnier-Cordon, Réunion départementale de l'environnement (September 17, 1981), CNR, AR0123, box 14, Réserves naturelles, Etude du massif forestier.

45. Cl. Gemaehling to M. le Rédacteur en Chef du Progrès (February 14, 1980), CNR, AR0123, box 14, Chute de Brégnier-Cordon, Divers—Relations publiques et interventions diverses (folder hereafter BC, Divers); J. Lecornu to M. le Préfet de l'Isère, de l'Ain, de Région (1985), CNR, AR0123, box 27, Réserves naturelles.

46. Hecht, *Radiance of France,* 2nd ed., afterword.

47. Special thanks go to my research assistant, Corinna Schlombs, for compiling data on France's post-1945 energy history, which suggested these conclusions.

48. Moulins to M. le Ministre de l'Industrie (April 21, 1978).

49. Legrand, note pour Monsieur le Ministre, Compagnie nationale du Rhône, Aménagement du Haut-Rhône, Chutes de Chautagne et de Belley (June 27, 1978); "Décret du 28 novembre 1978 déclarant d'utilité publique les travaux d'aménagement de la chute de Belley sur le Rhône (départements de l'Ain et de la Savoie)," *Journal Officiel* (December 1, 1978), in CNR, AR0121, box 10, Décret déclaration d'utilité publique.

50. The Ministry of Environment's demands are discussed in CNR, Aménagement de Brégnier-Cordon, Etat actuel du déroulement de la procédure administrative (January 16, 1979), CNR, AR0123, box 14, Impact. The CNR's pleas are from CNR, Mise en service des chutes (May 9, 1979), CNR, AR0121, box 2, Programme d'études et travaux. The DUP is Arnaud et al., Chute de Brégnier-Cordon, DUP, Dossier E, Rapport des ingénieurs (March 5, 1980).

51. J. Lecornu to M. le Directeur Général, Chute de Brégnier-Cordon, Note sur les aspects politiques de l'opération (July 18, 1980), CNR, AR0123, box 14, BC, Divers.

52. Paul Martin to M. le Chef de Cabinet du Président de la République (December 8, 1980). For two examples of this turn to Paris, see Le Bouchage to

M. le Préfet de l'Isère, Grenoble, Pétition concernant la chute de Sault-Brénaz (January 26, 1982); CNR, Une mise au point de la Compagnie nationale du Rhône sur les projets d'aménagement du Haut-Rhône (January 28, 1982). All in AR0126, box 2, SB, Enquête.

53. CNR, Une mise au point de la Compagnie nationale du Rhône sur les projets d'aménagement du Haut-Rhône (January 28, 1982); R. Bichet, note pour DT.E, Aménagement des chutes de Loyettes et Sault-Brénaz, Enquête d'utilité publique, Réunions d'information (January 25, 1982), CNR, AR0126, box 2, SB, Enquête.

54. Préfet de l'Ain, Procès-verbal de la réunion de la Commission Départementale des Sites, Perspectives et paysages (January 20, 1982), CNR, AR0126, box 37, Chute de Sault-Brénaz, Relations avec les administrations et les collectivités locales (mairies, etc.) (folder hereafter SB, Relations).

55. The petition is referenced in R. Bichet, note pour DT.E, Aménagement des chutes de Loyettes et Sault-Brénaz, Enquête d'utilité publique, Associations de défense de l'environnement (February 26, 1982); his quote is from R. Bichet, Section d'études des chutes de Loyettes et Sault-Brénaz, Note d'information (March 3, 1982). Both in CNR, AR0126, box 2, SB, Enquête.

56. Crépeau to CNR (?) (March 10, 1982).

57. "Michel Crépeau opposé à deux projets de barrages sur le Rhône," *Lyon Matin,* April 23, 1982, in BM-Lyon, AP3. For the CNR's denial, see R. Bichet to M. le Sous-Préfet, Belley, Aménagements de Loyettes et Sault-Brénaz, Enquête d'utilité publique (April 26, 1982), CNR, AR0126, box 2, SB, Enquête.

58. Ministère de l'Urbanisme et du Logement, Impact sur l'environnement des projets CNR, Chutes de Sault-Brénaz et Loyettes (1982), CNR, AR0126, box 37, SB, Relations.

59. Pierrick Eberhard, "Equipement, Aménagement du Haut-Rhône, M. Crépeau et le Haut Comité de l'Environnement se prononcent contre le projet CNR de barrage à Loyettes," *Le progrès,* November 13, 1982, in BM-Lyon, AP3.

60. An example of the CNR's defense is Bichet to M. le Sous-Préfet, Belley, Aménagements de Loyettes et Sault-Brénaz, Enquête d'utilité publique (April 26, 1982). The ongoing resistance of FRAPNA is illustrated by FRAPNA, Requête introductive d'instance (October 17, 1983), CNR, AR0126, box 2, Chute de Sault-Brénaz, Enquête d'utilité publique, Déclaration d'utilité publique.

61. Pierrick Eberhard, "Aménagement, Barrage de Loyettes: La CNR étudie une variante," *Le progrès,* December 2, 1982, in BM-Lyon, AP3. Moving the project may have placated some opponents, but this (new) location would still have had ecological consequences for the river downstream, including the confluence of the Ain and Rhône.

62. Cl. Gemaehling to Madame le Ministre de l'Environnement et de la Qualité de la Vie (January 18, 1984), CNR, AR0126, box 38, Sault-Brénaz, Brégnier-Cordon, Réserve naturelle (folder hereafter SB, BC, Réserve).

63. It is ironic that the CNR implemented its "integrated design" at its first two projects on the upper Rhône, since opposition increased over time. This opposition apparently did not persuade the agency to adopt this model at Brégnier-Cordon, Sault-Brénaz, or Loyettes (before its eventual demise).

64. J. Lecornu, note pour DT-E, DT-BT, Aménagement de Sault-Brénaz, Protection contre les crues des plaines de Brangues, Le Bouchage, Les Avenières, Saint-Benoit, Demande de DUP, Etude d'impact (September 21, 1984). For other examples of the competing objectives of flood protection and natural preserves, see Louis Mermaz to M. le Directeur Général (January 8, 1985); Préfet de l'Ain to M. le Directeur (March 18, 1985), CNR, AR0126, box 38, SB, BC, Réserve; Ch. Galvin to M. le Directeur de la CNR (June 11, 1986). Galvin is also explicit about agricultural concerns. Criticism of environmentalists can be found in FRAPNA to M. le Président du Conseil Général de l'Isère et al., Projet de digue sur le Rhône à Brangues (March 26, 1985); Pierre Duffe, Protection des plaines contre les crues du Rhône, Projet du Syndicat intercommunal de défense contre les eaux du Haut-Rhône, Réunion de conciliation du 11 juin 1985 (June 19, 1985), CNR, AR0126, box 52, Projet de protection des digues du Rhône, Liste des annexes. All but Préfet de l'Ain and Duffe in CNR, AR0126, box 34, Protection plaine de Brangues.

65. I have not statistically analyzed how often the terms "former," "dead," and "short-circuited" Rhône were used during the 1970s and 1980s. However, in surveying sources from this period, I observed a strong trend toward using "short-circuited."

66. Buschini to M. le Directeur de la SNR [sic], Aménagement du haut Rhône (suggestion) (August 3, 1976); Commerot and Javellas to M. le Ministre de l'Industrie, Chutes hydroliques [sic] et conséquences (April 4, 1977).

67. Porter, *Trust in Numbers*.

68. Cl. Gemaehling to M. le Chef du Service Interdépartemental de l'Industrie et des Mines Rhône-Alpes, Aménagement de la chute de Belley, Instruction préalable à la mise à l'enquête (November 9, 1976), CNR, AR0121, box 10, Cahier des charges.

69. M. Beslin, note pour DT.E, Aménagement du Haut Rhône, Usines de Chautagne et Belley, Débit réservé (March 11, 1977), CNR, AR0121, box 34, Etudes Générales; CNR, Aménagement de Chautagne et Belley sur le Haut-Rhône, Eléments de réponse à la note technique du Ministère de la Qualité de la Vie du 16 Février 1977 (March 14, 1977), CNR, AR0121, box 34, Convention CNR/SIVOM du Lac du Bourget et Département de la Savoie.

70. Beslin, note pour DT.E, Aménagement du Haut Rhône, Usines de Chautagne et Belley, Débit réservé (March 11, 1977). Quotation from CNR,

Aménagement de Chautagne et Belley sur le Haut-Rhône, Eléments de réponse à la note technique du Ministère de la Qualité de la Vie du 16 février 1977 (March 14, 1977). The declaration of unprofitability is from CNR, Aménagement du Haut-Rhône: Chutes de Chautagne, Belley et Brégnier-Cordon, Coefficients de valeur dans différentes hypothèses d'exploitation (June 8, 1978), CAC, 770764, box 23, Dossier I, Beige; Gemaehling, Letter to Service de l'Industrie et des Mines Rhône-Alpes, Aménagement du Haut-Rhône, Avis Fédération de Protection de la nature (May 16, 1977).

71. Beslin, note pour DT.E, Aménagement du Haut Rhône, Usines de Chautagne et Belley, Débit réservé (March 11, 1977).

72. The original agreement is described in CNR, Chute de Belley, Demande d'autorisation de travaux avec déclaration d'utilité publique, Dossier d'enquête (March 10, 1977). The revisions are outlined in M. Girard, Objet du rapport, Aménagement du Haut Rhône par la Compagnie nationale du Rhône, Avis du Conseil Général sur les chutes de Chautagne et de Belley, Rapport de M. Girard (June 27, 1977), CNR, AR0121, box 10, Réunions d'information avant l'enquête, pendant l'enquête. Persistent critics include Commerot and Javellas to M. le Ministre de l'Industrie, Chutes hydroliques [sic] et conséquences (April 4, 1977); CNR, Aménagement de Belley, Projet-variante proposé par le Comité de Défense du Rhône Savoyard-Bugiste (September 22, 1977); Louis Besson to M. le Directeur Général de la CNR, Aménagement du Haut-Rhône (March 2, 1978), CNR, AR0121, box 10, Résultats de l'enquête, Procès verbaux.

73. CNR, Aménagement de la Chute de Belley, Résultats de l'enquête d'utilité publique et des conférences (April 3, 1978), CNR, AR0121, box 5, Aménagement de la Chute de Belley, Résultats de l'enquête d'utilité publique et des conférences.

74. Notes de réunion, Groupe de travail Environnement-Industrie-CNR, Réunion du 22 juin 1978 (June 22, 1978), CAC, 770764, box 23, Dossier I, Beige.

75. Réunion du 7.7.78 (July 7, 1978), CNR, AR0121, box 10, Résultats de l'enquête, Procès verbaux; Jacques Darmon to M. le Ministre de l'Industrie, Aménagement du Haut-Rhône (July 27, 1978), CAC, 820752, box 23, Chautagne/Belley, Dossier DUP. Furthermore, when the *cahier des charges spéciales* for Belley were approved in December 1980, the final numbers were slightly higher than the Ministry's initial demands. See Ministère de l'Industrie, "Décret du 23 décembre 1980 relatif à l'aménagement de la chute de Belley, sur le Rhône," *Journal Officiel* (January 8, 1981), in CNR, AR0121, box 10, Décret déclaration d'utilité publique.

76. J. Thiellet to M. le Ministre de l'Industrie et de la Recherche (August 25, 1976), CAC, 770764, box 23, Dossier II, Agriculture; Arnaud to M. le Directeur Général de la CNR, Aménagement du Haut-Rhône, Chute de Brégnier-Cordon,

Demande d'autorisation de travaux avec déclaration d'utilité publique (August 1, 1977), CNR, AR0123, box 5, BC, Mise.

77. CNR, Aménagement de Brégnier-Cordon, Pertes de production en fonction du débit réservé, Solution de référence: Débit réservé 10 m3/s (September 8, 1977); M. Horgnies, Chute de Brégnier-Cordon, Demande d'autorisation de travaux avec déclaration d'utilité publique, La CNR pétionnaire, Conférence préliminaire à la mise à l'enquête, Clôture de la conférence ouverte le 13 mai 1976 entre l'Ingénieur en Chef de la Circonscription Electrique Sud-Est et l'Ingénieur en Chef du Service Régional de l'Aménagement des Eaux Rhône-Alpes (December 8, 1977). Both in CNR, AR0123, box 5, BC, Mise.

78. M. Pommier, DM-C, Aménagement de Brégnier-Cordon, Production de l'usine (January 5, 1978), CNR, AR0123, box 1, Aménagement de Brégnier-Cordon, Etudes générales et préliminaires. See also M. Pommier, DM-A, Aménagement de Brégnier-Cordon, Production de l'usine (July 3, 1978), CNR, Brégnier-Cordon, Production nette (GWh) (1978); CNR, Usine de Brégnier-Cordon, Productible net (GWh) (1978). All in CNR, AR0123, box 1, Aménagement de Brégnier-Cordon, Turbines, Usine: Productibilité, Alternateurs.

79. R. Pinatel to Conseil Supérieur de la Pêche, Aménagement de la chute de Brégnier-Cordon, Passage des poissons, Mesure du débit réservé (October 16, 1984), CNR, AR0123, box 60, Chute de Brégnier-Cordon, Pêche, Divers (folder hereafter BC, Pêche); Horgnies, Chute de Brégnier-Cordon, Demande d'autorisation de travaux avec déclaration d'utilité publique (December 8, 1977).

80. L. Cottin, Aménagement hydroélectrique du Haut Rhône, Chute de Brégnier-Cordon, Demande d'autorisation de travaux avec déclaration d'utilité publique, La CNR, pétitionnaire, Propositions complémentaires en vue de la mise à l'enquête, Rapport de l'Ingénieur des TPE (October 15, 1978), CNR, AR0123, box 5, BC, Mise; CNR, Chute de Brégnier-Cordon, DUP, Note de calcul de la puissance de la chute (April 25, 1979), CAC, 820752, box 22, Chute de Brégnier-Cordon, Demande d'autorisation de travaux avec déclaration d'utilité publique, Dossier B, Pièces complémentaires, pièce 1.

81. Anglers' demands are described in Alexandre Sogno to M. le Président de la Commission d'Enquête du barrage de Brégnier-Cordon (July 10, 1979), CAC, 820752, box 22, BC, II, Résultats, Dossier C. Quote from Points examinés chez M. Edou avec MM. Hirzman, Servat, Brochet, Tane, Roussel, Gaemeling, Savey, Pommier (May 7, 1980), CAC, 820752, box 23, Chautagne/Belley, Dossier II.

82. Contrast these debates with those over the reserved flow at Donzère-Mondragon discussed in Chapter 3.

83. Cl. Gemaehling to Direction Interdépartementale de l'Industrie, Aménagement de Brégnier-Cordon, Avis de la Commission des Sites de l'Isère (April 21, 1980), CNR, AR0123, box 5, BC, Résultats.

84. CNR to M. le Président (June 2, 1980), CNR, AR0123, box 31, Pink folder.

85. B. Lainez to M. Savey, Directeur de la CNR, Aménagement de Brégnier-Cordon, Débit réservé (July 11, 1980), CNR, AR0123, box 60, BC, Pêche; CNR, Haut-Rhône, Projet de marche pour l'aménagement du seuil de Yenne (Chute de Belley), Projet de marche pour les travaux de Génie Civil du barrage-usine de Champagneux (Chute de Brégnier-Cordon) (April 20, 1982), CNR, AR0121, box 39.

86. CNR, Aménagement du Haut-Rhône, Chute de Brégnier-Cordon, Note sur les conséquences financières des modifications apportées au projet sur la demande du Ministère de l'Environnement (March 19, 1981), CNR, AR0123, box 5, BC, Résultats.

87. P. Savey to Conseil Supérieur de la Pêche, Aménagement de la chute de Brégnier-Cordon, Contrôle du débit réservé (September 3, 1984), CNR, AR0123, box 60, BC, Pêche. This outcome is further supported by evidence in P. Savey to Direction Régionale de l'Industrie et de la Recherche, Division du Contrôle de l'Electricité, Grenoble, Aménagement de la Chute de Brégnier-Cordon, Débit réservé en aval du barrage, Essai de débit (March 22, 1985), CNR, AR0097, box 3, Brégnier-Cordon, Etudes Générales, Débits réservés.

88. CNR, Selon communication téléphonique de M. Bichet le 21 décembre 1981 (December 21, 1981); "Ain, Aménagement, Débat sans réserve pour un débit réservé," *Le Dauphiné libéré,* January 10, 1982, in BM-Lyon, AP3.

89. CNR, Comité technique du 12 décembre 1979, Aménagement énergétique du Haut-Rhône, Chutes de Sault-Brénaz et Loyettes, Notice (October 17, 1979), CNR, Comité technique, p.v.; CNR, Comité technique, Procès-verbal n. 161 de la réunion du 12 décembre 1979 (December 12, 1979); CNR, Comité technique du 12 décembre 1979, Annexe I au p.v. n. 161 du 12 décembre 1979, Aménagement énergétique du Haut-Rhône, Chutes de Sault-Brénaz et Loyettes, Notice (October 17, 1979). Both in CNR, Dossier for Comité technique n. 161 du 12 décembre 1979; CNR, Impact sur l'environnement (February 15, 1980); CNR, Chute de Sault-Brénaz, Cahier des charges spéciales (February 15, 1980). Both in CNR, AR0126, box 4, CNR, Chute de Sault-Brénaz, Demande d'autorisation de travaux avec déclaration d'utilité publique, Dossier d'enquête; CNR, Aménagement de Sault-Brénaz, Hypothèse de débit réservé: 20 m3/s pendant 9 mois; 60 m3/s pendant 3 mois, Coefficient de valeur du projet de base "aménagé" (January 10, 1983), CNR, AR0126, box 39, Chute de Sault-Brénaz, Production d'énergie, Rentabilité; J. Michel, note pour DT-A, Aménagement de la chute de Sault-Brénaz, Réponses aux observations de l'AEVR amont (May 4, 1984), CNR, AR0126, box 38, SB, BC, Réserve.

90. CNR, Aménagement du Haut-Rhône, Chute de Sault-Brénaz, Mémoire descriptif (February 15, 1980), CNR, AR0126, box 4, Chute de Sault-Brénaz,

Demande d'autorisation de travaux avec déclaration d'utilité publique, Dossier d'enquête; CNR, La chute de Sault-Brénaz, Note de présentation, Compagnie nationale du Rhône (1980).

91. The CNR's concerns are raised in CNR, Comité technique, Procès-verbal n. 161 de la réunion du 12 décembre 1979 (December 12, 1979); CNR, Comité technique du 12 décembre 1979, Annexe I au p.v. n. 161 du 12 décembre 1979, Aménagement énergétique du Haut-Rhône, Chutes de Sault-Brénaz et Loyettes, Notice (October 17, 1979). An example of local concern is H. Bonaque to M. le Directeur des Etudes et Travaux de la CNR, Projet d'aménagement de la chute de Sault-Brénaz sur le Haut-Rhône (May 22, 1980), CNR, AR0126, box 2, SB, Enquête.

92. EDF-EPEE, Modélisation thermique du Rhône (1981), CNR, AR0126, boxes 18 and 19, Aménagement des chutes de Loyettes et de Sault-Brénaz, Incidence des éclusées sur le régime thermique du Rhône, Graphiques.

93. R. Bichet, Note pour les cadres techniques de HR.B (February 20, 1987), CNR, AR0126, box 52, Consignes d'exploitation, Etiages et débits critiques. The official *consigne d'exploitation,* or statement of official operations procedures, was issued in J. F. Gros, Compagnie nationale du Rhône, Chute de Sault-Brénaz, Consigne d'exploitation (November 23, 1987), CNR, D30007, box 4, Consignes définitives et avenants.

Conclusion

1. CNR, "Une ambition nouvelle: Conjuguer Technique et Nature" (n.d.) (brochure); the CNR's headquarters had numerous copies available during the late 1990s; copy in author's possession.

2. Gabrielle Hecht, "Rupture-talk in the Nuclear Age: Conjugating Colonial Power in Africa," *Social Studies of Science* 32 (2002): 691–728.

3. Commission de Recherche et d'Information Indépendantes sur la Radioactivité, "Site nucléaire du Tricastin (Drôme): Fuite radioactive sur l'installation SOCATRI (filiale AREVA), Communiqué CRIIRAD du 9 juillet 2008," www.criirad.org/actualites/dossiers-08/tricastin-juil08/socatri/cpcriirad-socatri9juil.html (accessed August 5, 2009) (italics added; exclamation marks in original).

4. The events of July 2008 have been gleaned from the following: "Centrale de Tricastin: La fuite serait moins importante que prevue," *Le Point.fr* (July 9, 2008, updated July 10, 2008); Saget Estelle, "Les eaux troubles du Tricastin," *L'Express* (September 4, 2008), 94–95; Institut de Radioprotection et de Sureté Nucléaire, "Note d'information: Fuite d'une solution contenant de l'uranium à l'usine SOCATRI du Tricastin," (July 8, 2008); Christophe Labbé and Olivia Recasens, "Radioactivité—Le coupable est le tumulus," *Le Point.fr* (July 17, 2008, updated July 22, 2008); Marc Vignaud, "Tricastin: Un nouvel incident à la Socatri," *Le Point.fr* (August 8, 2008); Marc Vignaud, "Les sites nucléaires

provoquent une contamination des nappes phréatiques," *Le Point.fr* (October 14, 2008, updated October 16, 2008); Marc Vignaud, "Tricastin: de l'uranium militaire suspecté de contaminer la nappe," *Le Point.fr* (July 18, 2008); Olivier Vincent, "Le nucléaire sous surveillance," *L'Express* (October 30, 2008), 108–110, 112. All newspapers have been accessed through their online collections. Additional documents are available on the CRIIRAD website, www.criirad.org/. Djahane Salehabadi provided research assistance on the 2008 incident at Tricastin.

5. Charles Perrow, *Normal Accidents: Living with High-Risk Technologies* (Princeton: Princeton University Press, 1999).

6. Sara B. Pritchard and Thomas Zeller, "The Nature of Industrialization," in *The Illusory Boundary: Environment and Technology in History,* ed. Stephen Cutcliffe and Martin Reuss (Charlottesville: University of Virginia Press, 2010): 69–100.

7. Gabrielle Hecht, *The Radiance of France: Nuclear Power and National Identity in France after World War II,* 2nd ed. (Cambridge, MA: MIT Press, 2009), afterword.

8. Technologies do not inherently or necessarily obscure human-natural relations; some can illuminate them, at times quite purposefully. For instance, meters that calculate water consumption or Google Earth may facilitate a conception of the planet and human interactions with the nonhuman world in ways other technologies do not. In contrast, in *La trace du fleuve: La Seine et Paris, 1750–1850* (Paris: Editions de l'Ecole des hautes études en sciences sociales, 2000), Isabelle Backouche traces how technologies such as embankments physically impeded Parisians' connections to the Seine. From this, she concludes they had become distanced from the river. I would reframe her argument to maintain that Parisians were no more distanced from the Seine than they had been in the past; it was simply that these links were less immediate and visible. I am influenced here by Richard White, *The Organic Machine: The Remaking of the Columbia River* (New York: Hill and Wang, 1996), especially 182.

9. Wiebe E. Bijker, "Globalization and Vulnerability: Challenges and Opportunities for SHOT around Its Fiftieth Anniversary," *Technology and Culture* 50 (2009): 610.

10. This is an example of the ways nature can be technological. See the Introduction.

11. Ethanol is probably the best example of a recent energy bandwagon; Joan Fujimura discusses the concept of scientific bandwagon in "The Molecular Biological Bandwagon in Cancer Research: Where Social Worlds Meet," *Social Problems* 35 (1988): 261–283. Michael Pollan provides an overview of the critique of ethanol in *The Omnivore's Dilemma: A Natural History of Four Meals* (New York: Penguin, 2006). I discussed the implications of framing energy in

terms of envirotechnical analysis in "Towards an Envirotechnical Approach to Energy History and Policy," paper presented at the annual meeting of SHOT, Lisbon, October 11–14, 2008.

12. Patrick Guilhaudin, "Le Rhône: Quelques objectifs de gestion, présentation du Plan d'Action Rhône," *Revue de l'Agence de l'Eau Rhône-Méditerranée-Corse* (1994): 3–12, especially 4–7.

13. Quoted in Serge Alexis, Claude Amoros, Jean-Paul Chirouze, Patrick Guilhaudin, and Albert-Louis Roux, *Le Rhône: Histoire d'une évolution, du développement industriel au 'réaménagement durable'* (Agence de l'Eau Rhône Méditerranée Corse, 2002), in CNR, Documentation, AI0449, 6.

14. In April 2005, a "Charte de l'Environnement" was added to the French Constitution. It is unclear if the Rhône's "Charte d'Environnement du Rhône" served as a model. On the political valence of *charte*, see Lara Jennifer Moore, *Restoring Order: The Ecole des Chartes and the Organization of Archives and Libraries in France, 1820–1870* (Duluth: Litwin Books, 2008).

15. Michael Bess, *The Light-Green Society: Ecology and Technological Modernity in France, 1960–2000* (Chicago: University of Chicago Press, 2003), 112–113, 207; Le Ministre de l'Equipement, des Finances et de l'Industrie; Le Ministre de l'Economie, des Transports et du Logement; La Ministre de l'Aménagement du Territoire et de l'Environnement, *Directive à M. le Préfet de Région Rhône-Alpes, Préfet du Rhône, Coordonnateur du bassin Rhône-Méditérannée-Corse* (July 6, 1998), in CNR, Documentation, P0201-43. The policy and its implications are discussed extensively in Alexis et al., *Le Rhône*, 6–7.

16. DIREN and Agence de l'Eau Rhône-Méditeranné-Corse, *Programme décennal de restauration hydraulique et écologique du fleuve Rhône, Projet* (April 1999), in CNR, Documentation, P0201-43, 1–19.

17. Bess, *Light-Green Society*.

18. Hecht, *Radiance of France*, 2nd ed., afterword.

19. An overview of the proposal is described in Marlise Simons, "Water Scarce, Barcelona Plans Big Pipe To Tap Rhône," *New York Times*, July 19, 1999, www.nytimes.com/1999/07/19/world/water-scarce-barcelona-plans-big-pipe-to-tap-rhone.html (accessed May 20, 2009).

20. The cancellation of Loyettes in 1986 meant the only proposal the CNR had under consideration was the Rhine-Rhône liaison. This probably explains, at least partly, why CNR officials fought so hard for the project and were so frustrated by Voynet's cancellation of it in June 1997.

21. For the global reach of the CNR's activities, see the description on its website of its engineering division, www.cnr.tm.fr/fr/categorie.aspx?idcategorie=123, and a related brochure, www.cnr.tm.fr/medias_dynamique/ingenierie_fluviale_ok.pdf. The projects listed here come from an earlier version of this brochure. Similarly, an overview of the BRL's engineering division boasts projects worldwide:

www.brl.fr/index.php?page=7&rubrique=13&article=11&stat=no. (All accessed May 20, 2009.)

22. Neil Smith, *Uneven Development: Nature, Capital, and the Production of Space,* 3rd ed. (Athens: University of Georgia Press, 2008).

BIBLIOGRAPHY

Archives and Other Collections

Archives, Association des Amis du Vieux Donzère, Donzère
Books, pamphlets, newspaper and magazine articles, postcards, and photographs of Donzère-Mondragon and the town of Donzère.

Archives Départementales de la Drôme, Valence
Newspaper and magazine collections on CNR, Donzère-Mondragon.

Archives Départmentales de l'Hérault, Montpellier
Papers on BRL's and CNR's activities in the department; some regarding efforts in the Ardèche.

Archives Départmentales du Vaucluse, Avignon
Papers on CNR, Donzère-Mondragon.

Archives de la Direction Départementale de l'Equipement de la Drôme, Valence
Papers on CNR, Donzère-Mondragon.

Archives Municipales, Pierrelatte
Minutes of meetings, Conseil Municipal.
Papers on Canal de Pierrelatte, Association foncière de remembrement, ASA pour l'irrigation des régions Pierrelatte-Orange.

Archives Municipales, Saint-Paul-Trois-Châteaux
Minutes of meetings, Conseil Municipal.
Papers on Canal de Pierrelatte, CNR, Donzère-Mondragon.

Registre des délibérations de l'Association Foncière Syndicale de Saint-Paul-Trois-Châteaux. Series AF8, AF9, AF10, AF11, AF12, AF14, AF15, AF16, AF17, AF18, 5.0.3, 5.0.15, 5.0.16, 1AV.3bis, 1D8, 3F8.

Archives Nationales, Paris

Papers on CNR, Donzère-Mondragon, Rhône River.
Series F10, boxes 2671, 2672, 3761, 3762.
Series F14, boxes 12608, 12609, 12610, 12611, 14643, 14647, 14649, 14696, 14697, 15367.

Bibliothèque Municipale de Lyon

Books and local publications on CNR, Donzère-Mondragon, Rhône River; compiled newspaper articles (L'aménagement du Rhône, Suivi de la liaison du Rhône au Rhin, vols. 1–5).

Centre des Archives Contemporaines, Fontainebleau

Papers on Belley, Brégnier-Cordon, Chautagne, CNR, Donzère-Mondragon, Loyettes, Rhône River, Sault-Brénaz.
Series 770614, box 85.
Series 770764, boxes 17, 19.
Series 780467, box 24.
Series 810205, box 34.
Series 820752, boxes 22, 23.
Series 850575, box 4.

Archives, Compagnie Nationale du Rhône, Lyon

Papers from the Comité technique (1935–present); various documents pertaining to the following projects.
  Belley
    Series AR0044, box 4.
    Series AR0098, boxes 1, 2, 3, 5, 9, 12.
    Series AR0121, boxes 2, 3, 4, 5, 6, 8, 9, 10, 18, 21, 26, 27, 33, 34, 35, 37, 38, 39, 40, 45, 47, 48, 50, 53, 55, 57, 59, 61.
    Series AR0123, boxes 60, 61.
    Series AR0126, box 37.
    Series D30012, boxes 1, 2, 3, 7.
  Brégnier-Cordon
    Series AR0097, boxes 1, 3, 4, 7, 8.
    Series AR0099, box 3.
    Series AR0123, boxes 1, 2, 3, 5, 6, 12, 13, 14, 26, 27, 31, 36, 39, 40, 50, 53, 60, 61.

Series AR0126, boxes 38, 39, 52.
Series D30008, boxes 1, 4, 5, 7.

Chautagne

Series AR0044, box 4.
Series AR0097, box 3.
Series AR0098, box 12.
Series AR0121, boxes 6, 10, 18, 26, 34, 38.
Series AR0123, boxes 5, 6, 60, 61.
Series AR0126, box 38.

Donzère-Mondragon

Series AR0086, boxes 3, 20, 44, 45, 46, 48, 49.
Series AR0103, boxes 3, 4, 5, 6, 10, 11, 12, 13, 15, 17, 18, 20, 21, 26, 28, 29, 49, 53, 61, 62, 65, 67, 69, 70, 71, 75, 76, 77, 78, 79, 80, 81, 82, 84, 87, 89, 94, 135, 136, 155, 158, 175.
Series AR0106, boxes 23, 43, 44, 46, 47, 49, 51, 54, 56, 59, 61, 62, 70, 72, 73, 93, 94, 96.
Series D30020, box 3.

Sault-Brénaz

Series AR0029, boxes 1, 3, 4, 5.
Series AR0099, boxes 1, 2, 3, 4.
Series AR0123, box 6.
Series AR0126, boxes 2, 3, 4, 5, 6, 7, 8, 11, 16, 20, 21, 34, 37, 38, 39, 49, 52, 53, 55, 56, 60, 61, 85, 91.
Series D30007, boxes 2, 4, 5, 6, 10.

Documentation, Compagnie Nationale du Rhône, Lyon

Newspapers, magazines, pamphlets, books, photographs, films, and other materials on Belley, Chautagne, Brégnier-Cordon, Donzère-Mondragon, and Sault-Brénaz published by non-CNR sources; similar materials on the history of the CNR and its general program.

La Documentation Française, Paris

Government publications and studies of CNR, Donzère-Mondragon, hydroelectric development in France, Rhône River.

Archives, Fédération Rhône-Alpes de Protection de la Nature, Lyon

Newspaper and magazine coverage of upper Rhône; internal and public memos, minutes, studies, and other documents pertaining to the development of the upper Rhône and the organization's activities.

Photothèque, Compagnie Nationale du Rhône, Lyon
Various photographs of the CNR's projects.

Interviews

Conducted by Sara B. Pritchard

Current and former CNR administrators, engineers, technicians, and employees (1998–1999): M. Bourgad; J.-M. Bouvard; M. Carbonari; M. Charbonnier; E. Doutrieux; M. Gueret; H. Laydier; C. Jimenez; M. Moussa; M. Paris; R. Pinatel; J. Ponce; C. Terrier.
Residents of Donzère (1999): M. Nicholas; M. Reynard (daughter); M. Reynard (father)
Residents of Pierrelatte (1999): Y. Guéret.
FRAPNA cofounders (2003): M. Coulet; P. Lebreton.

Conducted by Olivier Mondon; recording available in Archives Municipales de Saint-Paul-Trois-Châteaux

J. Laurent; A. Stemmelin.

Primary Sources

A.G. "Aménagements du Rhône, La CNR devra remonter le courant." *Voix de l'Ain,* January 15, 1982.
"Ain, Aménagement, Débat sans réserve pour un débit reservé." *Le Dauphiné libéré,* January 10, 1982.
Allix, André. "Le Rhône." *Réalisations industrielles,* June-July 1951.
"L'aménagement de la vallée du Rhône-Moyen." *Liaisons Rhône-Alpes* 35 (1969): 1–3.
"L'aménagement du fleuve." *Bulletin du PCM* 10 (1976): 22–26.
"L'aménagement du Haut-Rhône." *Les nouvelles annales de l'Ain* (1983–1984): 147–170.
"L'aménagement du haut Rhône, une commission sénatoriale a étudié sur le terrain la future chute de Belley." *Le progrès,* March 4, 1975.
"L'aménagement du Rhône." *Bulletin AFC,* n.d.
"Aménagement du Rhône, un confluent tout neuf." *Métropole,* October 18, 1978.
Aubert, Jean. "La canalisation du Bas-Rhône." *Annales des Ponts et Chaussées* (1947).
"Les automobilistes pêcheurs sportifs: Un grand fleuve est mort." *Le progrès soir,* December 24, 1981.
Barron, Louis. *Le Rhône.* Paris: Librairie Renouard, 1891.

Bayard, P. "Le vrai visage de l'aménagement de Pierre-Bénite sur le Rhône." *L'ingénieur* 73 (1967): 25.
Bernard, A. "La récente industrialisation de la région de Pierrelatte et ses incidences géographiques." *Association universitaire d'études drômoises* 3–4 (1974): 29–46.
———. "Le Rhône en 1981." *Etudes drômoises* (1982): 43–52.
———. "Sur l'histoire du Tricastin." *Association universitaire d'études drômoises* 2 (1973): 19–24.
Bideau, Daniel. *De Lyon à Valence au gré du Rhône.* Lyon: Elie Bellier, 1982.
Boher, L. "Aménagement de la chute de Donzère-Mondragon, L'usine André Blondel de la Compagnie nationale du Rhône." *Revue d'électricité de mécanique* (1953).
Bonnefoy, Georges. "L'usine génératrice André Blondel." *Revue générale de l'électricité* 65 (1956): 5–33.
Bordeaux, Louis. "L'aménagement du Rhône: Étude d'économie politique." Ph.D. diss., Université de Lyon, 1919.
Breittmayer, Albert. *Le Rhône: Sa navigation depuis les temps anciens jusqu'à nos jours.* Lyon: Henri Georg, 1904.
Bret, Paul-Louis. "La formule du Rhône." *Rapports: France–Etats Unis,* November 1952.
Bueb, F. "La bataille du canal à grand garabit, Le canal Rhin-Rhône: Tous les coups sont permis, Le rapport du professeur Linder qui concluait à une 'catastrophe écologique' avait été réécrit.'" *Libération,* May 9, 1977.
C.H. "Aménagement du Rhône: Peu de conséquences sur la qualité des eaux du fleuve." *Le Journal,* September 2, 1983.
"Le canal à grand garabit contesté par les scientifiques." *Ecologie hebdo,* November 18, 1977.
Carrière, Marcel. "Découverte de la France, pays des merveilles, IX: Donzère-, capitale du Rhône." *Détective, l'hébdomadaire des secrets du monde,* 1951.
Castelnau, André. "A Brens, près de Belley, on a remis le Rhône dans son lit... pour le faire travailler! (455 millions de kWh par an)." *Le progrès,* March 21, 1982.
———. "Equipement, Le nouveau visage du Rhône canalisé." *Le progrès,* April 14, 1982.
Cédié, Roger. "Deux expériences d'économie mixte dans l'industrie électrique." Ph.D. diss., Université de Paris, 1943.
Chabert, Louis. "Introduction à l'étude du nucléaire rhodanien." *Revue de géographie de Lyon* 62 (1987): 141–148.
———. "Les transformations des communes nucléaires de la vallée du Rhône." *Revue de géographie de Lyon* 62 (1987): 161–191.
Chaintreuil, Hildebert. "Il y a 40 ans, Histoire d'hommes, à l'ombre d'un grand barrage." *Cahiers Rhône-Alpes d'histoire sociale* 16 (1990): 14–25.

Chambre de Commerce de la Drôme. *La Drôme.* Romans: J.A. Domergue, 1950.

Champion, M. Maurice. *Recherches historiques sur les inondations du Rhône et de la Loire.* Paris: Imp. de Panckoucke, 1856.

Chignol, Jean. "Samedi, Donzère-Mondragon a été inauguré officiellement par le président de la République ... et 2.000 CRS." *Les Allobroges,* October 27, 1952.

Clavel, Bernard. "Témoignage, J'ai vu mourir le Rhône, je ne veux pas voir mourir le Doubs." *Le Monde,* June 22, 1978.

Cluzea, Marc. "Rhône, merveilleuse mine d'énergie." *Le pèlerin,* February 16, 1958.

"CNR: Crise de l'energie et aménagement du Haut-Rhône." *Bref Rhône Alpes* 365 (December 5, 1973).

Collin, M. "Etude des incidences de l'installation de l'usine atomique de Pierrelatte sur les structures et la démographie d'un petit village voisin: La Garde-Adhémar." *Association universitaire d'études drômoises* 2 (1975): 15–17.

Compagnie Nationale du Rhône. *L'aménagement du Rhône et l'agriculture.* Lyon: Imprimerie Rivet, 1981.

———. "Publi-information. La CNR: Un aménageur fluvial au service des collectivités." *Vie publique,* October 1991.

———. *Son programme, ses réalisations.* Villeurbanne: Studio Villeurbannais, 1971.

"Comurhex: La main dans le fût." *Hebdo-Pays* 19 (July-August 1979).

Coordination des Associations de la vallée du Rhône, Mairie de Givors. "Une association de lutte contre la pollution." *Stop Pollution/Vallée du Rhône* 1 (1975): 1.

———. "Centrales nucléaires: Pour ou contre?" *Stop Pollution/Vallée du Rhône* 1 (1975): 1.

———. "Livre blanc de la pollution du Rhône." *Stop Pollution/Vallée du Rhône* 8 (1982).

Cros, M. "Un petit domaine agricole du Tricastin." *Association universitaire d'études drômoises* 4 (1973): 21–24.

"Défense du fleuve Rhône et de la rivière d'Ain: Création d'un comité de coordination." *Le progrès,* December 26, 1981.

Delattre, Pierre. "L'aménagement du Rhône." *Travaux* 286 (August 1958): 695.

———. *La chute de Donzère-Mondragon sur le Rhône.* Paris: Société des ingénieurs civils de France, 1952.

Delettrez, J. M. *Le Rhône de Genève à la Méditerranée.* France: B. Arthaud, 1974.

Delmas, Claude. *L'aménagement du territoire.* Paris: Presses universitaires de France, 1962.

Deriol, Claudius. "Entre Donzère et Mondragon, Le Rhône va sauter une marche de vingt-deux mètres." *France Magazine,* June 25, 1950.

"Le développement du Rhône Moyen: À la recherche d'une stratégie." *Liaisons Rhône-Alpes* 24 (March 1975): 43–51.

La Direction de la Documentation, ed. *Donzère-Mondragon et l'aménagement du Rhône.* Paris: La Documentation française, 1950.

La Documentation française. "L'équipement hydroélectrique de la France." *Notes, documentaires et études* 524, Série française 122 (1947): 496–587.

Donzère-Mondragon. Grenoble: La Houille Blanche, 1955.

Dubesset, Pierre. "Une retombée agricole du nucléaire rhodanien: Les serres chauffées." *Revue de géographie de Lyon* 62 (1987): 193–217.

Eberhard, Pierrick. "Aménagement, Barrage de Loyettes: La CNR étudie une variante." *Le progrès,* December 2, 1982.

———. "Equipement, Aménagement du Haut-Rhône, M. Crépeau et le Haut Comité de l'Environnement se prononcent contre le projet CNR de barrage à Loyettes." *Le progrès,* November 13, 1982.

"Ecologie: Construction du barrage de Loyettes, la CNR rassure les opposants. . . ." *Le progrès,* January 16, 1982.

Economic Cooperation Administration. *3 Years of the Marshall Plan.* Washington, DC: Government Printing Office, 1951.

"Econotes, Chautagne et Belley: Les barrages pour 900.000 habitants." *Objectif Savoie-Léman,* June 18, 1977.

"Environnement, La CNR contestée." *Le progrès,* January 24, 1982.

Fangeat, Jean. "Devant les ambassadeurs de 45 nations, M. Vincent Auriol a mis en marche l'usine géant de Donzère-Mondragon." *Le Dauphiné libéré,* October 27, 1952.

Faucher, Daniel. "'Du Rhône sauvage au Rhône discipliné.'" *Association universitaire d'études drômoises* 3–4 (1975): 48–52.

———. *L'homme et le Rhône.* Paris: Editions Gallimard, 1968.

Faure, Gabriel. *The Banks of the Rhone from Lyons to Arles.* Trans. Frank Kemp. Grenoble: J. Rey, 1922.

Faure, Raoul. "Le Rhône (II, Au fil de son histoire)." *Bibliothèque de travail* (1959).

Fioravante, J. *Aménagement hydro-agricole Rhône Moyen.* Romans-sur-Isère: Ets J-A Domergue SA, May 1969.

Fontaine, J. P., and J. P. Roux. "Les problèmes d'implantation soulevés par la construction du complexe EDF-Eurodif." *Revue générale nucléaire* (May-June 1977): 184–189.

Fournier, Bruno. "Inauguration: L'usine hydraulique de Chautagne, premier maillon de l'aménagement du haut Rhône." *Le Dauphiné libéré,* February 13, 1981.

Fournier, Martine, and Pierre-Claude Tracol. *Hommage aux mariniers du Rhône.* Valence: Imp. réunies, 1980.

Fritsch, Robert. "Un paysage à sauvegarder: La vallée du Rhône jurassien de Bellegarde à Lagnieu." *Bulletin de la société d'histoire naturelle de la Savoie* 185 (1977): 7.

Gemaehling, Cl. "Vers l'achèvement de l'aménagement du Rhône." *Techniques et sciences municipales* 3 (1974): 93–103.

Gemaehling, Cl., and P. Savey. "The Multipurpose Development of the River Rhône Valley," *Water Power: IPC Electrical and Electronic Press* (November 1972).

George, Pierre. *Les pays de la Saône et du Rhône.* Paris: Presses universitaires de France, 1946.

GEPAR. *Etude des perspectives agricoles et rurales rhodaniennes.* Lyon: Impression Super Plan, May 1972.

Gilbert, Alain. "Les chantiers de la CNR à Bregnier et à Belley, Bétonner, exploiter . . . et respecter le Rhône." *Voix de l'Ain,* December 26, 1980.

"Giscard: 'La France au rendez-vous de l'Europe.'" *Le quotidien de Paris,* November 25, 1975.

"Le Grand Delta: Un rendez-vous avec l'Europe . . . des journées internationales 1972 à l'horizon 1980." *La vie française* 1452 bis (1973).

Gravier, J. F. *L'aménagement du territoire et l'avenir des régions françaises.* Villeneuve-Saint-Georges: Imp. l'Union Typographique, 1964.

———. *Paris et le désert français.* Paris: Flammarion, 1947.

Gres, G. "Les chantiers de Donzère-Mondragon." *Travaux* 195 (January 1951): 25.

Grosjean, Thierry. "Les méandres d'un canal." *Ecologie* (September–October 1987): 31–33.

Grosrichard, François. "L'inauguration du barrage de Vaugris, 'La liaison Rhin-Rhône se réalisera si elle constitue vraiment une grande ambition collective.'" *Le Monde,* October 5-6, 1980.

Grosso, René, Pierre Nicolas, and Lucien Perret. "Donzère-Mondragon: Les chantiers (1947–1951)." *Bibliothèque de travail* 166 (1951).

———. "La peine des hommes à Donzère-Mondragon." *Bibliothèque de travail* 167 (1951): 13.

Guex-Rolle, Henriette. *Rhône.* Lausanne: Librairie Marguerat, 1956.

Guichard, Olivier. *Aménager la France: Inventaire de l'avenir.* Paris: Robert Laffont, 1965.

Guilhaudin, Patrick. "Le Rhône: Quelques objectifs de gestion, présentation du Plan d'Action Rhône." *Revue de l'Agence de l'Eau Rhône-Méditerranée-Corse* (1994): 3–12.

Guinier du Vignard, Roland. "Rhin-Rhône: Un trait d'union qui divise." *Le quotidien de Paris,* November 25, 1975.

Holtz-Bonneau, Françoise. "Des barrages à contre-sens." *Ecologie* 344 (March 15–April 15, 1982): 19–20.
"Invité par la Compagnie nationale du Rhône, M. Vincent Auriol accompagné de 53 ambassadeurs et diplomates étrangers inaugurera samedi prochain le Canal Donzère-Mondragon et l'Usine André Blondel." *Le Dauphiné libéré,* October 18, 1952.
Jourdan, Guy. "Aménagement du Haut-Rhône, la FRAPNA pas d'accord." *Le progrès,* January 22, 1982.
———. "Remous autour du projet de barrage de Loyettes." *Le progrès,* January 14, 1982.
Labadie, J. "L'aménagement du Rhône de Donzère à Mondragon." *Science et vie* 381 (1949): 380–385.
Lachat, R. L. "Alors que l'on s'apprête à inaugurer avec éclat le gigantesque ouvrage rappelons que ce sont 2 Grenoblois MM. Giroud et Perrin qui, en 1811, et pour la somme de 40 000 fr. lancèrent les premiers travaux de Donzère-Mondragon." *Le Dauphiné libéré,* October 23, 1952.
Lamour, Philippe. *L'aménagement du territoire: Principes, éléments directeurs, méthodes et moyens.* Paris: Les Editions de l'épargne, 1964.
———. *Histoire des canaux du Rhône.* Paris: Editions Sodirep, 1961.
———. *La réforme du régime foncier.* Paris: Les Editions de l'épargne, 1964.
———. "Rhin-Rhône: Une liaison pour l'Occident." *Le Monde,* July 11, 1975.
———. *60 millions de Français.* Paris: Buchet, Chastel, 1967.
Largeron, D. "Les lônes remplacées par des zones." *Le progrès,* October 3, 1980.
de Latil, Pierre. "Donzère-Mondragon, chantier n. 1 de France." *Science et vie,* February 1952.
Lebreton, Philippe. "Non aux barrages de Loyettes et de Sault Brenaz." *Le courrier de la nature,* May-June 1982, 31–33.
Lejeune, Yves, and Louis Chabert. "La complémentarité du nucléaire et de l'hydraulique: L'exemple de Rhône-Alpes." *Revue de géographie de Lyon* 62 (1987): 149–160.
Lenthéric, Charles. *Du Saint-Gothard à la mer, Le Rhône, Histoire d'un fleuve.* Paris: Libraire Plon, 1905.
———. *La région du Bas-Rhône.* Paris: Librairie Hachette et Compagnie, 1881.
M.J.G. "Nos lecteurs nous écrivent, Le Haut-Rhône." *Le progrès,* August 6, 1976.
M.-R. G. "Economie et écologie, Les questions auxquelles il faut réfléchir pour que la liasion Rhône-Rhin nous apporte un maximum d'avantage et un minimum de nuisances." *Le progrès,* January 31, 1975.

"M. Paul Ribeyre: 'Cette grande idée peut constituer pour la jeunesse le symbole des ambitions nouvelles de la France.'" *Le progrès,* November 25, 1975.

Le Masson, Henri. "L'aménagement du Rhône de Genève à Donzère-Mondragon." *France Illustration,* June 19, 1948.

Mauron, Marie. *Au fil du Rhône: Des glaciers à la mer.* Paris: Horizons de France, 1957.

Mayer, René. "Le Rhône, Principal axe de transport français." *Delta: Revue d'action régionale* 19 (1968): 28–35.

Michel, Jean. "Le problème du Rhône: L'aménagement intégral du fleuve. Son triple point de vue: navigation, forces motrices, irrigations." Ph.D. diss., Ecole des Sciences Politiques, 1932.

Michel, R. "Derniers aménagements hydrauliques du haut Rhône." *Le progrès,* February 26, 1982.

"Michel Crépeau opposé à deux projets de barrages sur le Rhône." *Lyon Matin,* April 23, 1982.

Ministère de l'Equipement et du Logement et du Ministère des Transports. "Equipement et développement du sillon rhodanien et du sud-est français." *Equipement, Logement, Transports* (1969).

Moulins, Max. "Le département de l'Ardèche et l'aménagement du Rhône." *L'Ardèche aujourd'hui et demain* 6 (1974): 26–34.

Office de Publicité Générale. *Donzère-Mondragon.* Paris: Editions Maurice André, 1953.

Osterman, Jean. "Le Rhône dompté." *La vie française,* March 17, 1980.

Pelletier, Jean. "Les centrales nucléaires rhodaniennes: Essai de sitologie." *Revue de géographie de Lyon* 62 (1987): 241–260.

Perronnet, R. *La société d'économie mixte d'aménagement régional: L'exemple de la Compagnie nationale d'aménagement de la région du Bas-Rhône et du Languedoc.* Montpellier: Paysan du Midi, 1962.

"Philippe Lebreton: 'L'Ain c'est ma patrie.'" *Le Dauphiné libéré,* November 16, 1982.

Pinay, Antoine. "Une alliance d'avenir à l'échelle de l'Europe: Le grand delta." *Delta, Revue d'action régionale* (1968): 7–13.

Plécy, Albert. "Donzère-Mondragon, à la poursuite de 2 milliards de kilowatts." *Paris match,* November 1, 1952.

"Point final de l'équipement du Bas-Rhône, Le barrage de Vaugris inauguré par Raymond Barre." *Lyon Matin,* October 4, 1980.

"La pollution du Rhône au banc des accusés." *Information municipale* 6 (1977): 45–47.

Portnoff, A-Y. "Grands chantiers: Le Tricastin, banc d'essai social." *L'usine nouvelle,* November 6, 1975, 82–83.

"Pour Condrieu, L'aménagement du Rhône pose des problèmes." *Le Journal,* September 21, 1978.

Pour un environnement "à la française": Textes et déclarations de M. Valéry Giscard d'Estaing, Président de la République (Paris: La Documentation française, 1977).

"Les réalisations françaises, Donzère-Mondragon, second Génissiat rhodanien." *Cahiers français d'information: La Documentation française* (December 1, 1948).

Résidence du conseil. *Le barrage de Génissiat et l'aménagement du Rhône, Notes documentaires et études, N. 839, 1948.* Paris: Direction de la Documentation, 1948.

"Le Rhône à grand garabit de Lyon à la Mediterranée." *Le moniteur: Des travaux publics et du bâtiment,* October 13, 1980, 28–33.

"The Rhône Valley." *Water Power* (1960): 214–223.

Ritter, Jean. *Le Rhône.* Paris: Presses universitaires de France, 1973.

Rognani, Jean-Yves. "Tricastin, premier chantier d'Europe." *Regards sur l'actualité* 29 (1977).

Rousseau, Pierre. *Glaciers et torrents, énergie et lumière.* Paris: Hachette, 1955.

Roussilhon, André. "L'utilisation du Rhône au point de vue de la navigation et de l'irrigation." Ph.D. diss., Faculté de Droit, Université de Paris, 1914.

Sabran, Jacques. "Le nouveau Rhône est né." *Le Dauphiné libéré,* October 24, 1952.

———. "Le plus puissant barrage d'Europe." *Le Dauphiné libéré,* April 3, 1952.

Savey, P. "Les aspects systémiques de l'aménagement du Rhône." *Revue de l'énergie* 316 (1979): 497–504.

———. *La CNR à 60 ans, Une brève histoire de la CNR à l'occasion de son anniversaire.* Lyon: CNR, 1993.

Sécretariat d'état à la présidence du conseil et à l'information. *Le Rhône, son rôle économique (aménagement, navigation fluviale, énergie hydroélectrique, irrigation: Crau et Camargue), Notes documentaires et études, N. 309, 1946.* Paris: Direction de la Documentation, 1946.

"Seurre: La liaison Rhône-Rhine-Danube grande entreprise européenne."*Le Monde,* November 6, 1973.

Simon, Vincent. "Le Rhône, fleuve-dieu, fleuve-béton." *Métropole: Magazine d'informations de Lyon et Rhône-Alpes,* June-July 1979, 17–21.

Souchon, Marie-Françoise. *La Compagnie nationale d'aménagement de la région du Bas-Rhône-Languedoc.* Grenoble: Editions Cujas, 1968.

Soulié, J. L., and M. Mandrin. "Le Rhône assassiné." *La gueule ouverte: Combat non-violent, Hebdomadaire d'écologie politique et de désobéissance civil* 215 (June 21, 1978): 2–3.

Soustelle, Jacques. *L'avenir du "Grand Delta" et l'Europe.* Paris: IMP, n.d.

Terrin, Pierre. "Le Grand Delta." *Transports* 147 (1962): 391–394.

Thomas, Abel. *Sillon Rhodanien: Axe-Rhin Méditerranée: Rapport du Commissaire à l'aménagement du territoire.* Paris: R. Lacer, 1960.

Thomas, Claude. "L'équipement hydro-électrique en France, Du plus grand chantier au plus haut barrage." *France Illustration,* April 1, 1950.

Thompson, Ian. *The Lower Rhône and Marseille.* Oxford: Oxford University Press, 1975.

Tournier, Gilbert. *En descendant le cours du Rhône.* Lyon: Chambre de Commerce de Lyon, 1960.

———. "L'équipement énergétique du Rhône." *Urbanisme: Revue française* 61 (1958).

———. *Rhône, dieu conquis.* Paris: Librarie Plon, 1952.

———. *Le Rhône, fleuve dieu, vous parle.* Paris: Librairie Arthème Fayard, 1957.

———. *La vallée du Rhône: Son passé, son présent, son avenir. Exposition de documents historiques, projets et plans. Catalogue.* Sommières: Atelier Antoine Demontoy, 1954.

———. *La vallée impériale: Couloir de l'Europe.* Lyon: Audin, 1977.

Tournier, Gilbert, Michel Laferrère, and François Tornas. *Lyonnais: Beaujolais, Forez, Vivarais.* Paris: Librairie Larousse, 1973.

U.S. Congress, House Committee on Foreign Affairs, *Report to the President of the United States by Samuel I. Rosenman on Civilian Supplies for the Liberated Areas of Northwest Europe (Prepared Pursuant to Letter of Franklin D. Roosevelt Dated January 20, 1945),* 79th Cong., 1st sess., 1945.

U.S. Congress, Senate Hearings before Committee on Foreign Relations, 82nd Cong., 2nd sess., March/April 1952.

Vallée, L.-L. *Du Rhône et du Lac de Genève: Ou, des grands travaux à exécuter pour la navigation du Léman à la mer.* Paris: Librairie Scientifique-Industrielle de L. Mathias, 1843.

Varaschin, Alain and Denis. "Energie et mutations industrielles du Haut-Rhône." *Visages de l'Ain* 198 (1985): 7–10.

Virenque, Jean-François. *Nouveaux destins du Rhône: Navigation, électricité, irrigation.* Paris: Services d'information de la mission spéciale en France de l'ECA, n.d.

Vivian, H. "Le Rhône supérieur franco-suisse: Fluctuations naturelles et artificielles du régime fluvial." *Revue de géographie alpine* 1–2 (1986): 157–165.

Weckel, Marcel. "L'aménagement du Rhône, exposé par M. l'Ingénieur général, lors du 796e déjeuner-conférence de la Société de Géographie Commerciale le 12 décembre 1973." *Revue économique française* 2 (1974): 43–58.

Wohrer, Henri. "L'aménagement du Rhône et la loi du 27 mai 1921." Ph.D. diss., Université d'Aix-Marseille, 1925.

Visual Materials

Barrage de Donzère-Mondragon. Les Actualités Françaises (October 5, 1950), www.ina.fr. Accessed August 12, 2008.
Crin-Blanc: Cheval sauvage. Films Montsouris, 1953. Film.
Demeurisse, René. *Donzère-Mondragon, Pointes-sèches originales.* Lyon: Compagnie Nationale du Rhône, 1952.
Inondations dans la vallée du Rhône. Les Actualités Françaises (November 29, 1951), www.ina.fr. Accessed August 12, 2008.
L'or du Rhône. Les Films Caravelle, 1950. Film.
Le plus formidable barrage d'Europe occidentale. Les Actualités Françaises (January 29, 1948), www.ina.fr. Accessed August 12, 2008.
Le problème d'aujourd'hui: La sécheresse et le charbon. Les Actualités Françaises (December 7, 1945), www.ina.fr. Accessed August 12, 2008.
Tracol, Michel-André. *Quand le Rhône était un fleuve: Album de cartes postales, de photographies, et de gravures composé et légendé par l'auteur.* 3rd ed. Tain l'Hermitage: MAT Editeurs, 1985.
Les travaux de barrage sur le Rhin et le Rhône. Les Actualités Françaises (June 23, 1949), www.ina.fr. Accessed August 12, 2008.

ACKNOWLEDGMENTS

I have been asked countless times how I ended up writing about the Rhône; or, as several French colleagues put it, "our little river." My circuitous path to the Rhône began with undergraduate coursework in environmental history. Then my French professor, Michel Rocchi, encouraged me to put my language skills to work and to investigate the country's environmental history. When I undertook a senior thesis with Drew Isenberg, I decided to examine the entwined social, legal, and environmental histories of Commencement Bay, near Tacoma, Washington. These two independent research projects inspired me not only to pursue graduate studies in history but also to spend more time in France. During the summer between my junior and senior years in college, I had traveled for several weeks there. A train ride from Nice to Paris first sparked my interest in the Rhône. Picturesque Roman ruins, manicured vineyards, and nuclear reactors might appear to have little in common, but each is linked to the river. It was these connections, and how and why they were forged, that made me want to study the Rhône and its history. This book is the result of this meandering intellectual and personal history.

All scholars incur debts, and mine are numerous. It is an honor and a pleasure to recognize the many individuals, professional communities, institutions, and foundations that all helped to make this book possible.

This project could not have been undertaken without the cooperation of the Compagnie Nationale du Rhône (CNR). As the CNR is not part of the public domain in France, its archives are private, not public. The CNR was therefore under no obligation to grant me access to its collections, first during a preliminary research trip in 1997 and then in 1998, 1999, and 2003. I extend my sincere thanks to the CNR's administration, especially Christian Terrier, for agreeing to open their archives. The CNR's Communications division, then headed by Jean-Michel Bouvard, gave me a tour of the upper Rhône during my first research trip. That tour, not to mention the CNR's basically untapped archives, sold me on the project. I am grateful to current and former employees who granted interviews: M. Bourgad, J.-M. Bouvard, M. Carbonari, M. Charbonnier, E. Doutrieux, M. Gueret, H. Laydier, C. Jimenez, M. Moussa, M. Paris, R. Pinatel, and C. Terrier. In 1999, H. Laydier also gave me a fascinating tour of the CNR's facilities along the central Rhône.

Many thanks also go to the CNR's archival and documentation *équipes,* who made my research life much easier. Claire du Besset and Isabelle Manevy-Sellier oversee the Archives and Documentation divisions respectively. F. Alery, C. Beguet, A. Fombonne, S. Martin, M.-A. Narmand, N. Pacorel, J. Ponce, and L. Roger provided additional help. Isabelle kindly facilitated the logistical arrangements for my repeated trips to the CNR; Marie-Agnès made sure I had all the documents I wanted—needless to say, a historian's dream.

This book also tapped archives and other repositories across France. I thank the archivists and staff at the Archives Nationales; Centre des Archives Contemporaines; La Documentation Française; Archives Départmentales de la Drôme, de l'Hérault, and du Vaucluse; Archives de la Direction Départementale de l'Equipement de la Drôme; Archives Municipales de Pierrelatte and de Saint-Paul-Trois-Châteaux; and Bibliothèque Municipale de Lyon.

I thank Mado Nicholas and Michèle Reynard of the Association des Amis du Vieux Donzère for granting me access to the association's materials and their personal collections on the history of the town of Donzère and arranging interviews with several longtime residents. The Rhône valley's main environmental organization, Fédération Rhône-Alpes de Protection de la Nature, allowed me to consult their archives in 2003; its founders, Monique Coulet and Philippe Lebreton, also shared their valuable perspectives during interviews that summer.

During my studies I received financial support from the Andrew C. Mellon Foundation and National Science Foundation (NSF), as well as Stanford University's Department of History, Program in History and Philosophy of Science, Dean's Office, and Institute for International Studies. A Science, Technology, and Society Scholar's Award from the NSF, Bernadotte Schmitt Grant for Research in European History from the American Historical Association, and the Brooke Hindle Postdoctoral Fellowship from the Society for the History of Technology funded new research that proved central to the book. This project also received generous support from several programs at Montana State University and the Dean of Arts and Sciences at Cornell University. A faculty fellowship at Cornell's Society for the Humanities—on the timely theme "Water: A Critical Concept in the Humanities," no less—allowed me to bring the book to fruition. Thanks to various grants, I was able to hire Angie Boyce, Djahane Salehabadi, Corinna Schlombs, and Jennifer Sessions, all capable research assistants, at different points in this book's history.

I have had the privilege of working with Harvard University Press over the past few years. I would especially like to thank Patrice Higonnet for his confidence in this project and for allowing it to be published as part of the Harvard Historical Studies series. Kathleen McDermott was a wonderful editor. It was a joy working with Joe Stoll from the Department of Geography at Syracuse University on the maps. Connie Hsu Swenson, Robert Kulik, and Martha Ramsey provided excellent editorial assistance.

Previous versions of some material appeared in *La technologie au risque de l'histoire*, ed. Robert Belot, Michel Cotte, and Pierre Lamard (Belfort: l'Université de Technologie de Belfort-Montbéliard, 2000); *French Historical Studies*; and *The Nature of Cities: Culture, Landscape, and Urban Space*, ed. Andrew C. Isenberg (New York: University of Rochester Press, 2006).

As any scholar knows, archives, funding, and presses matter, but so do people. I am delighted to recognize the numerous mentors, colleagues, and friends who have contributed to this book. I am happy to start with Gabrielle Hecht and Richard White, who were exceptional coadvisers and proved quite a team. Both provided practical advice, insightful critiques, and unyielding support, not to mention inspiration through their own pathbreaking scholarship. Stellar mentors are rare. I was extremely fortunate to have had not one but two. These few words don't do either of them justice. A supportive team of readers at Stanford, both official and unofficial, including Lou Roberts, Paula Findlen, Tim Lenoir, and the late John Wirth, also strengthened the book.

Since beginning this research, my intellectual community has widened, and I am grateful for the reception I have received from several professional communities. The manuscript benefited from feedback on papers presented at conferences of the American Society for Environmental History, International Committee for the History of Technology, Society for French Historical Studies, Society for the History of Technology, Society for Social Studies of Science, Society for the Study of French History, and Western Society for French History. Talks based on parts of this book were also given to thoughtful audiences at Colby College, Cornell, Georgia Institute of Technology, Harvard, MIT, Princeton, Rutgers, University of California–Davis, and the University of Pennsylvania. A timely NSF-sponsored workshop, "Envirotech," at the University of Maryland in 2006 helped me crystallize several key ideas that I develop at length in the book.

I owe special thanks to a number of senior scholars for their encouragement: Herrick Chapman, Mark Cioc, Bill Cronon, Steve Cutcliffe, Deborah Fitzgerald, Drew Isenberg, Mott Greene, John Krige, Paul Josephson, Marty Melosi, Marty Reuss, Harriet Ritvo, John Staudenmaier, Jonathan Steinberg, Jeffrey Stine, and Joel Tarr.

Colleagues working at the intersection of environmental history and the history of technology have particularly influenced my work. It has been a pleasure thinking with Brian Black, Hugh Gorman, Ann Greene, Dolly Jørgenson, Finn Arne Jørgenson, Tim LeCain, Betsy Mendelsohn, Ed Russell, Fred Quivick, Paul

Sutter, Peter Thorsheim, Frank Uekoetter, Jim Williams, and Tom Zeller. I also acknowledge my compatriots who work on science, technology, and the environment in modern France: Michael Osborne, Tamara Whited, and especially Michael Bess.

This project could not have been completed were it not for conversations with scholars in France. I extend sincere thanks to Christophe Bonneuil, Gabrielle Bouleau, Jean-Paul Bravard, Florian Charvolin, Alexandre Giandou, André Micoud, Benoît Vernière, and André Vincent.

I am indebted to my former colleagues in the Department of History and Philosophy at Montana State University, including Prasanta Bandyopadhyay, Gordon Brittan, Rob Campbell, David Cherry, Susan Cohen, Kristen Intemann, David Large, Tim LeCain, Dale Martin, Michelle Maskiell, Mary Murphy, Michael Reidy, Bob Rydell, Lynda Sexson, Billy G. Smith, and Brett Walker.

Most recently, I have received encouragement from my colleagues in the Department of Science and Technology Studies at Cornell, including Peter Dear, Steve Hilgartner, Ron Kline, Bruce Lewenstein, Michael Lynch, Trevor Pinch, Rachel Prentice, Judith Reppy, Margaret Rossiter, Phoebe Sengers, Suman Seth, and Kathleen Vogel. It was a pleasure being a faculty fellow at Cornell's Society for the Humanities in the fall of 2008. I particularly appreciated conversations with Jenny Gaynor, Christine Marran, Aaron Sachs, and Eric Tagliocozzo. The students in my Society-associated course helped me articulate several key issues; in fact, after much hair-pulling, I realized that the title of the course made the perfect book title.

I would also like to thank the members of my recent writing groups: Johanna Crane, Durba Ghosh, TJ Hinrichs, Stacey Langwick, Sherry Martin, Rachel Prentice, Jessica Ratcliff, Kathleen Vogel, and Marina Welker. In addition, current and former graduate students have shaped my thinking and thus this book: Daniel Ahlquist, Jaime Allison, Angie Boyce, E. Jerry Jessee, Amy Kohout, Anto Mohsin, Megan Raby, Djahane Salehabadi, Tyson Vaughn, Benjamin Wang, and Michael Wise.

I salute other colleagues and friends for engaging conversations, participating in conference panels, and wading through early drafts: Karl Appuhn, Tuska Benes, Matthew Booker, Kip Curtis, Sarah Elkind, Karin Ellison, Malick Ghachem, Jeanne Haffner, Marc Hall, Toby Jones, Arn Keeling, Ari Kelman, Matt Klingle, Tom Lekan, Sarah Leonard, Neil Maher, Jennifer Milligan, Suzanne Moon, Georgina Montgomery, Carla Nappi, Tara Nummedal, Jared Orsi, Karen Oslund, Cindy Ott, Chris Pearson, Lise Sedrez, Sarah Phillips, Jennifer Siegel, Paul Sutter, Marsha Weisiger, Bob Wilson, and everyone already mentioned above. Ellen Stroud read an early draft of the manuscript and helped me see the forest for the trees. Rob Martello read most of a later version and helped me clarify several issues with not only insight but also much needed humor. Long before that, Lara Moore endured the earliest drafts. I thought of Lara so many times as I ventured on the arduous process of writing a book. I sorely miss her brilliance, collaboration, and friendship.

I appreciate all the support I have received from my families, both biological and adopted, American and French. I will always be grateful to Claude Bernard, Marie Acevedo, and their son Rémi for opening their home to a stranger. Hervé and Michèle Chaine offered fine meals and good conversation during my stays in Lyon. Most of all, my aunt and uncle, Marilyn and Maxime Laville, have taken me in as their fifth child during my frequent trips to France. I am also happy to acknowledge a family of friends. Although we are spread around the globe, the University of Puget Sound Honors crew, their spouses, and their kids have been good friends for a long time. In particular, Lisa Kozleski and Rebecca Page deserve thanks for too many things to name here.

My parents, Barbara Pritchard and Robert Pritchard, have given me steadfast, unquestioning support for as long as I can remember. Both of them have shaped this book in many invisible ways. Without my recognizing it until recently, my father showed me the rewards of life as a researcher and scholar. Through quiet example, my mother pointed me toward intellectual curiosity, a lifetime of learning, and a commitment to writing history that matters.

Last but certainly not least, Ron LeCain came into my life through an unusual confluence of events. I am so happy and grateful that he did. Ron's unyielding support of all facets of my life, personal and professional, has grounded me over the past few years. His confidence, patience, and love have sustained me through challenging times. For these reasons and many more, I dedicate this book to him.

INDEX

Actor-network theory (ANT), 12
Administration des Eaux et Forêts, 30, 33, 96
"Affaire Bresson," 143, 227
Agencies de l'Eau, 33, 247
Agency, of nature, 11–12, 14–15
Agriculture, xv–xvi, xvii, 26–27; 19th century, 37; CNR reconstitution/improvement of, 117–130; competing goals of, 5; crop choices in, 122–123, 129, 130, 174; diversion approach and, 82–83; electricity in transformation of, 128–129; environmentalism and, 205–208; hydroelectricity vs., 38–39, 50, 51–52; industrialization and, 113–116, 123–124, 188, 189; in multipurpose development, 39–45; politics of, 122–124; in postwar development, 171–174; postwar food shortages and, 7; rationalization of, 124–130; regionalization and, 171–174, 175–177, 188, 189; Rhône formula on, 79–85; upper Rhône projects and, 227–228, 230–231; Vichy government on, 48. *See also* Génie Rural; Irrigation
Aigues-Mortes, 28
Alatout, Samer, 10–11
Albin, Jules, 138
Allix, André, 65
Aménagement. See Development

Aménagement du territoire et l'avenir des régions françaises, L' (Gravier), 170–171
Amis de la Nature, 223
Anderson, Benedict, 7, 8, 9, 10
André Blondel hydroelectric plant, 69, 87–88, 94–95. *See also* Donzère-Mondragon
Antinuclear movement, 201
Ardèche River, xv
Arrighi de Casanova, J., 126
Association pour le Développement Économique et Social du Sud-Est Français, L'. *See* Grand Delta, Le (group)
Association pour l'Environnement de la Vallée du Rhône, 222
Aubert, Jean, 82, 94
Audebrand, G., 106
Auriol, Vincent, 55, 67, 74
Avène dam, 176
Avignon, xvi
Aymon, J. P., 62–63, 72–73

Barcelona aqueduct, 27, 250
Barre, Raymond, 197, 198
Bas-Rhône, 177–178
Bassin Rhône-Méditerranée-Corse (RMC), 247–248
Bastet, Henri, 135
Baviskar, Amita, 2
Bayard, P., 112
Beaucaire, xvi

359

Beauchastel, 121
Belley, 216–217, 219, 220, 230; delays in, 225; reserved flows at, 231–234
Bess, Michael, 210, 211, 249
Bibliothèque de travail, 61, 64
Bichet, R., 228
Bijker, Wiebe, 8
Bio-territoriality, 10–11
Blackbourn, David, 9
Bollaert, Emile, 55, 65, 98
Bonnier, J., 83, 154, 156
Boudrant, M., 111
Bourgin, M., 154, 156
Bourg-Saint-Andéol, 115, 134, 143
Braudel, Fernand, 193, 199
Brégnier-Cordon, 216–217, 219; delays in, 226–227; opposition to, 222, 224; reserved flows at, 234–237
Bresson, M., 143
Bret, Paul-Louis, 64
BRL. *See* Compagnie National de l'Aménagement de la Région du Bas-Rhône et du Languedoc (BRL)
Bugey, 237–238

Camargue, xvi, 205
Canal de Bourne, 127
Canal de Pierrelatte, 34, 116, 127
Canal de Provence, 174
Canal des Deux Mers. *See* Canal du Midi
Canal du Midi, 30, 31, 133, 153
Canals. *See* Channelization; Counter-canal network; Recharge programs
Carrière, Marcel, 66
Casiers, 176
Castelnau, André, 195
CEA. *See* Commissariat à l'Energie Atomique (CEA)
Centrale Nucléaire de Tricastin, 105, 106–107, 242–243
Centralization, 164, 165; critiques of, 166–171; fragmented, 33; regional, 188–190
Centre Nucléaire de Pierrelatte, 242; Donzère-Mondragon and, 100–102; electricity consumed by, 110; hydrology for, 102–103; water discharge at, 107–109; water supplies for, 104–106
CGA. *See* Confédération Générale de l'Agriculture (CGA)
Channelization, xvii; diversion approach vs., 80–85; navigability and, 35–39; political authority and, 30–31; water transfer system and, 153–159
Charlemagne, 34, 57
Charte d'Environnement du Rhône, 248
Châteauneuf, 111
Chautagne, 216–217, 219, 220; delays in, 225; reserved flows at, 231–234
Claudius-Petit, Eugène, 165, 170
Clavel, Bernard, 204
Clemenceau, Georges, 167
Climate change, 246, 249–250
CNR. *See* Compagnie Nationale du Rhône (CNR)
Coal, 49–50
Code Napoléon (1804), 31–33
CODERA. *See* Coordination Pour la Défense du Fleuve Rhône et de la Rivière d'Ain (CODERA)
Colbert, Jean-Baptiste, 30
Cold War, 9–10, 58, 77, 112, 199, 209
Collaboration, 4
Comité de Défense du Rhône Savoyard-Bugiste, 223–224
Comité Ecologie, 222
Commerce, channelization and, 35–36. *See also* Industrialization
Commissariat à l'Energie Atomique (CEA), 78; Donzère-Mondragon and, 100–102; on nuclear power flow control, 102–103; water demands of, 104–106, 109–113. *See also* Centre Nucléaire de Pierrelatte
Commissariat Général au Plan (the Plan), xi–xvii. *See also* Plan de Modernization et d'Equipement
Commission Départementale des Sites, Perspectives, et Paysages, 227–228
Commission de Recherche et d'Information Indépendantes sur la Radioactivité (CRIIRAD), 242

Common Market. *See* European Economic Community (EEC)
Communist Party, 74
Communities, imagined, 7–8
Compagnie National de l'Aménagement de la Région du Bas-Rhône et du Languedoc (BRL), 129–130, 171–180; on Barcelona aqueduct, 250; expertise marketing by, 251; irrigation and, 172, 174, 175–177; mission of, 173–174; in regional planning, 179; workforce demographics in, 180
Compagnie Nationale du Rhône (CNR), 4, 24; *actionnaires* in, 46; agricultural modernization and, 117–130; BRL conflict with, 178–179; creation of, 45–48; declarations of public utility, 218–220; diversion approach in, 80–85; economic viability of, 215–217, 232–234, 236; electricity rates of, 128–129; environmental impact statements and, 203–204, 213; environmentalism and, 221–225; expanded authority of, 161; expertise marketing by, 250–251; fear of litigation, 142–143; flood management policy of, 110–112; gigantomania of, 207–208; greening of, 240–242; groundwater changes and, 132–162; hydroelectric paradigm of, 48–54; legal responsibility of for groundwater changes, 152–159; on locals, 85, 87, 88–89; in modernization, 73–77; multipurpose mandate of, 50–51, 79, 82–83, 113–116, 125; nationalization and, 52–54; naturalization by, 196–197; nuclear power and, 78–79, 102–113; on nuclear power flow control, 102–103; in postwar national identity, 67–73; projects completed by, 47; protests against, 213; public vs. private statements by, 140–152; regionalization and, 174–175; in regional planning, 179; restoration projects, 247–250; rhetorical strategies of, 58, 59, 65–66, 230, 240–242; Rhône formula of, 79–85,

86; share of blame for groundwater changes, 146–152; technical committee of, 46; upper Rhône projects, 194–239; water guarantees and, 109–113; WWII projects of, 80, 82
Confédération Générale de l'Agriculture (CGA), 171
Conseil d'Etat, 46
Conseil Général des Ponts et Chaussées, 80, 144–146
Constituencies: in the CNR, 46–48; competing goals of, 2–3, 5, 24, 37, 102–103; conflict among, 25–26, 51–52; in development, 203–204; engineers, 48–49; in environmentalism, 205–208; multipurpose development and, 38–45; nuclear power and, 102–103, 107–109; postwar, 78–79
Constructivist perspective, 10, 12, 14, 244
Controversy studies, 133
Convention agricole, 121–122
Cooperative des Irrigants de la Basse Vallée de l'Arc, 128
Coordination Pour la Défense du Fleuve Rhône et de la Rivière d'Ain (CODERA), 222, 228
Co-production, 8–9, 17–18, 244
Corps des Ponts et Chaussées, 4
Correia, Francisco Nunes, 32
Côte d'Azur, 181–191
Counter-canal network, 152–159. *See also* Groundwater
Coustou, Guillaume, 60–61
Crépeau, Michel, 212, 228–229
Creys-Malville, 214
Crin-Blanc: Cheval sauvage (film), 60
Cronon, William, 11

Dagand, A., 145
Daladier, Édouard, 144–145
DATAR. *See* Délégation à l'Aménagement du Territoire et d'Action Régionale (DATAR)
David, A., 116, 119, 156
Davis, Diana, 164–165
Debré, Michel, 168

Declarations of public utility (DUPs), 218–220, 225–226
Decolonization, 9–10, 163–164; expertise marketing and, 250–251; population growth and, 184–185; regionalization and, 169, 188, 192
De Gaulle, Charles, 57, 102, 208
Delattre, Pierre, 56; on diversion approach, 84; on Donzère-Mondragon, 91, 98; on groundwater changes, 132, 140–152, 156; on high-chute dams, 83; on nuclear energy, 100, 101, 107; on reserve flows, 98
Délégation à l'Aménagement du Territoire et d'Action Régionale (DATAR), 170, 179, 200
Délégation Générale de l'Equipement National, 169–170
Delettrez, J. M., 197
Delta, Revue d'action régionale (journal), 181–191
Deriol, Claudius, 62
Descendant le cours du Rhône, En (Tournier), 73
Determinism, technological, 12, 245
Develop France: Inventory of the Future (Guichard), 189–190
Development, 163–192; constituencies in, 203–204; in developing nations, 27; diversion approach in, 38–39; environmentalism and, 5, 193–194, 202–204; Europeanization of, 200–201; expertise marketing and, 250–251; flood management and, 35, 36; gender ideology and, 195, 196; geography and, 167; gigantomania in, 207–208; inequality in, 163–192; institutions and power in, 4–6; integrated, 230; legitimation of, 195–197; multipurpose, 2–3, 38–45; national identity and, 198–201, 208–209; naturalization of, 172–173; opposition to, 26, 222–225; postwar, 78–131; redefinitions of, 192–211; regional, 163–192; regional companies for, 173; rhetorical strategies for, 50, 57–59, 194–197, 204–205; Rhône formula for, 79–85, 230; state building and, 3; state role in, 40–42; before the Third Republic, 34–36; urban vs. rural, 169–171
Direction de la Protection de la Nature, 234
Diversion approach, 80–85; in Donzère-Mondragon, 85–100; flow control and, 89–100; pollution control and, 100–101
Donzère-Mondragon, 49, 54–77; agricultural impacts of, 117–130; canal placement for, 85–89; counter-canal network, 152–159; diversion approach, 80–85; drainage issues at, 126–127; in flood management, 110–112; flow control in, 89–100; former/dead Rhône at, 93–99; "Frenchness" of, 72–73; geology of, 87–88; "greatness" of, 66–73; hydrology impacts of, 105–106; inauguration of, 55–57; industrialization and, 113–116; local opposition to, 138–140; local responses to, 132–162; naturalization of technology and, 63–66; nuclear facilities and, 100–102, 102–113, 114; pollution control and, 100–101; promise of, 73–77; reservoir height at, 91, 92; restoration projects and, 249; rhetorical strategies for, 50, 57–59, 64–66; topography around, 83–84; tributaries and, 87. *See also* Groundwater
Donzère-Mondragon (Herriot), 61
Drôme River, xv
Drought, groundwater changes and, 135–136, 142, 145, 148
Dumas, G., 137
DUPs. *See* Declarations of public utility (DUPs)
Durance River, xvi

Ecogovernmentality, 11
Ecole des Ponts et Chaussées, 48
Ecole Polytechnique, 48
Ecological restoration, 247–250
Ecology, 6; of Donzère-Mondragon, 96–97, 99–100; flow control and,

99–100, 231; groundwater changes and, 159–260; nuclear power and, 103

Economic inequality, 74–76; European integration and, 200; in "Le Grand Delta," 181–291; postwar, 124; urban/rural, 166–171

EDF. *See* Electricité de France (EDF)

Edict Concerning the General Regulation of Waters and Forests (1669), 30

Eiffel Tower, 68

Electricité de France (EDF), 52, 78; Donzère-Mondragon and, 100–102; on economic viability, 215, 216, 232; electricity supplied by, 128; on nuclear power flow control, 102–103; water demands of, 106–107, 109–113. *See also* Centrale Nucléaire de Tricastin

Elites, xi, 29; 19th century, 35; BRL, 174; CNR, 79; CNR public vs. private statements and, 140–152; in engineering, 48; environmentalism and, 238–239; groundwater debates and, 132–133, 135; in regional planning, 168, 178–179; on reserved flows, 233; scientists, 203–204; in state building, 8–9

Energy. *See* Coal; Hydroelectric power; Nuclear power

Energy crises, 43, 49–51, 194, 198, 214–217

Environmental history, 3–4, 4–6, 243–251; constructivism in, 14, 15–16; in envirotechnical analysis, 11–18; on human-natural relations, 2–3; methods from, 5; sources in, 15; STS compared with, 14–17

Environmental impact statements (EISs), 203–204, 213, 219–220

Environmentalism: bureaucratic hurdles in, 218–221; development vs., 5, 193–194; legitimation of, 236–237; modernism and, 210–211, 251; national identity and, 205–208, 208–210; nationalization of, 212–213; rhetorical strategies of, 202–203, 204–205; romanticism in, 204–205, 224; in society, 221–225; in the state, 218–221, 238–239; upper Rhône and, 201–204, 227–231

Environmental management, 3; political territories and, 10–11; politics and technology in, 29–33; postwar modernization and, 73–76; state building and, 8–9

Environmental movement, 26–27

Environmental protection, xvii

Envirotechnical analysis, 11–18, 244–247; regime concept in, 23–24

Envirotechnical landscapes: creation of Rhône, 28–29; definition of, 1–2, 13; technopolitics of, 9

Envirotechnical regimes, 3–4, 18, 23–24; 1970s to 1980s, 213; BRL, 177–179; change in, 238–239; definition of, 23

Envirotechnical systems, 3–4, 18; BRL, 177–179; definition of, 19; environmentalism and, 209–211; nature as technological in, 22–23; politics of, 133–134; postwar development and, 78–131; river flow control and, 89–100, 233–234; technology as natural in, 21–22

Eurodif. *See* European Gaseous Diffusion Uranium Enrichment Consortium (Eurodif)

European Economic Community (EEC), 123, 163, 186–187

European Gaseous Diffusion Uranium Enrichment Consortium (Eurodif), 100

European integration, 9–10, 163–164; agriculture and, 123; anxiety about, 168; energy policies and, 249–250; industrialization and, 186–87; national identity and, 165, 199; regionalization and, 186–187, 192; rhetorical strategies for, 58

Europeanization, 200–201

European Union, 249–250

Faure, Raoul, 64
Fédération Française des Sociétés de Protection de la Nature, 222
Federation of Those Expropriated by the Compagnie Nationale du Rhône in Bollène, 138
Fédération Rhône-Alpes de Protection de la Nature (FRAPNA), 202, 204, 206, 222, 224, 229–230
Feyzin industrial site, 174–175
Fifth Republic (1958–present), 199
Financing, 223; government role in, 42; Marshall Plan in, 70, 72; postwar, 82; *rentabilité* and, 215–217, 232–234, 236
Fioravante, J., 107, 120–121, 123–124
Fish and Fishing, xvi, 205; Donzère-Mondragon and, 96–97, 99; restoration projects and, 249; upper Rhône projects and, 222, 224, 235, 236
Flooding, xiv, xv; in 1840 and 1856, 34–35, 36; benefits of, 59; constructivist categorization of, 16; former dead Rhône and, 99; diversion approach and, 84–85; flow control and, 91, 99; history of, 28; management policies, nuclear power and, 105–106, 110–112, 230–231; thousand-year, 34–35; upper Rhône projects and, 230–231
Flow control: at Donzère-Mondragon, 89–100; former/dead/short-circuited Rhône, 93–99, 110, 204, 231; guarantees in, 109–113; hydroelectricity and, 94–95; nuclear power and, 102–113; reserved flow, 96–99; restoration projects and, 249; in upper Rhône projects, 223–224, 230–238; variability in, 90, 94–96
Fos, xvii
Foucault, Michel, 11
Fourques, 176–77
Fourth Republic (1946–1958), 49, 199
FRAPNA. See Fédération Rhône-Alpes de Protection de la Nature (FRAPNA)
French Revolution, 31–32, 34, 166

Freycinet, Charles de, 199–200
Furka glacier, xii

Gaffière canal, 107–108
Galley, R., 102–3, 104–105, 108
Gard River, xvi
Gascar, Pierre, 205
Gaujac, P. de, 111
Gemaehling, Claude, 101, 109, 198, 224; on reserved flows, 231–232, 235–236; on upper Rhône projects, 230
Gender ideology, 60–61, 68–69, 195, 196
Génération Ecologie, 202
Génie Rural, 4, 78; on agriculture reconstitution, 117, 119, 121; Canal de Bourne and, 127; engineering elites and, 48–49; on groundwater changes, 154, 156, 157; groundwater changes and, 144–146; on impact of industrialization, 115
Génissiat, xiv, 48, 82, 113
Geologic discontinuities, 149–150
Geology issues, 87–88, 148–150
Germany, 7, 64, 186–187
Germinal (Zola), 63
Giandou, Alexandre, 54
Gigantomania, 207–208, 223
Giguet, R., 51–52
Gilbert, Alain, 207
Gilles, Simon, 136
Giscard d'Estaing, Valéry, 200, 219–220, 221
Government: centralization in, 26, 33, 164, 165, 166–171; development roles of, 40–42; in hydroelectric development, 42–45, 212, 214–217; postwar instability of, 74; public utility laws, 35–36; rationalization by, 124–130; regionalization and, 26, 165–192; river management authority of, 29–33
Governmentality, 11
Grand Coulee Dam, 45
Grand Delta, Le (group), 163, 181–183, 185–186

Grand Delta, Le (region), 166, 180–191; as crossroads, 190–191; population growth in, 184–185
Grandeur, 57, 198–201; development and, 201, 209; of Donzère-Mondragon, 67–70; Donzère-Mondragon and, 57–59, 77; Germany compared with, 64; hydroelectricity and, 49; Upper Rhône and, 198–199; U.S. as threat to, 70, 72. *See also* National identity
Gravier, Jean-François, 124, 163–164, 166, 171, 170, 188–189; on inequality, 200; on laissez-faire economics, 167; on urban vs. rural development, 170–171
Great Britain, 45
Green Party, 193, 202
Groundwater, 132–162; changes in as improvements, 151; CNR share of blame for changes in, 146–152, 151–152; drought blamed for changes in, 135–136, 142, 145, 148; fear of litigation over, 142–143; geology and, 148–150; legal status of, 152, 153; locals' perspectives on, 134–140; nuclear power and, 105–109, 242–243; public vs. private CNR statements on, 140–152; recharge programs, 105–106, 107, 153–159, 161; solutions to changes in, 152–159; underground barrier to, 157–158; water transfer system and, 153–159. *See also* Counter-canal network; Recharge programs
Groupe Ain Nature, 222
Groupe de Nature d'Isère, 223
Groupe d'Etude des Perspectives Agricoles et Rurales Rhodaniennes, Le (GEPAR), 125–126
Groupement des Pêcheurs Sportifs, 224
Guichard, Olivier, 189–190

Hainard, Robert, 224
Hanotaux, Gabriel, 64
Harvey, David, 66
Haussmannization, 169
Hecht, Gabrielle, 9, 17, 23, 245
Hechter, Michael, 164
Henry, Marc, 98, 110, 111; on groundwater changes, 148, 151, 152–153
Herriot, Édouard, 41, 56, 199; on conquest of nature, 61; on Donzère-Mondragon, 67; in historicization, 65; on regionalism, 165, 167
High-chute model, 80, 82, 83
Historicization development and, 196–197; of former/dead Rhônes, 93–99; of technology, 65–66
Holtz-Bonneau, Françoise, 205–206, 207
Homme et le Rhône, L' (Faucher), 197
Hughes, Thomas, 12, 19
Hybridity, 246–247
"Hydro-Agricultural Development of the Mid-Rhône, The" (Fioravante), 123–124
Hydroelectric power, xiv, xvii; 19th century, 37–39; in 1970s and 1980s, 26–27; agriculture transformed by, 128–29; agriculture vs., 122; CNR paradigm for, 48–54; diversion approach and, 82–83; diversion approach in, 38–39; Donzère-Mondragon, 85–89; economic viability of, 215–17, 232–234, 236; energy crises and, 43, 49–51, 214–217; environmental constraints on, 19–20, 90; envirotechnical analysis on, 13; envirotechnical systems in, 21–22; flow control and, 94–95; government role in developing, 42–45, 214–215; industrialization and, 113–116; in modernization, 49–54; nationalism and, 216–217; nationalization of, 52–53; nuclear power demands on, 110; nuclear power vs., 225; in regional development, 172; rhetorical strategies for, 50–51; Rhône formula on, 79–85, 86, 230; river flow control and, 89–100; upper Rhône projects, 214–217
Hydrology, 243; definition of, 103; groundwater changes and, 132–162; nuclear power and, 102–104, 105–106

Identity. *See* National identity
Identity of France, The (Braudel), 193, 199
Imagined Communities (Anderson), 7
Immigration, 118, 123
Imperialism, 9–10, 163–164; internal, 164–165; population growth and, 184–185; regional planning and, 180, 191
Industrialization, xvii, 6–11, 113–116; of agriculture, 124–130; agriculture vs., 113–116, 123–124; CNR constituency for, 46; geography of European, 186–187; groundwater changes and, 162; hydroelectric development and, 43–44; in multi-purpose development, 39–45; regionalization and, 188, 189; upper Rhône, 198, 200
"Influence of the CNR's work at Donzère-Mondragon on the behavior of groundwater," 149–150
Inquiry commissions, 218
Irrigation, xv–xvi, xvii; crop choices and, 129; Donzère-Mondragon and, 115–116, 117–118, 126–128; electrification of, 128–129; groundwater changes and, 153–159; navigability vs., 37; rational, 125; in regional development, 172, 174, 175–177. *See also* Agriculture; Génie Rural
Isère River, xv
Italy, 45

Jacobins, 166
Jonage, xiv, 38–39, 43
Josephson, Paul, 207

Kirchner M., 132
Knowledge, politics of, 133–134; in CNR public vs. private statements, 140–152; groundwater changes and, 135, 137, 140–152
Konvitz, Josef, 31

Labadie, J., 63, 70
Labor strikes, 74–75
Lachaux, Emile, 136

La Garde-Adhémar, 135, 142
Lamorisse, Albert, 60
Lamour, Philippe, 171–173, 179, 187, 188
Land reform, 117–118, 129
Languedoc, 177–178, 181–191
Larmand, André, 136
Latil, Pierre de, 64, 73
Latour, Bruno, 14–15, 149
Lauragais Audois, 176
Lauzon canal, 107–108
Lebreton, Philippe, 206
Lecornu, J., 226, 230–231
Legitimation: of development, 195–197; of environmental concerns, 236–237; historicization in, 65–66; political, 9; rhetorical strategies in, 195–197
Legrand, M., 225–226
Léman Lake, xi, xii
Le Pen, Jean-Marie, 173
Lilienthal, David, 172
Locals, xvi, 25–26; 19th century, 37; as barrier to progress, 75–76; CNR representation of, 46; conflicts with, 37; credibility of, 147; Donzère-Mondragon effects on, 85, 87, 88–89; environmentalism and, 205–208, 221–225; on flow control, 96–97; on groundwater, 134–140; groundwater and, 132–162; on property rights and development, 44; protests by, 213; in regional development, 173; upper Rhône projects and, 227–231
Louis XIV, 30–31, 57
Louvel, Jean-Marie, 56
Low-chute model, 80
Loyettes, 193, 194, 212, 216–217; opposition to, 205, 227–231
Lugudunum, xiv, 28
Lyon, xiv–xv, xvii; chambers of commerce meeting in, 39; floods in, 34–35; population in, 184; regionalization and, 174–175

Marcoule, 110
Marseille, 184, 185

Marshall Plan, 70, 72, 171–172
Masson, Henri le, 59
Materiality/materialization: of communities, 7–9; cultural processes vs., 17–18; of technology, 14–15
Mathian, J., 147, 148–149
McEvoy, Arthur, 13
Mediterranean, 190
Messmer Plan (1974), 214, 215, 225
Michel, R., 206
Michelet, Jules, 59, 195
Ministère de l'Aménagement du Territoire et de l'Environnement, 249
Ministry of Agriculture, 125–126, 128, 179
Ministry of Industry, 225–226
Ministry of the Environment, 213, 218–219, 220, 221, 228–229; legitimation of environmentalism and, 238–239; on reserved flows, 231, 234
Ministry of Urbanism and Housing, 229
Mistral winds, 62, 101
Mitchell, Timothy, 16
Mitterrand, François, 200
Modernism, 58–59, 130–131, 250–251; contradictions in, 76; environmentalism and, 210–211; historicization and, 65–66; limits of, 133
Modernization, xi; after WWII, 49–54; of agriculture, 117–130; definitions of, 25; dislocation and disruption from, 73–76; Donzère-Mondragon in, 66–73; economic stability and, 186; electricity in, 49–54; environmentalism and, 238–239; groundwater changes and, 153; national identity and, 3, 6–11; promise of, 73–77; regionalization and, 172–173; rhetorical strategies for, 50, 57–59, 65–66; unequal, 163–164; upper Rhône, 198; water in, 174
Monnet, Jean, xi, 171–172
Montélimar, 91
Morand, L., 138
Morel, Gilles, 207

Moulins, Max, 217, 220, 225
Mukerji, Chandra, 8–9
Murphy, Michelle, 23

Napoléon Bonaparte, 33
Napoléon III, 35–36
National identity, 3, 6–11; after WWII, 57–59, 66–73; cultural constructions of, 9; development and, 198–201; Donzère-Mondragon and, 57–59; energy crises and, 49–51, 216–217; environmentalism and, 205–208; European integration and, 165; greatness in, 66–73; hydroelectricity and, 49; nuclear power and, 212; rural life in, 169–170; technology in, 3, 6–11, 57–59, 72–73
Nationalization, 52–53, 247–248
Natural flow theory, 32–33
Nature: agency of, 11–12, 14–15; blurred boundaries of, 99–100, 196, 242–243; definitions of, 18–19; environmentalist rhetoric on, 204–205; in envirotechnical analysis, 11–18; gender ideology and, 60–63; groundwater changes as, 149–150; human relations with, 2–3, 5, 13; national identity and, 3, 6–11, 58–59, 77, 199; naturalization of development and, 172–173; naturalization of nuclear power and, 103; naturalization of technology and, 63–66, 93–94; renewable energy and, 246–247; rhetorical strategies on, 240–242; social and political shaping of, 14; state building and, 10–11; as technological, 22–23. *See also* environmentalism; Envirotechnical landscapes; Envirotechnical systems
Naturecultures. *See* Envirotechnical analysis; Environmental landscapes
Nature preserves, xvii
Navigability, xiv, xvi; agricultural goals vs., 37; channelization and, 35–39; competing goals of, 5; diversion approach and, 82–84; flow control and, 92; hydroelectricity vs., 38–39, 52; industrialization and, 113–116;

Navigability (*continued*)
  in multipurpose development, 39–45; politics and, 30; Rhône formula on, 79–85
New Destinies of the Rhône (Virenque), 65
"New Goal, A: Uniting Technology and Nature," 240, 241, 251
Nourrit, L., 115
Nuclear power, xvii, 4, 212–239; accidents in, 242–243; "all nuclear" policy on, 111–112, 207, 214–217; antinuclear movement and, 201, 214; current production of, 214; Donzère-Mondragon and, 100–102; energy crises and, 214–217; environmental constraints on, 20; environmentalism on, 207; flow control for, 103–113, 237–238; groundwater and, 105–109; hydroelectric development vs., 225; hydrology for, 102–104; national identity and, 57; naturalization of, 103; pollution control and, 100–101, 108–109, 237–238; priorities of, 78–79; technopolitics of, 245; uranium dependency and, 216; water discharge in, 107–109; water guarantees for, 109–113
Nye, David, 1

Office des Transports, 39
Oil crisis (1970s), 194, 198
Oil refineries, xvii, 28
O'Neill, Karen, 9
OPEC, 214
Orb River, 176
Or du Rhône, L' (film), 61, 73–74
Ornano, Michel d', 216, 217

Pampelonne, M. de, 96
Panama Canal, 70
Paris: CNR representation of, 46; dominance of in development, 163–64, 166, 170–171; electricity demands of, 41, 42; regional metropoles and, 188
Paris et le désert français (Gravier), 124, 163–164, 166, 167, 170, 200
Paris-Lyon-Marseille (PLM) railroad, 36–37, 46
Paul, Marcel, 49
Perrier, Léon, 40; in historicization, 65; on hydroelectricity, 44; on regional development, 41–42; on regionalism, 165, 167
Pfahl, M., 144
Pierre-Bénite, 174–175, 248–249
Pierrelatte. *See* Centre Nucléaire de Pierrelatte
Pinay, Antoine, 56, 163, 164, 186, 191; on regional centralization, 188; on regional planning, 190
Pintat Commission, 216–217
Plan d'action Rhône, 248
Plan Migrateurs Rhône-Méditerranée, 249
Plan de Modernisation et d'Equipment, xi, 53; First, 53, 170; Second, 165, 179; Fifth, 122; Seventh, 219–220. *See also* Commissariat Général au Plan
Plan National d'Aménagement du Territoire (1950), 170
Plécy, Albert, 70, 72
Political geographies. *See* Europeanization; National identity; Regionalization
Politics, 243; in environmental management, 29–33; of envirotechnical systems, 23; of knowledge, 133–135, 140–152; of water, 29–33, 133–134
Pollution: environmentalism and, 208; nuclear energy and, 100–101, 105–108; thermal, 237–238
Pommeret, Henri, 180
Population, 184–185
Poujadisme, 140, 173
Power, 213; colonialism and, 163–165; in development decisions, 4–6; in envirotechnical landscape creation, 2; in state building, 9; urban, 188
Prior appropriation doctrine, 32
Property rights, 32–33, 44
Prospective, 189–190
Provence, 174, 181–191
Public utility laws, 35–36

Radiodiffusion Française, 55
Railroads, 35–37, 46, 52
Rastoin, Édouard, 145
Rationalization, 124–130
Recharge programs, 105–106, 107, 153–159, 161. *See also* Counter-canal network; Groundwater
Recreation, xvii, 26–27, 224
Regimes. *See* Envirotechnical regimes
Regionalization, 26, 163–192; agriculture in, 171–174; balance in, 188–189; challenges in, 188–191; CNR mission and, 174–175; desert metaphor in, 182–183; economic disparities in, 124; economic stability and, 185–186; environment as asset in, 182–184; envirotechnical possibilities in, 184; European integration and, 186–87; geography in, 167–168; Gravier and Pinay on, 166–171; irrigation and, 172, 174, 175–177; large-scale planning in, 179; naturalization in, 172–173; population growth and, 184–185; regional development companies in, 173–174; resistance to, 168–169; rhetorical strategies for, 58; Rhône in, 171–180, 180–191; systems thinking in, 189–190; tourism and, 177–178
Remembrement, 117, 118–122, 126, 129
Rendement, 215
Renewable energy, 246–247, 249–250
Rentabilité, 215–216, 232–234, 236
Restoration projects, 247–250
Rhetorical strategies, 240–241; environmentalist, 202–203, 204–205; on groundwater changes, 134–140, 148–149; national identity and, 57–59; on social and economic change, 74–75; on upper Rhône, 194–198; against upper Rhône projects, 230
Rhine-Rhône canal, 193–194, 199–200, 207, 248
Rhône, dieu conquis (Tournier), 60, 65, 67

Rhône, fleuve dieu, vous parle, Le (Tournier), 63, 64
Rhône-Alpes region, 181–191
Rhône de Genève à la Méditerranée, Le (Delettrez), 197
Rhône River: before 1945, 24; Barcelona aqueduct from, 27, 250; average flow of, 94–95; delta of, xvi–xvii, 180–191; depictions of, 59–63; destiny of, 65, 197; diversion approach in, 80–85; flow control, 89–100; former/dead/short-circuited, 93–99, 110, 204, 231; as nation's river, 58–59; regional development and, 171–191; restoration of, 27, 247–250; Rhine liaison with, 193–194, 199–200, 207, 248; Rhône formula for, 79–85, 86, 230; saving the, 204–208; symbolic centrality of, 2; as technology, 183–184, 192; tributaries of, xiii, xiv–xv, 87; upper, 194–239; volume of, xvi; as wilderness, 210–211. *See also* Flooding; Flow control
Ribèyre, Paul, 200
Rights, public vs. private, 32–33
Riparian rights, 32
Roman Empire, xiv, 28, 34
Romanticism, 204–205, 224
Roosevelt, Franklin, 40
Rostagni, J., 138, 142, 147–148
Rougier, Roger, 145, 146
Roussin, Joseph, 136
Rupture-talk, 240–242
Rural areas: electrification of, 128–129, 172; idealization of, 169–171; Le Grand Delta, 182–183; postwar exodus from, 118, 123; regionalization and, 179; unequal development in, 164–192; urban areas vs., 37, 167–171
Russell, Edmund, 21

Sabran, Jacques, 62, 75
SACTARD. *See* Société Anonyme de Coordination des Travaux d'Aménagement du Rhône à Donzère (SACTARD)

Sahlins, Peter, 7, 8
Salenc, Pierre, 40, 119; on agriculture, 121, 122, 125, 126, 127, 129
Sanitation, xvi
Saône River, xiv, xv
Sault-Brénaz, xiv, 194, 212, 216–217; opposition to, 205, 227–231; reserved flows at, 237–238
Science and technology studies (STS), 243–251; constructivism in, 14, 15–17; controversy studies in, 133; on co-production, 8–9, 17–18; environmental history compared with, 14–17; methods from, 5; on nature, 12; naturecultures in, 1–2
Scott, James C., 8, 58–59
Second Empire (1852–1870), 35–36
Seeing Like a State (Scott), 8, 58–59
Service de la Pêche, 78, 97, 99
Service du Rhône, 35
Simon, Vincent, 198–199, 201
Site Nucléaire du Tricastin, 100
6ème Circonscription Electrique, 144, 151, 154, 156
60 millions de Français (Lamour), 179
Social constructivist approaches, 15, 16
Socialist Party, 168, 202
Social mobility, 189
Social structure, 245–247; agriculture modernization and, 120–121; development inequalities and, 167–168; Donzère-Mondragon and, 85, 87; groundwater changes and, 161–162; imperialism and, 163–165; postwar modernization and, 74–76, 138
Société Anonyme de Coordination des Travaux d'Aménagement du Rhône à Donzère (SACTARD), 144
Société d'Etudes des Canaux de la River Droite du Rhône, 129, 173
Société du Sud-Electrique, 128
Soil erosion/degradation, 122
Solétanche, 157–158
Soustelle, Jacques, 180, 183, 185, 189, 190
Soviet Union, 45
Spain, 45

State building, 3, 6–11; cultural constructions in, 9; extension of jurisdiction and, 30–33; groundwater changes and, 162; naturalization in, 31–32; politics and nature in, 10–11; regionalization and, 168–169, 187–188, 192; spatial/environmental contexts in, 7–8; Third Republic, 36. *See also* National identity
Stop Pollution: Vallée du Rhône (magazine), 208
STS. *See* Science and technology studies (STS)
Suez Canal, 70
Superphénix, 214
Switzerland, 9
Systems theory, 12, 19
Systems thinking, 189–190

Tacitus, 34
Tanguy-Prigent, François, 172
Tazieff, Haroun, 205
Technology: aesthetics of, 68–69; artifacts of, 245; blurred boundaries of, 99–100, 196, 242–243; definitions of, 18–19; depoliticization of, 5–6; determinism of, 12, 245; environmental constraints on, 19–20; environmental management and, 3, 29–33; in envirotechnical analysis, 11–18; gender ideology and, 61–63; history of, 11–18; life cycles of, 13; materiality of, 14–15; as mediator with nature, 13; national identity and, 3, 6–11, 57–59, 72–73, 77; as natural, 21–22; naturalization of, 63–66, 93–94, 240–242; Rhône as, 183–184, 192; social and political shaping of, 14. *See also* Envirotechnical landscapes; Envirotechnical systems
Technopolitics, 8
Tennessee Valley Authority (TVA), 40, 45, 172
Third Republic (1870–1940), 7, 36
Those Affected by the Canal of Donzère-Mondragon, 138
Tocqueville, Alexis de, 166

Tourism, xvii, 130, 177–178
Tournier, Gilbert, 60, 67, 143, 200; on collaboration, 75; on destiny, 65; on development, 195; on groundwater changes, 147–148; naturalization of technology by, 63, 64; on promise of modernization, 73
Train à Grande Vitesse, 223
Transportation. *See* Navigability
Tricastin. *See* Centrale Nucléaire de Tricastin; Site Nucléaire du Tricastin

United States: France compared with, 183–184; French *grandeur* threatened by, 70, 72; gigantomania of, 207; hegemony of, 58; hydroelectric development in, 45; prior appropriation doctrine in, 32; Tennessee Valley Authority, 40, 45, 172; wilderness mythology in, 11–12
Uranium, 216
Urban areas: critiques of, 166–71; electricity demands in, 41, 42; regionalization and, 169–171, 188–189; rural areas vs., 37, 169–171; unequal development in, 164–192
Usine Eurodif, 100, 105, 107

Vallabrègue, 178–179
Varennes, Charles, 132, 137

Vichy regime, 7, 48–49, 169–170, 189
Virenque, Jean-François, 61, 65, 68–69, 197
Vollant, Jacques, 132–133
Voynet, Dominique, 193

Water law of 1992, 247–248
Water Power (journal), 65
Water rights, 32–33
Water supplies, xvi. *See also* Groundwater; Irrigation
"When the Rhône Ran Free" (Hainard), 224
White, Richard, 20
Wilderness imagery, 11–12, 210–211
Working Goup on the Upper Rhône, 222
World War I, hydroelectric development and, 43–44
World War II: dislocation caused by, 73–76; Donzère-Mondragon and, 55–57; hydroelectric paradigm after, 48–54, 76; national identity after, 57–59, 66–73; reconstruction after, xi, xv, 6–11, 24–25, 73–76, 164, 171–174

Yom Kippur War (1973), 214

Zola, Emile, 63

Harvard University Press is a member of Green Press Initiative (greenpressinitiative.org), a nonprofit organization working to help publishers and printers increase their use of recycled paper and decrease their use of fiber derived from endangered forests. This book was printed on recycled paper containing 30% post-consumer waste and processed chlorine free.